Undergraduate Topology

Undergraduate Topology

Robert H. Kasriel

Dover Publications, Inc.
Mineola, New York

Bibliographical Note

This Dover edition, first published in 2009, is an unabridged
republication of the work originally published in 1971 by W. B.
Saunders Company, Philadelphia.

Library of Congress Cataloging-in-Publication Data

Kasriel, Robert H. (Robert Herman)
    Undergraduate topology / Robert H. Kasriel.
      p. cm.
    Originally published: Philadelphia : W.B. Saunders, 1971.
    Includes bibliographical references and index.
    ISBN-13: 978-0-486-47419-9
    ISBN-10: 0-486-47419-4
    1. Topology. I. Title.

QA611.K34 2009
514—dc22

2009015895

Manufactured in the United States by Courier Corporation
47419401
www.doverpublications.com

To Ernestine, Sarita, and David

Preface

General topology can be a valuable tool to the graduate student of mathematics in such courses as complex analysis, real analysis, and functional analysis. Since he is likely to take one or more courses in analysis during his first year of graduate work, it is to his advantage to have taken a course in general topology before beginning his graduate program. This text is intended as an introduction to general topology for upper division undergraduates who intend to continue their study of mathematics in graduate school. It is essentially self-contained except for elementary calculus.

The reader who has had sufficient experience with elementary set theory can skip Chapter 1, except for whatever review of it he feels necessary. He should, of course, become familiar with the notation introduced in that chapter. The reader who can already do the review exercises at the end of Chapter 1 certainly has sufficient background to skip that chapter. The next chapter introduces enough machinery to establish the completeness of Euclidean n-space and a few other special properties of that space, for example, the nested interval theorem. Thus, the reader who has had a quarter or a semester's work in advanced calculus should be able to begin with Chapter 3, using Chapter 2 for reference when the need arises. Chapters 3 and 4 present a detailed study of metric spaces and mappings on metric spaces. The next chapter deals with several important metric spaces and applications to analysis. It is quite likely that by the time the reader gets to Chapter 5 he will have had some exposure to an elementary course in differential equations. If he has not, he can defer reading the application to differential equations without interfering with the continuity of the study. Beginning with Chapter 6, the book considers general topological spaces and mappings on such spaces.

Almost every one of the 112 sections in the book ends with a set of exercises. Some of the exercises are designed to help the reader to become acquainted with the concepts just defined, whereas others are intended to help to prepare the reader for what is to follow. The reader is also asked to prove a number of theorems; in other exercises he is given a proposition that calls for a proof or a counterexample. There are also several sets of review exercises which are designed to help the reader to gain an overview of large portions of the subject matter. Throughout, he is given an opportunity to take an active part in the development of the subject matter.

I owe a debt of gratitude to all the professors who taught me at the University of Virginia. In particular, I continue to feel the influence of G. T. Whyburn in my mathematical interests, teaching, and now in the writing of this book. I also recognize in this text the influence of various books from which I taught, particularly J. L. Kelley's *General Topology*. I am grateful to Norman Levine, Albert Novikoff, and Arlo Schurle for their critical reading of the original manuscript and for their detailed and very helpful reviews. Their valuable suggestions aided me greatly in the revision of the original manuscript and resulted in many improvements. I wish to thank George Cain, William McKibben, and Sanford Wiener for their helpful suggestions and for their careful reading of the galleys. I am indebted to my editor George Fleming of W. B. Saunders for encouraging me to undertake this project and for his continued interest and help all through it. In addition, all the other members of Saunders with whom the book brought me in contact were always cooperative and helpful. Also, during the early stages of the writing I was fortunate in having several students who studied from the manuscript as I was preparing it. I found their comments quite helpful. In this respect I wish to thank W. G. Christian, C. J. Holland, W. P. McKibben, H. B. Overton, G. Redd, R. J. Schaffer, and N. Warsi. I am grateful to my many colleagues who were willing to discuss various parts of the manuscript with me and especially to J. M. Osborn, E. J. Pitts, and W. R. Smythe, who made valuable suggestions on the basis of their classroom experience with parts of the original manuscript. I appreciate the conscientious job of typing done on various parts of the manuscript by Mrs. Frances Fowler and Mrs. Virginia Wilson and also by my daughter Sarita and brother Victor. Finally, I want to thank my wife Ernestine and my son David for helping me with the tedious task of proofreading from manuscript to galleys and with the indexing.

Contents

5

METRIC SPACES: SOME EXAMPLES AND APPLICATIONS 143

6

GENERAL TOPOLOGICAL SPACES AND MAPPINGS ON TOPOLOGICAL SPACES 162

7

COMPACTNESS AND RELATED PROPERTIES 202

8

CONNECTEDNESS AND RELATED CONCEPTS 218

9

QUOTIENT SPACES 230

10

NET AND FILTER CONVERGENCE 248

11

PRODUCT SPACES 261

To the Instructor

Note: Each of the following exercises is referred to in the text or in a subsequent exercise. Therefore it is suggested that they be included among those assigned.

Page 6: 1

Page 8: 2

Page 9: 2

Page 10: 1(a)

Page 11: 3

Page 13: 2, 3, 5

Page 17: 1(b)

Page 18: 5(e), 5(f)

Page 23: 9, 10

Page 35: 10

Page 53: 5

Page 65: 1

Page 66: 3

Page 67: 4, 5

Page 70: 3

Page 71: 3, 7

Page 77: 2, 5

Page 87: 3

Page 102: 10

Page 105: 6

Page 110: 4

Page 112: 2(d)

Page 113: 5

Page 114: 7

Page 116: 4

Page 118: 2, 6

Page 121: 3

Page 122: 3

Page 123: 4, 5, 7

Page 126: 5

Page 128: 2

Page 142: 15

Page 146: 3

Page 165: 4, 5

Page 169: 6

Page 183: 5

Page 185: 4, 6

Page 186: 9, 10

Page 187: 16, 17

Page 196: 5

Page 197: 4

Page 207: 4, 6

Page 210: 3

Page 211: 4

Page 225: 14

Page 225: 4

Page 234: 5

Page 255: 4

Notation for Some Important Sets

$\mathbf{R}$ = the set of all real numbers
$\mathbf{Q}$ = the set of all rational numbers
$\mathbf{Z}$ = the set of all integers
$\mathbf{P}$ = the set of all positive integers
$\mathbf{P}_n = \{1, 2, 3, \ldots, n\}$
$\mathbf{R}_+$ = the set of all positive real numbers
$\mathbf{R}^n$ = the set of all n–tuples of real numbers

1

Sets, Functions, and Relations

A fundamental concept in mathematics is that of function, a relation that assigns to each element in one set, the domain, a unique element in a second set, the range of the function. For functions considered in elementary courses, the domains and ranges are very often subsets of real numbers. However, even in elementary courses the student comes in contact with functions whose domains are sets of objects other than real numbers. For example, the multiplication operation for numbers is a function p which relates each ordered pair of real numbers (x, y) with a unique real number $p(x, y) = x \cdot y$. The domain of this product function p is the set of all ordered pairs of real numbers, and the range of p is the set of all real numbers. In elementary calculus, the definite integral $\int_a^b$ is a function whose domain is the set of all real-valued functions that are Riemann integrable on the closed interval $[a, b]$, and whose range is the set of all real numbers. For, given an integrable function f, there corresponds to it a unique real number $\int_a^b f$, usually written $\int_a^b f(x)\, dx$. Because there are a great many different kinds of sets of mathematical objects, and functions defined on these sets, that are important in mathematics, a study of elementary set theory and of functions defined on abstract sets can provide a unifying foundation for a variety of mathematical subjects.

In this chapter we give a nonaxiomatic treatment of some important topics from set theory. The notions of function from one set into another and of relations between two sets will be discussed in detail. These two notions will then play a critical role in almost every topic discussed in the remainder of the chapter.

1. SETS AND MEMBERSHIP

By a set we shall mean a collection of objects. Each object of a particular set will be called an element of the set, a point of the set, or a member of the set. If

A is a set and x is an element of A, we shall use the notation $x \in A$ to indicate this fact. For example, if A is the set of all real numbers x such that $x^2 > 2$, then $5 \in A$. If an object x is not a member of the set A, we use the notation $x \notin A$ to indicate this fact. Thus, if A is the same set of numbers described previously, $\sqrt{2} \notin A$. If A is a set and B is a set and we write $A = B$, we mean that each element that is in A is also in B and that each element that is in B is also in A. Thus A and B are two names for the same set.

Now suppose that x is an object and we consider the set whose only element is x. It will be convenient to designate such a set with the symbol $\{x\}$. Thus $\{2\}$ is the set whose only element is 2. Note that $x \in \{x\}$. We make a clear distinction between the object x and the set $\{x\}$ whose only element is this object. This is the same distinction one makes, for example, between Mr. Jones and a committee of one whose only member is Mr. Jones.

Sometimes we can designate a set by listing all of its elements. If x and y are objects, we shall designate the set whose only elements are x and y by $\{x, y\}$. This notation is consistent with the notation $\{x\}$ described in the previous paragraph. Let us emphasize that both $\{x, y\}$ and $\{y, x\}$ denote the same set. Note that there is no notion of order involved here. Next, let n be a positive integer. Using the same kind of notation, we may write $\{1, 2, 3, \ldots, n\}$ to designate the first n positive integers in the set of all positive integers. Sometimes it is not convenient or even possible to designate a set by listing its elements explicitly. However, it might be possible to describe a set by some defining property. For example, consider the set of all real numbers x such that $0 \leq x \leq 1$. We may use $\{x:x$ is a real number and $0 \leq x \leq 1\}$ to designate this set. Actually, since it is clear from the statement $0 \leq x \leq 1$ that we are dealing with real numbers, the notation may be shortened to $\{x:0 \leq x \leq 1\}$.

More generally, the notation $\{x:x \ldots\}$ will be used to designate the set of all x such that $x \ldots$, where "$x \ldots$" will be replaced by a description of a property that defines the set. For example, "$\{x:x$ is a number and $-1 < x < 1\}$" reads "the set of all x such that x is a number and $-1 < x < 1$."

We shall also assume that there is a set called the *empty set* which has no elements. The empty set is designated by the symbol $\varnothing$. Thus, there is no y such that $y \in \varnothing$. Because of the availability of the empty set, it would be perfectly proper for someone to say, "Let S be the set of all real solutions of the equation $x^2 + x + 100 = 0$," even though this equation has no real solutions. The set S is the empty set.

EXERCISES: **SETS AND MEMBERSHIP**

In each of the following exercises we will assume that the sets involved are subsets of real numbers; thus, no explicit mention is made of the fact that the elements are real numbers.

1. List explicitly the elements of the set
$$\{x:x < 0 \text{ and } (x - 1)(x + 2)(x + 3) = 0\}.$$

2. List the elements of the following set.
$$\{x:3x - 1 \text{ is a multiple of } 3\}$$
(A number a is a multiple of 3 provided that $a = 3k$ for some integer k.)

3. Sketch on a number line each of the following sets.

(a) $\{x : |x - 1| \leq 3\}$
(b) $\{x : |x - 1| \leq 3 \quad \text{and} \quad |x| \leq 2\}$
(c) $\{x : |x - 1| \leq 3 \quad \text{or} \quad |x| \leq 2\}$

Note that in (b) each of the elements in the set must satisfy both of the inequalities, and that in (c) an element x is in the set if it satisfies at least one of the inequalities. We shall always use the connective *or* in that sense.

2. SOME REMARKS ON THE USE OF THE CONNECTIVES *and, or, implies*

First of all, we assume that each statement p is either *true* or *false* (but not both). We sometimes speak of a statement as having *true* or *false* as its *truth value*. Thus, a false statement has *false* as its truth value. This assumption would be a real restriction in ordinary language. For example, consider the statement, "Professor Robinson is a good lecturer." A classification for this statement which is more appropriate than *true* or *false* might be *never, seldom, usually, almost always, always*.

If p is a statement and q is a statement, the compound statement *p and q* is assigned the truth value as indicated by the following table. Notice that the assignment below agrees with common usage in ordinary conversation.

TRUTH VALUES FOR *p and q*

p	q	*p and q*
True	True	True
True	False	False
False	True	False
False	False	False

If p, q are statements, we will call the compound statement *p or q* true provided that at least one of the statements is true. Otherwise we will give the statement *p or q* a truth value of false. We indicate this agreement schematically as follows:

TRUTH VALUES FOR *p or q*

p	q	*p or q*
True	True	True
True	False	True
False	True	True
False	False	False

The reader is already familiar with the fact that the conditional *if p, then q* is used frequently in mathematics. In an *if p, then q* statement, the p is called the *hypothesis* or the *assumption*, and the q is called the *conclusion*. In mathematics, when the truth of the conditional *if p, then q* is known, the truth of q is inferred, provided that the truth of p is known. However, *nothing can be inferred about q if it is known that p is false*. For example, if one starts with the incorrect assumption that $-1 = 1$, one can conclude, by squaring both sides, that $1 = 1$, a true statement. On the other hand, by beginning with $-1 = 1$, one can obtain the false statement $0 = 2$ by adding 1 to both sides. Consistent with mathematical usage, we will label the conditional *if p, then q* as false in case q is false when p is true. Otherwise, the conditional is labeled as true. Thus we have the following truth table for the conditional, *if p, then q*.

TRUTH VALUES FOR THE CONDITIONAL *if p, then q*

p	q	*if p, then q*
True	True	True
True	False	False
False	True	True
False	False	True

We must distinguish carefully between the truth of an *if p, then q* statement and the truth of the conclusion q. For we see from the truth table that when the hypothesis p is false, the statement *if p, then q* is true for an arbitrary q.

We shall use *p implies q* to be synonymous with the statement *if p, then q*. Other common ways of saying the same thing are *p is a sufficient condition for q*, and *q is a necessary condition for p*. The reader should try to convince himself that the language of the synonymous statements for the conditional is reasonable.

From usage in previous mathematics courses, the reader should be familiar with the term *converse*. By the *converse* of an implication *if p, then q*, we mean the implication *if q, then p*.

It is apparent that the truth values for the *negation* of a statement p, abbreviated *not p*, should be as indicated in the next table.

TRUTH VALUES FOR *not p*

p	*not p*
True	False
False	True

In the previous discussion we have considered a statement p and have called its negation *not p*. In actual situations in mathematics, it is often not enlightening for one to negate a statement simply by saying that it is false. It is often better to write the negation in a more useful form. For example, consider the following statement:

For each x, if x is a real number, then $(x + 1)^2 = x^2 + 2x + 1$.

For the negation of this statement, we may write

For at least one real number x, it is not true that $(x + 1)^2 = x^2 + 2x + 1$.

The point here is that the negation of the statement that something always happens is not that it never happens, but rather that there is at least one instance for which it does not happen.

From the truth values assigned in the preceding paragraphs, the truth values of more complicated statements can be determined. For example, consider the following table for the compound statement indicated. The columns are numbered in the order in which they were completed.

$$(p \text{ implies } q) \text{ and } (q \text{ implies } p)$$

(1)	(2)	(3)	(4)	(5)
p	q	p implies q	q implies p	$(p$ implies $q)$ and $(q$ implies $p)$
True	True	True	True	True
True	False	False	True	False
False	True	True	False	False
False	False	True	True	True

We see that the compound statement *(p implies q) and (q implies p)* is given a value of *true* when p and q both have the same truth values. Otherwise it is given the value *false*.

2.1. Definition. Equivalent statements. *Suppose that p and q are statements for which the compound statement*

2.1(a). $\qquad\qquad (p \text{ implies } q) \quad \text{and} \quad (q \text{ implies } p)$

is true for all values under consideration. We then say that p and q are equivalent statements.

In mathematical texts and papers, the language

p if and only if q

is often used for the longer statement in 2.1(a). Note that the *p if q* simply means *q implies p*; *p only if q* refers to the fact that *p implies q*. Further, in view of alternate ways of expressing the conditional, the statement *(p implies q) and (q implies p)* may be expressed by the statement *p is a necessary and sufficient condition for q*.

Suppose that p and q are statements and that we wish to prove that *p implies q*. One of the methods of proof with which the reader is familiar is the so-called indirect proof. Essentially, that method of proof consists of assuming that the conclusion q is false, and then proving that this assumption leads to a statement which is inconsistent with the hypothesis, thus arriving at a contradiction. In using this method, we are essentially assuming that the statement *(not q implies not p)* implies the statement *(p implies q)*. Moreover, it can be shown that *(p implies q) implies (not q implies not p)* is true for all statements p and q. Thus, the statements *p implies q* and *not q implies not p* are equivalent for all statements p and q. We shall discuss this further after the next definition.

2.2. Definition. Contrapositive. *The statement*

not q implies not p

is called the contrapositive of the statement

p implies q.

To prove that the statement *p implies q* is equivalent to its contrapositive, we need to show that the following two statements are both true in all cases.

(a) (*p implies q*) *implies* (*not q implies not p*)

(b) (*not q implies not p*) *implies* (*p implies q*)

The truth of (a) is shown schematically by the following table. As in the previous table, the columns are numbered in the order in which they were completed.

(p implies q) implies (not q implies not p)

(1)	(2)	(3)	(4)	(5)	(7)	(6)
p	*q*	*not q*	*not p*	(*p implies q*)	*implies*	(*not q implies not p*)
T	T	F	F	T	T	T
T	F	T	F	F	T	F
F	T	F	T	T	T	T
F	F	T	T	T	T	T

The verification of part (b) is left as an exercise for the reader. See Exercise 1, page 6 .

Note that the contrapositive of *p implies q*, just discussed, should not be confused with the converse of *p implies q*. Recall that the converse of *p implies q* is *q implies p*, which is not equivalent to *p implies q*.

In the previous discussion we have indicated that the statement (*p implies q*) *implies* (*not q implies not p*) is true for all possible cases involving the truth values of statements *p* and *q*. This is an example of a *tautology*. Suppose *S* is a sentence involving statements $p_1, p_2, \ldots, p_n$. Suppose furthermore that *S* is true for all possible cases involving the truth values of $p_1, p_2, \ldots,$ and p_n. *S* is then called a *tautology*.

EXERCISES: THE USE OF CONNECTIVES *and, or, implies*

1. Demonstrate by means of a table showing truth values that the following is a true statement for any choice of *p* and *q*. Thus, show that it is a tautology.

    (not q implies not p) implies (p implies q)

2. Show by means of a truth table that the statement ((*p implies q*) *and* (*q implies r*)) *implies* (*p implies r*) is a tautology.

3. Show by means of a truth table that (*p and q*) *implies* (*p or q*) is a tautology.

4. Suppose that *p* and *q* are statements such that (*p and q*) is a false statement. Does it follow that the statement

    (p is false) or (q is false)

    is a true statement?

5. Negate the following statement: *If two angles of a triangle have equal measure, then the length of two sides of that triangle are equal.*

6. Write the contrapositive of the statement in Exercise 5.

7. Write the converse of the statement in Exercise 5.

8. Write the contrapositive of the following statement: *If a person belongs to Committee A, then he must be a member of Committee B and he must be a member of Committee C.*

9. Write the contrapositive of the following statement. If $x \in A$ and $x \in B$, then $x \in C$.

3. SUBSETS

3.1. Definition. Subset of a set. *Suppose that A and B are sets such that each element of A is an element of B. We shall then say that A is a subset of B, or that A is contained in B (written $A \subset B$).*

If $A \subset B$ we shall also say that *B contains A* $(B \supset A)$. It is to be noted that if $A \subset B$ and $B \subset A$, then $A = B$.

Notice that if A is a set, then $A \subset A$. All other subsets of A are referred to as *proper* subsets of A.

3.2. Definition. Proper subset. *If A and B are sets such that $B \subset A$ and $A \neq B$, then B is said to be a proper subset of A. If B is a proper subset of A, then B is said to be properly contained in A.*

(It should be pointed out to the reader that some texts use the notation $A \subseteq B$ to indicate that A is a subset of B and $A \subset B$ is used to indicate that A is a proper subset of B. We shall *not* follow this practice in this book.)

In the previous section we noted that if p and q are statements and if p is false and q is true or false, then the statement p *implies* q is a true statement. This fact is used in proving the following statement.

3.3. *Remark.* The empty set $\varnothing$ is a subset of every set.

PROOF. Suppose that A is a set. Then from the previous paragraph, $x \in \varnothing$ implies that $x \in A$ is a true statement, since $x \in \varnothing$ is a false statement. Then from the definition of subset it follows that $\varnothing$ is a subset of A.

In our previous discussion we referred to $\varnothing$ as *the* empty set. This would not be appropriate unless there were only one empty set. We show next that by virtue of the definitions that we have agreed to accept, there is indeed only one empty set. To see this, notice that if $\varnothing_1$ and $\varnothing_2$ are empty sets, then the previous argument in 3.3 would apply to each. This fact would imply that $\varnothing_1 \subset \varnothing_2$ and $\varnothing_2 \subset \varnothing_1$. However, it would then follow that $\varnothing_1 = \varnothing_2$.

4. UNION AND INTERSECTION OF SETS

We define next what is meant by the union and the intersection of two sets. The reader should note the critical role that the connectives—*and, or*—play in the definition.

4.1. Definition. Union of sets. *Suppose that A and B are sets. Then the union of A and B (written A $\cup$ B) is defined to be the set of all objects x such that x $\in$ A or x $\in$ B. Using set notation, we may write*

$$A \cup B = \{x : x \in A \text{ or } x \in B\}.$$

4.2. Definition. Intersection of sets. *Suppose that A and B are sets. Then the intersection of A and B (written A $\cap$ B) is the set of all objects x such that x $\in$ A and x $\in$ B. In set notation, we may write this as*

$$A \cap B = \{x : x \in A \text{ and } x \in B\}.$$

Thus, if A is the set of all solutions of the equation $(x - 1)(x - 2) = 0$ and B is the set of all solutions of the equation $(x - 1)(x - 3) = 0$, then $A \cup B = \{1, 2, 3\}$ and $A \cap B = \{1\}$.

If $A \cap B = \varnothing$ we say that A and B are *disjoint*.

EXERCISES: UNION AND INTERSECTION OF SETS

The reader is already familiar with the fact that the points in the plane can be thought of as ordered pairs of real numbers. All sets in the following exercises are subsets of the plane in which a rectangular coordinate system has been introduced.

1. Let G_1 be the graph of the equation $x^2 + y^2 = 16$, and let G_2 be the graph of the equation $x^2 - y^2 = 1$. Sketch the sets $G_1 \cup G_2$ and $G_1 \cap G_2$.

2. We define the sets A, B, and C as follows: $A = \{(x, y) : x^2 + y^2 \leq 9\}$, $B = \{(x, y) : x + y \geq 3\}$, $C = \{(x, y) : x \geq 0\}$.

    Draw sketches of each of the following sets: (a) $A \cup (B \cup C)$, (b) $A \cap (B \cup C)$, (c) $(A \cap B) \cup (A \cap C)$, (d) $(A \cup B) \cup C$, (e) $A \cup (B \cap C)$, (f) $(A \cup B) \cap (A \cup C)$.

3. Let $A = \{(x, y) : x + y \leq 5\}$, $B = \{(x, y) : x + y \geq 3\}$, $C = \{(x, y) : x \geq 3\}$, and $D = \{(x, y) : y \geq 3\}$.

    Draw a sketch of each of the following sets: (a) $(A \cap B) \cap C$, (b) $[(A \cap B) \cap C] \cap D$.

5. COMPLEMENTATION

5.1. Definition. Complement of a set. *If A and B are sets, then we define A $-$ B (A minus B) as follows:*

$$A - B = \{x : x \in A \text{ and } x \notin B\}.$$

A $-$ B is said to be the complement of B with respect to A. Note that if B $\cap$ A $= \varnothing$, then A $-$ B $=$ A. Also, if B $\supset$ A, then A $-$ B $= \varnothing$.

Often in a particular study in mathematics, the entire discussion is centered around some "universal" set. For example, the setting of a discussion might be the set of all real numbers, **R**. In a situation such as this, rather than referring to a set **R** — S as the complement of S relative to **R**, we may simply refer to **R** — S as the complement of S, the phrase "relative to **R**" being understood. In that case, the shorter notation $\sim S$ would be used instead of **R** — S. Thus, if our universal set in a particular discussion is the set of all real numbers **R** and I is the set of all irrational real numbers, then $\sim I$ is the complement of I, or the set of all rational real numbers.

EXERCISES: COMPLEMENTATION

In this set of exercises we take our universal set to be the plane. Thus $\sim A$ will stand for the complement of A with respect to the plane. Further, A, B, and C are taken to be the same sets as sets A, B, and C in Exercise 2, page 8.

1. Sketch each of the following sets: (a) $\sim(A \cap B)$, (b) $(\sim A) \cup (B)$, (c) $\sim(A \cup B)$, (d) $(\sim A) \cap (B)$, (e) $C - A$, (f) $\sim(A \cap C)$, (g) $(\sim A) \cup (\sim B)$, (h) $(\sim A) \cap (A)$, (i) $C - (A \cup B)$, (j) $(C - A) \cap (C - B)$, (k) $\sim(\sim A)$.

2. On the basis of some of the sketches made in the previous exercise, formulate a proposition about relations that exist concerning complementation, union, and intersection. Try out your conjecture on other examples. In subsequent exercises you will be asked to try to prove such conjectures.

6. SET IDENTITIES AND OTHER SET RELATIONS

In Exercise 2, this page, you might have conjectured that if A and B are subsets of a set X, then $(X - A) \cup (X - B) = X - (A \cap B)$. Assuming that this statement holds in general, we call such a statement a *set identity*.

Suppose that X and Y are sets and we wish to prove that $X = Y$. One way of doing this is to show that $X \subset Y$ and $Y \subset X$. Thus, we would let $x \in X$ and show that $x \in Y$. Further, we would let $x \in Y$ and show that $x \in X$. Often, of course, a set identity can be established by making use of previously proved identities. As an illustration, let us consider the following statement.

6.1. Theorem. *Let X, A, and B be sets. Then*

$$(X - A) \cup (X - B) = X - (A \cap B).$$

PROOF. To prove this identity, we will first show that $(X - A) \cup (X - B) \subset X - (A \cap B)$. Toward this end, let $x \in (X - A) \cup (X - B)$. Then $x \in (X - A)$ or $x \in (X - B)$. Suppose that $x \in (X - A)$. Then $x \in X$ and $x \notin A$. But if $x \notin A$, then $x \notin (A \cap B)$. Thus $x \in X - (A \cap B)$. Similarly we can show that if $x \in (X - B)$, then $x \in X - (A \cap B)$. Thus we have shown that $(X - A) \cup (X - B) \subset X - (A \cap B)$.

Next, we show that $X - (A \cap B) \subset (X - A) \cup (X - B)$. To see this, let $x \in X - (A \cap B)$. Then $x \in X$ and $x \notin A \cap B$. Since $x \notin (A \cap B)$, it must be true that $x \notin A$ or that $x \notin B$. Suppose first that $x \notin A$. Then $x \in X - A$ and, hence, $x \in (X - A) \cup (X - B)$. If the other possibility is true, i.e., if $x \notin B$, then $x \in X - B$ so that $x \in (X - A) \cup (X - B)$. Hence, we have verified that $X - (A \cap B) \subset (X - A) \cup (X - B)$, and the proof has been completed.

In Exercise 2(b) and (c), page   8 , the reader might have been led to make the conjecture that $A \cap (B \cup C) = (A \cap B) \cup (A \cap C)$ is an identity. We show next that this is indeed an identity.

6.2.   Theorem.   *Let A, B, and C be sets. Then,*

$$A \cap (B \cup C) = (A \cap B) \cup (A \cap C).$$

PROOF.   We show first that $A \cap (B \cup C) \subset (A \cap B) \cup (A \cap C)$. In order to do so, let $x \in A \cap (B \cup C)$. Then $x \in A$, and (i) $x \in B$ or (ii) $x \in C$. If (i) holds, recalling that $x \in A$, it follows that $x \in A \cap B$. Similarly, if (ii) holds, then $x \in A \cap C$. In either case, $x \in (A \cap B) \cup (A \cap C)$.

To show that $(A \cap B) \cup (A \cap C) \subset A \cap (B \cup C)$, let $x \in (A \cap B) \cup (A \cap C)$. Then $x \in A \cap B$ or $x \in A \cap C$. Since $A \cap B \subset A \cap (B \cup C)$ (see Exercise 1(a), this page) and $A \cap C \subset A \cap (B \cup C)$, it follows that in either case $x \in A \cap (B \cup C)$. This completes the proof.

Summarizing our discussion, if we want to show that $X \subset Y$, where X and Y are sets, we consider an arbitrary element x in X and show that x is also in Y. We may show that $X = Y$ by showing that $X \subset Y$ and $Y \subset X$.

As pointed out previously, once an identity has been proved, it can be used to prove others. Or, we may use previously proved identities to see whether we may rewrite expressions in a way that will perhaps suit our purposes better. For example, consider the set $X - [A \cap (B \cap C)]$. Using Theorem 6.1, we may write $X - [A \cap (B \cap C)]$ as $(X - A) \cup [X - (B \cap C)]$. Then using 6.1 again, we have $X - [A \cap (B \cap C)] = (X - A) \cup [(X - B) \cup (X - C)]$.

EXERCISES: SET IDENTITIES AND OTHER SET RELATIONS

In the following exercises, all sets are assumed to be subsets of some universal set **U**. Thus $\sim A$ will stand for $\mathbf{U} - A$.

1.   Prove that if $A \subset B$, then:
(a) $A \cap C \subset B \cap C$
(b) $\sim B \subset \sim A$
(c) $A \cap B = A$
(d) $A \cup C \subset B \cup C$

2.   Verify that each of the following is an identity.
(a) $A \cup \varnothing = A$
(b) $A \cap \varnothing = \varnothing$
(c) $A \cap A = A$
(d) $A \cup A = A$
(e) $(A \cup B) \cup C = A \cup (B \cup C)$
(f) $(A \cap B) \cap C = A \cap (B \cap C)$
(g) $A \cup (B \cap C) = (A \cup B) \cap (A \cup C)$
(h) $X - (A \cup B) = (X - A) \cap (X - B)$
(i) $A \cap \sim A = \varnothing$
(j) $A \cup \sim A = \mathbf{U}$

3. Prove that if $A \subset C$ and $B \subset C$, then $A \cup B \subset C$.

4. Prove that if $A \subset B$ and $A \subset C$, then $A \subset B \cap C$.

Note that in Theorem 6.1 and Exercise 2(h), if X happens to be a universal set, then these identities take the form

$$\sim(A \cap B) = (\sim A) \cup (\sim B)$$
$$\sim(A \cup B) = (\sim A) \cap (\sim B).$$

7. COUNTEREXAMPLES

In a normal working situation a mathematician is often led to a general statement from some particular examples. If he cannot prove the general statement, he then looks for an example to show that the statement is not always true. Such an example is called a *counterexample* to the statement. Often the counterexample gives him insight into what additional hypothesis needs to be added to force the conclusion to be valid. After he has a correct version of a statement, he often wonders if he has as "good" a theorem as he is able to get. Once again, counterexamples come into play. The mathematician might then try to change various parts of the hypothesis and seek counterexamples to show that he may not change the hypothesis in the way in which he tried. Sometimes at first one is not able to prove a statement which in fact turns out to be correct. Often in attempting to construct the counterexample, the mathematician obtains insight into how to prove the statement. The role of construction of counterexamples in creative mathematics should be emphasized to the student of mathematics. Although some individuals seem much more adept at the technique than others, the skill can be developed by practice. In many of the excercises in the remainder of the text the reader will be asked to determine whether or not the particular statement is necessarily true, and to justify his answer with a proof of the statement or a counterexample.

EXERCISES: COUNTEREXAMPLES

In each of the following exercises state whether the statement is necessarily true. Assume that A, B, and C are subsets of a universal set **U**. Justify with a proof or a counterexample.

1. If $A \cup C = B \cup C$, then $A = B$

2. $(A \cup B) - B = A$

3. $(A - B) \cup B = A$

4. $\sim(A - B) = \sim(A \cap \sim B)$

5. $\sim(\sim(\sim A)) = \sim A$

6. $A \cup (B - C) = (A \cup B) - C$

7. $\sim(A - B) = (\sim A) \cup B$

8. If $A - B = C - B$, then $A = C$

9. $A - (B \cap C) = (A - B) \cap (A - C)$

8. COLLECTIONS OF SETS

So far we have defined what is meant by the union of two sets and the inter-section of two sets. We shall have need to form the union and intersection of many (rather than just two) sets. To do this, we shall extend the definition of *union* to arbitrary collections of sets, and the definition of *intersection* to arbitrary non-empty collections of sets. Note that if we say that $\mathcal{K}$ is a nonempty collection of sets, then $\mathcal{K}$ has at least one element in it and each element in $\mathcal{K}$ is itself a set.

8.1. Definition. Union and intersection of collections of sets. *If $\mathcal{K}$ is a collection of sets, then the union of elements of $\mathcal{K}$ (written $\bigcup \mathcal{K}$) is defined by*

$$\bigcup \mathcal{K} = \{x : x \in K \text{ for at least one } K \in \mathcal{K}\}.$$

If $\mathcal{K}$ is a nonempty collection of sets, then the intersection of elements of $\mathcal{K}$ (written $\bigcap \mathcal{K}$) is defined by

$$\bigcap \mathcal{K} = \{x : x \in K \text{ for each } K \in \mathcal{K}\}.$$

These definitions extend the notions of union and intersection given previously. For example, suppose that $\mathcal{K} = \{A, B\}$, where A and B are sets; then $\bigcup \mathcal{K} = A \cup B$ and $\bigcap \mathcal{K} = A \cap B$. Suppose that $\mathcal{K} = \{A, B, C\}$. We will also agree in that case to write $\bigcup \mathcal{K}$ as $A \cup B \cup C$ and $\bigcap \mathcal{K}$ as $A \cap B \cap C$. It is easy to see that $A \cup B \cup C = (A \cup B) \cup C = A \cup (B \cup C)$, and that $A \cap B \cap C = (A \cap B) \cap C = A \cap (B \cap C)$.

Suppose that $\mathcal{K}$ is a collection of sets $K_1, K_2, \ldots, K_n$. Then using the nota-tion introduced previously, we may write $\mathcal{K} = \{K_i : i = 1, 2, \ldots, n\}$. Thus, we can write

$$\bigcup \mathcal{K} = \bigcup \{K_i : i = 1, 2, \ldots, n\}$$

and

$$\bigcap \mathcal{K} = \bigcap \{K_i : i = 1, 2, \ldots, n\}.$$

Sometimes $\bigcup \{K_i : i = 1, 2, \ldots, n\}$ is also written as $K_1 \cup K_2 \cup \cdots \cup K_n$, and $\bigcap \{K_i : i = 1, 2, \ldots, n\}$ is written as $K_1 \cap K_2 \cap \cdots \cap K_n$.

When we consider the collection of sets $\mathcal{K} = \{K_i : i = 1, 2, \ldots, n\}$, the sub-script i is known as an index. In the example just considered, i "runs through" the first n positive integers. The set of the first n positive integers in that case is called an *indexing set* for the collection $\mathcal{K}$. Sometimes convenient indexing sets occur when we consider infinite collections. For example, suppose that for each positive real number r, we let S_r be the circle $\{(x, y) : x^2 + y^2 = r^2\}$ in the plane, with center at $(0, 0)$ and radius equal to r. Let $\mathcal{K}$ be the collection of all such circles. Then a convenient notation for $\mathcal{K}$ is $\{S_r : r \in \text{set of all positive real numbers}\}$. Here the indexing set for $\mathcal{K}$ is the set of all positive real numbers. We now give a definition of an indexing set for a collection of sets.

8.2. Definition. Indexed collection. *Suppose that Δ is a set, and that for each α in Δ, there corresponds a set K_α. Then the collection of sets $\mathcal{K} = \{K_\alpha : \alpha \in \Delta\}$ is an indexed collection of sets, and Δ is said to be an indexing set for the collection $\mathcal{K}$. (Note that $\alpha \neq \beta$ does not imply that $K_\alpha \neq K_\beta$.)*

8.3.   Definition.   The power set of a set.   *Let S be a set. By $\mathscr{P}(S)$, the power set of S, we shall mean the collection of all subsets of S.*

8.4.   Notation for some important sets.   *At this point it will be convenient to introduce notation for some sets that will occur frequently in the text.*

$$\mathbf{R} = \text{the set of all real numbers}$$
$$\mathbf{Q} = \text{the set of all rational numbers}$$
$$\mathbf{Z} = \text{the set of all integers}$$
$$\mathbf{P} = \text{the set of all positive integers}$$
$$\mathbf{P}_n = \{1, 2, 3, \dots, n\}$$
$$\mathbf{R}_+ = \text{the collection of all positive real numbers.}$$

Also, recall from calculus that if a and b are real numbers such that $a < b$, then (a, b) is the open interval $\{x : a < x < b\}$ and $[a, b]$ is the closed interval $\{x : a \leqq x \leqq b\}$.

EXERCISES: COLLECTIONS OF SETS

1.   Suppose that A, B, and C are the following subsets of the plane:
$A = \{(x, y) : x^2 + y^2 \leq 16\}$, $B = \{(x, y) : x \geq 0$ and $y \leq 0\}$,
$C = \{(x, y) : y \leq x\}$. If $\mathscr{H}$ is the collection of sets $\{A, B, C\}$,
sketch each of the following sets:
   (a) $\bigcap \mathscr{H}$
   (b) $\bigcup \mathscr{H}$
   (c) $\bigcup \mathscr{H} - \bigcap \mathscr{H}$

2.   Recall that $\mathbf{P}$ is the symbol for the set of positive integers.
Suppose that for each $n \in \mathbf{P}$, we let $A_n = \{x : x \geq n\}$. Describe
the sets $\bigcup \{A_n : n \in \mathbf{P}\}$ and $\bigcap \{A_n : n \in \mathbf{P}\}$.

3.   Suppose that for each $n \in \mathbf{P}$, K_n is a nonempty set such that
$K_{n+1} \subset K_n$. Let $\mathscr{H} = \{K_n : n \in \mathbf{P}\}$.
      In each of the following, if the statement is necessarily
true, say so and justify your answer. If the statement is not
necessarily true, give a counterexample to justify your answer.
   (a) $\bigcup \mathscr{H} = K_1$
   (b) $\bigcap \{K_i : i = 1, 2, 3, \dots, n\} = K_n$
   (c) $\bigcap \mathscr{H} \neq \varnothing$

4.   For each real number $r > 0$, let $L_r = \{x : x \geq r\}$. Sketch the
sets $\bigcup \{L_r : r > 0\}$ and $\bigcap \{L_r : r > 0\}$ on a number line. If a
set happens to be empty, say so.

5.   Let U be a set and let $\mathscr{H}$ be a nonempty collection of subsets of
U. $\sim$ will signify the complement with respect to U. Prove the
following set identities. The identities are quite important and
are known as De Morgan's Laws.
   (a) $\sim (\bigcup \{K : K \in \mathscr{H}\}) = \bigcap \{\sim K : K \in \mathscr{H}\}$
   (b) $\sim (\bigcap \{K : K \in \mathscr{H}\}) = \bigcup \{\sim K : K \in \mathscr{H}\}$

6.   Let $S = \{1, 2, 3, 4, 5\}$ and let $\mathscr{P}(S)$ be the power set of S.
List the elements in $\mathscr{P}(S)$.

9. CARTESIAN PRODUCT

The student is already familiar with the notion of an ordered pair from analytic geometry.   When the pair of so-called Cartesian coordinates (x, y) is used to represent a point in the plane, there are distinct roles played by the "first" co-ordinate and the "second" coordinate.   We note that $(a, b) = (c, d)$ if and only if $a = c$ and $b = d$.   Thus, one represents the plane as the collection of all ordered pairs (x, y), where x and y are real numbers.   This notion is the parent of the more general notion of the Cartesian product of two sets to be discussed in this section.

9.1.   Definition.   Cartesian product of two sets.   *Let A and B be sets.   Then by $A \times B$, the Cartesian product of A and B, we shall mean the set of all ordered pairs (a, b) for $a \in A$ and $b \in B$.   Thus*

$$A \times B = \{(a, b) : a \in A \text{ and } b \in B\}.$$

Furthermore A^2 will denote the set $A \times A$.

We emphasize that $A \times B$ is not, in general, the same as $B \times A$.   For example, let $A = \{1, 2, 3\}$ and $B = \{1, 2\}$. $A \times B = \{(1, 1), (2, 1), (3, 1), (1, 2), (2, 2), (3, 2)\}$.   On the other hand, $B \times A = \{(1, 1), (1, 2), (1, 3), (2, 1), (2, 2), (2, 3)\}$.

[Note that the notation (a, b) for the ordered pair is the same as that used for an open interval with endpoints a and b, if a and b are real numbers.   This is not likely to cause trouble, since it will be clear from the context as to which is meant.   If it is not clear, the words "ordered pair" or "interval" should be used to modify (a, b).]

9.2.   Theorem.   *Suppose that A, B, and C are sets.   Then*

$$A \times (B \cap C) = (A \times B) \cap (A \times C).$$

PROOF.   First we show that $A \times (B \cap C) \subset (A \times B) \cap (A \times C)$.   To see this, suppose that $(x, y) \in A \times (B \cap C)$.   Then $x \in A$ and $y \in B \cap C$.   Hence, $x \in A$ and $(y \in B$ and $y \in C)$.   Thus, $(x \in A$ and $y \in B)$ and $(x \in A$ and $y \in C)$.   It then follows that $(x, y) \in A \times B$ and $(x, y) \in A \times C$.   Consequently, $(x, y) \in (A \times B) \cap (A \times C)$.   We have shown that $A \times (B \cap C) \subset (A \times B) \cap (A \times C)$.

We leave as an exercise the proof that $(A \times B) \cap (A \times C) \subset A \times (B \cap C)$.

EXERCISES: CARTESIAN PRODUCT

1.   Suppose that $A \subset B$ and C is a set.   Prove that $A \times C \subset B \times C$.

2.   Let $A = \{1, 2, 3\}$, $B = \{a, b\}$, and $C = \{\alpha, \beta\}$.   List the ele-ments in each of the following sets: (a) $A \times (B \cup C)$, (b) $(A \times B) \cup (A \times C)$, (c) $(A \cup B) \times C$, (d) $(A \times C) \cup (B \times C)$.

3.   Are any of the sets in Exercise 2 the same?   If so write the set identities that are suggested by your observations.   Try to prove your conjectures.

4. Suppose that A is a set consisting of five elements and B is a set consisting of three elements. How many elements does the set $A \times B$ have? The set $B \times A$?

5. Suppose that A is a set consisting of m elements and B is a set consisting of n elements, where m and n are positive integers. How many elements are there in $A \times B$?

6. Suppose that A consists of three elements, B consists of four elements and C consists of two elements. How many elements are there in the set $(A \times B) \times C$?

10. FUNCTIONS

The reader is already familiar with the notion of function from elementary texts. As often used in calculus texts, for example, a function f is regarded as a correspondence between one set of numbers called the domain of the function and another set of numbers called the range of the function such that to each x in the domain there corresponds one and only one number $f(x)$ in the range. In that approach, the word correspondence is generally not defined. However, the notion of function is so primitive that the concept is probably very clear in the reader's mind.

With the notion of function in mind, the reader should recall that one can define the graph of a function f as the collection $\{(x, y) : x \in (\text{domain of } f)$ and $y = f(x)\}$. (Do not confuse this abstract notion of the "graph of a function" with the "sketch" of the graph on paper.)

Notice that if we know the correspondence, i.e., if we know what $f(x)$ is for each x, we can tell whether (x, y) is in the graph. Furthermore, if we know that (a, b) is a point in the graph, we know that $f(a) = b$. This suggests that we might define function in terms of what we have been speaking of as its graph. There are certain advantages to this approach, and we shall use it in this text. However, the reader should keep in mind that the essential feature behind the concept is the notion that "to each x there is a uniquely determined $f(x)$."

10.1. Definitions. Function, domain, range. *A function f is a set of ordered pairs such that if $(x, y_1) \in f$ and $(x, y_2) \in f$, then $y_1 = y_2$. (Thus, no two distinct pairs making up the function have the same first element.) If f is a function, then the set of all x such that x is the first coordinate of a pair $(x, y) \in f$ is called the domain of f. The set of all y such that y is the second coordinate of a pair $(x, y) \in f$ is called the range of f.*

The *domain of f* will be denoted by Dom f and the *range of f* will be denoted by Range f. For each $x \in \text{Dom } f$, $f(x)$ will denote the unique element in Range f such that $(x, f(x)) \in f$.

Although to us a function is defined as a collection of ordered pairs with a certain additional property, "single-valuedness," we will often find it convenient to define a particular function f by giving Dom f and a rule that specifies explicitly the functional value for each $x \in \text{Dom } f$. For example, let f be the function whose domain is $\{x: -1 \leqq x \leqq 1\}$, and whose functional value $f(x)$ for each x in this closed interval $[-1, 1]$ is given by the rule $f(x) = -x^2 + 1$. Thus, $f = \{(x, y): -1 \leqq x \leqq 1 \text{ and } y = -x^2 + 1\}$.

It should be emphasized that, from our point of view, if two different function-defining rules give the same set of ordered pairs, the functions so defined are the same. For example, let f^* be the function whose domain is $\{x: -1 \leqq x \leqq 1\}$ and whose functional value $f^*(x)$ for each $x \in [-1, 1]$ is given by the rule $f^*(x) = \sqrt{x^4 - 2x^2 + 1}$, where $\sqrt{}$ is the symbol for nonnegative square root. From a computational standpoint this rule is different from the rule defining f in the last paragraph. However, if we note that for $x \in [-1, 1]$, $-x^2 + 1$ is the positive square root of $x^4 - 2x^2 + 1$, we see that the two rules define the same set of ordered pairs for the domain specified. Thus, $f = f^*$.

Frequently in elementary texts, a problem is posed something like this: "Consider the function $y = x^2$." What is meant by such a statement is this: "Consider the function f given by the rule $y = f(x) = x^2$ with the domain being the real number system." A statement such as "Consider the function $y = 1/x$" is probably intended to convey that a function f is to be considered for which $f(x) = 1/x$ with Dom f being the set of real numbers x for which $1/x$ is meaningful.

We should note that the function f given by the rule $f(x) = x^2$ for each real x is not the same as the function f^* given by $f^*(x) = x^2$ for each real $x \in [-1, 1]$. From our point of view f^* is a proper subset of f. In a case like this, we shall also say that f^* is the *restriction* of f to $[-1, 1]$. This concept of the restriction of a function will be defined formally later.

We note abstractly that if $x \in \text{Dom } f$, then $f(x)$ is simply the element in Range f such that $(x, f(x)) \in f$. However, much of the language currently used concerning functions conveys the notion that "f *transforms* x into $f(x)$," or that "f *maps* x onto $f(x)$." In some of our investigations it will be helpful to emphasize the ordered pair aspect of function. At other times it will be more helpful to think in terms of the mapping or transformation point of view. For this reason we will use the words *mapping*, *map*, and *transformation* as synonyms for *function*.

10.2. Definition. Function from X into Y. *Suppose that f is a function whose domain is X and whose range is contained in but is not necessarily all of Y. Then f is said to be a function (mapping, map, or transformation) from X into Y.*

If f is a function from X into Y we also say that f *maps* or *carries* X into Y. Rather than writing "f is a function from X into Y," we often write "$f: X \to Y$ is a function," which we read "f from X into Y is a function."

10.3. Definitions. Onto mappings or surjections. *Suppose that the function $f: X \to Y$ is such that the range of f is all of Y. Then f is said to be a function from X onto Y or, alternatively, f is said to be a surjection from X to Y. In such a case we will write more briefly, "$f: X \to Y$ is an onto mapping" or "$f: X \to Y$ is a surjection."*

The sine function defined on the real line $\mathbf{R}$ maps $\mathbf{R}$ into $\mathbf{R}$ but maps $\mathbf{R}$ onto $[-1, 1]$. Thus, we would say that sine$: \mathbf{R} \to [-1, 1]$ is a surjection, although sine$: \mathbf{R} \to \mathbf{R}$ is not. Thus, we see that whether a function f on X into Y is a surjection depends not only on the function f but also on the set Y.

10.4. Definitions. One-to-one or injective function; bijection. *A function f is said to be a one-to-one or injective function, or an injection, provided that for all x and y in Dom f,*

$$f(x) = f(y) \text{ implies that } x = y.$$

A map $f: X \to Y$ is said to be a bijection if it is both an injection and a surjection.

If f is a one-to-one function from X onto Y, it is sometimes referred to as a *one-to-one correspondence* between X and Y.

10.5. Examples. ⁷ Let f, g, and h be the functions defined on **R** and given by $f(x) = x \sin x$, $g(x) = x^3$, and $h(x) = \dfrac{x}{1 + |x|}$. The function f maps **R** onto **R** ($f: \mathbf{R} \to \mathbf{R}$ is a surjection). Note also that $g: \mathbf{R} \to \mathbf{R}$ is a one-to-one and onto mapping ($g: \mathbf{R} \to \mathbf{R}$ is a bijection). Moreover, the mapping $h: \mathbf{R} \to \mathbf{R}$ is not a bijection, but $h: \mathbf{R} \to (-1, 1)$ is a bijection.

10.6. Definition. The image of a subset of the domain. *Suppose that $f: X \to Y$ is a function and $A \subset X$. The set $f[A]$, the image of A under f, is defined to be the following subset of Y:*

$$f[A] = \{y : y = f(a) \text{ for at least one } a \in A\}.$$

Often "$f[A]$" is read as "f of A" rather than the longer "the image of A under f."

10.7. Example. Let $f: \mathbf{R} \to \mathbf{R}$ be given by $f(x) = x^2$. Let A be the closed interval $[-2, 2]$, and $B = [0, 2]$. Note that $f[A] = f[B] = [0, 4]$. Furthermore, Range $f = f[\mathbf{R}] = \{y : 0 \leq y\}$.

10.8. Remark. Let $f: X \to Y$ be a function. We make the following observations concerning the image of subsets of X.

10.8(a). $f[\varnothing] = \varnothing$
10.8(b). $f[X]$ is the range of f
10.8(c). $f[\{x\}] = \{f(x)\}$ for each $x \in X$
10.8(d). $f[A] = \cup \{f[\{a\}] : a \in A\}$, where $A \subset X$
10.8(e). $f[A] = \{y : (a, y) \in f \text{ for at least one } a \in A\}$, where $A \subset X$
10.8(f). If $f[X] = Y$, then $f: X \to Y$ is a surjection.

EXERCISES: FUNCTIONS

1. In each of the following a set of ordered pairs Γ is given. In each case, determine whether Γ is a function and, if it is, determine if it is a one-to-one function.

   (a) Let $\Gamma = \{(x, y) : -1 \leq x \leq 1 \text{ and } x^2 + y^2 = 1\}$.
   (b) Let $\Gamma = \{(x, y) : -1 \leq x \leq 1, y \geq 0, \text{ and } x^2 + y^2 = 1\}$.
   (c) Let $\Gamma = \{(x, y) : 0 \leq x \leq 1, y \geq 0, \text{ and } x^2 + y^2 = 1\}$.
   (d) Let $\mathscr{F}$ be the collection of all real-valued differentiable functions defined on the open interval (a, b). Let $\Gamma = \{(f, f') : f \in \mathscr{F} \text{ and } f' \text{ is the derivative of } f\}$.
   (e) Let X be the collection of all continuous real-valued functions defined on the closed interval $[a, b]$. Let

$$\Gamma = \left\{ \left(f, \int_a^b f(x)\, dx \right) : f \in X \right\}.$$

(The reader should have determined that the collections of ordered pairs in (d) and (e) are functions. Thus, we see that there are functions dealt with in calculus whose domains are sets other than subsets of reals. This fact is generally not emphasized in elementary courses in calculus.)

2. Let $f:\mathbf{R} \times \mathbf{R} \to \mathbf{R} \times \mathbf{R}$ be the function defined as follows: For each $(x, y) \in \mathbf{R} \times \mathbf{R}$, let $f((x, y)) = (a, b)$ where

$$a = x + 2y$$

and

$$b = 2x + 4y.$$

Which of the following terms applies to $f:\mathbf{R} \times \mathbf{R} \to \mathbf{R} \times \mathbf{R}$? (a) surjective, (b) bijective, (c) injective.

3. Repeat the question in Exercise 2 for the system

$$a = 3x + 2y$$
$$b = 6x - 2y.$$

4. Let f be a map from the set of all reals $\mathbf{R}$ into $\mathbf{R}$. Suppose furthermore that if x_1 and x_2 are in $\mathbf{R}$ and $x_1 < x_2$, then $f(x_1) < f(x_2)$. Is it necessarily true that f is one-to-one? Is it necessarily true that $f[\mathbf{R}] = \mathbf{R}$? Justify your answers.

5. Consider the function $f:X \to Y$. Suppose that A and B are subsets of X. Decide which of the following statements are necessarily true. Justify your answers.
    (a) If $A \cap B = \varnothing$, then $f[A] \cap f[B] = \varnothing$.
    (b) If $f[A] \cap f[B] = \varnothing$, then $A \cap B = \varnothing$.
    (c) If $A \subset B$, then $f[A] \subset f[B]$.
    (d) $f[A - B] = f[A] - f[B]$.
    (e) $f[A \cup B] = f[A] \cup f[B]$.
    (f) $f[A \cap B] \subset f[A] \cap f[B]$.
    (g) $f[A \cap B] = f[A] \cap f[B]$.

11. RELATIONS

We are all familiar with the term relation when it is used with respect to certain sets of human beings. For example, let M be the collection of all male human beings living at this moment, and let F be the collection of all female human beings now alive. Let us write

$$m \, R \, f \text{ if and only if } m \in M, f \in F \text{ and } m \text{ is the husband of } f.$$

Thus, the relation that we are considering is the relation *is the husband of*. Note that, in this example, an element of M is written first and an element of F is written last. Also, in this example, since monogamy is not universal, there are elements of M related to more than one element of F, and similarly more than one element of *M* may be related to a particular element of *F*. Furthermore, there are elements of *M* that are not married and are hence related to no element of *F*.

Another familiar example is the relation *is greater than* ($>$) between real numbers.

Since there are a multitude of different relations that are useful in mathematics, it is economical to abstract certain properties possessed by various relations. The abstraction of the notion of relation itself turns out to be quite simple. Notice that in each of the two examples given (*is the husband of* and *is greater than*), when one object is related to a second object, there is a specific role played by the first object and a specific role played by the second. This observation immediately suggests the use of the notion of an ordered pair. Suppose that in the first example we consider the set *M* X *F* and we let *R* be that subset of *M* X *F* such that $(m, f) \in R$ if and only if *m is the husband of f*. Then *R* could be thought of as a listing of all married couples for which the husband's name is always written first. In the second example, *is greater than*, we may think of the relation $>$ as the subset of **R** X **R** such that $(x, y) \in >$ provided that $x > y$.

Thus, abstractly we may define a relation *R* as a collection of ordered pairs. However, in applying this abstract definition to particular cases, often we will want to revert to the former notation. We can accomplish this very easily by making the agreement that if $(x, y) \in R$, then we may write $x \, R \, y$.

In defining a relation as a set of ordered pairs, we see that we may also look at the notion of relation as a generalization of the notion of function. Thus, a function *f* can be thought of as a relation with an additional single-valuedness property, namely,

$$(x, y_1) \in f \quad \text{and} \quad (x, y_2) \in f \text{ implies that } y_1 = y_2.$$

Accordingly, we shall extend the definitions of domain of *R*, range of *R* and *R*[*A*] to an arbitrary relation *R* that is not necessarily a function. Furthermore, we shall define the *inverse* of a relation. The notion of the inverse of a relation will be of special interest to us when the relation is a function.

11.1. Definitions. Relation, domain, and range of a relation. *A relation is defined to be a set of ordered pairs. If R is a relation, then the domain of R (Dom R) and the range of R (Range R) are defined as follows*:

$$\text{Dom } R = \{x : (x, y) \in R \text{ for some } y\}$$
$$\text{Range } R = \{y : (x, y) \in R \text{ for some } x\}$$

If R is a relation and $(x, y) \in R$, *we also write* $x \, R \, y$.

11.2. Definition. Relation between sets. *Suppose that X and Y are sets and* $R \subset X \times Y$. *Then R is said to be a relation between X and Y.*

We should observe that if *f* is a function from *X* into *Y*, then *f* is a relation between *X* and *Y* with the following additional property:

$$\text{For each } x \in X, \text{ there is a unique } y \in Y$$
$$\text{such that } (x, y) \in f.$$

We see from the definition of relation, that a relation need not have the "single-valued" property. Furthermore, if *R* is a relation between *X* and *Y*, then

Dom $R \subset X$ but Dom R need not be all of X. Note that in Exercise 1(a), page 1 7 , Γ is a relation but not a function.

Recall that for a function $f: X \to Y$ and $A \subset X$, we defined (in 10.6)

$$f[A] = \{y : y = f(a) \text{ for at least one } a \in A\}.$$

We noted in 10.8 (e) that

$$f[A] = \{y : (a, y) \in f \text{ for at least one } a \in A\}.$$

We can similarly define sets $R[A]$, where $A \subset X$ and R is a relation between X and Y.

11.3.   Definition.   The set $R[A]$ for a relation R.   *Let R be a relation between sets X and Y. For $A \subset X$, we define*

$$R[A] = \{y : (a, y) \in R \text{ for at least one } a \in A\}.$$

When A is a set consisting of exactly one point, we make the following abbreviation:

$$R[\{x\}] = R[x].$$

In Definition 11.3, it should be noted that A need not be a subset of Dom R, since Dom R need not be all of X.

11.4.   Definition.   Inverse of a relation.   *Let R be a relation between X and Y. Then R^{-1} is the following relation defined between Y and X.*

$$R^{-1} = \{(y, x) : (x, y) \in R\}.$$

On the basis of Definition 11.4, we note that

$$\text{if } x \, R \, y, \quad \text{then } y \, R^{-1} \, x.$$

We also note that

$$\text{Range } R = \text{Dom } R^{-1}$$

and

$$\text{Range } R^{-1} = \text{Dom } R.$$

Furthermore,

$$(R^{-1})^{-1} = R.$$

The inverse of a function need not be a function (see Exercise 1(b), page 17). However, if $f: X \to Y$ is a function, then f^{-1} is always defined and is a relation between Y and X. Consequently the set $f^{-1}[B]$ is also defined for each $B \subset Y$.

The reader should verify the following remark.

11.5.   *Remark.*   Consider the function $f: X \to Y$. For $B \subset Y$,

$$\begin{aligned}
f^{-1}[B] &= \{x : (y, x) \in f^{-1} \text{ for at least one } y \in B\} \\
&= \{x : (x, y) \in f \text{ for at least one } y \in B\} \\
&= \{x : f(x) \in B\}.
\end{aligned}$$

We point out for emphasis that the equalities in 11.5 are correct for any $B \subset Y$ and that B need not be contained in $f[X]$, the range of f.

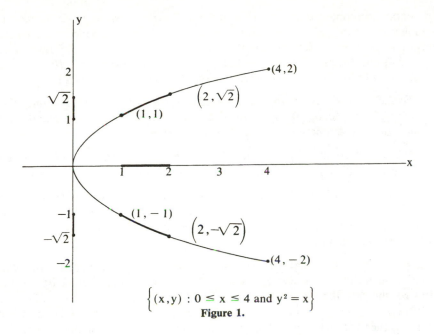

$$\left\{(x,y) : 0 \leq x \leq 4 \text{ and } y^2 = x\right\}$$
Figure 1.

11.6. *Example.* Let $R = \{(x, y): 0 \leq x \leq 4 \text{ and } y^2 = x\}$.
From the sketch, note that

$$\text{Dom } R = \{x: 0 \leq x \leq 4\}, \text{ Range } R = \{y: -2 \leq y \leq 2\},$$
$$R[2] = \{-\sqrt{2}, \sqrt{2}\}, \ R[\{x: 1 \leq x \leq 2\}] =$$
$$\{y: 1 \leq y \leq \sqrt{2}\} \cup \{y: -\sqrt{2} \leq y \leq -1\}.$$

Note that R is not a function but R^{-1} is a function.

11.7. *Example.* Let $R = \{(x, y): 0 \leq x \leq 1 \text{ and } 0 \leq y \leq 5\}$. Then $R^{-1} = \{(x, y): 0 \leq x \leq 5 \text{ and } 0 \leq y \leq 1\}$. Note that, in this example, for each $a \in$ Dom R, $R[a] = \{y: 0 \leq y \leq 5\}$, and for each $b \in$ Range R, $R^{-1}[b] = \{x: 0 \leq x \leq 1\}$.

11.8. *Example.* Let R be the following subset of $\mathbf{R} \times \mathbf{R}$:
$$R = \{(x, y): x < y\}.$$
Then for each $a \in \mathbf{R}$,
$$R[a] = \{y: a < y\}$$
and
$$R^{-1}[a] = \{x: x < a\}.$$

As we have seen, if F is a function, F^{-1} is a relation but not necessarily a function, for it is possible that for an element $y \in$ Range F, $F^{-1}[y]$ may consist of more than one element. However, for a certain class of functions, it can be proved that for each F in the class, F^{-1} is a function. These are the one-to-one functions defined in 10.4.

11.9. **Theorem.** *If F is a one-to-one function, then F^{-1} is a function. Further, F^{-1} is one-to-one.*

PROOF. Suppose that F is a one-to-one function. Since F is a function, it is a relation. Hence, F^{-1} is a relation. To show that F^{-1} is a function we need show only that for $(a, b) \in F^{-1}$ and $(a, d) \in F^{-1}$, $b = d$. But $(a, b) \in F^{-1}$ implies that $(b, a) \in F$, and $(a, d) \in F^{-1}$ implies that $(d, a) \in F$. However, since $(b, a) \in F$ and $(d, a) \in F$, it follows that $b = d$ since F is one-to-one. Thus F^{-1} is a function.

To show that F^{-1} is one-to-one, we need to show that if $(a, b) \in F^{-1}$ and $(c, b) \in F^{-1}$, then $a = c$. But $(a, b) \in F^{-1}$ and $(c, b) \in F^{-1}$ imply that $(b, a) \in F$ and $(b, c) \in F$, and since F is a function, $a = c$. This completes the proof.

11.10. Theorem. *Suppose that F is a one-to-one function. Then for each $x \in$ Dom F, $F^{-1}(F(x)) = x$. Further, for each $y \in$ Range F, $F(F^{-1}(y)) = y$.*

PROOF. By 11.9, F^{-1} is a function. Let $x \in$ Dom F. Then $(x, F(x)) \in F$ and, thus, $(F(x), x) \in F^{-1}$. Hence, $F^{-1}(F(x)) = x$. Similarly, let $y \in$ Range F. Then $(y, F^{-1}(y)) \in F^{-1}$ so that $(F^{-1}(y), y) \in F$. From this it follows that $F(F^{-1}(y)) = y$. This completes the proof.

We can now show that the converse of 11.9 is also true. Suppose that F is a function such that F^{-1} is also a function. Note that the proof of 11.10 depended only on the fact that F and F^{-1} were functions. Hence, if $F(x) = F(y)$, then $F^{-1}(F(x)) = F^{-1}(F(y))$, and by the proof of 11.10, $x = y$ so that F is one-to-one. This together with 11.9 gives the following characterization of one-to-one functions.

11.11. Theorem. *A function F is a one-to-one function if and only if F^{-1} is a function.*

EXERCISES: RELATIONS

In Exercises 1 through 5, all relations are subsets of the plane. In each case, draw a sketch of R, and give Dom R, Range R, $R[0]$ and $R^{-1}[0]$.

1. Let $(x, y) \in R$ provided that (x, y) satisfies each of the following inequalities: $x + y \leq 3$, $y - x \geq 0$, $x \geq -3$.

2. Let R be the set of all (x, y) that satisfy $x^2 - y^2 \leq 1$ and $y^2 - x^2 \leq 1$.

3. Let R be the set of all (x, y) such that $x - y$ is a multiple of 3.

4. Let R be a subset of the plane such that $(x, y) \in R$ provided that $x - y \leq \frac{1}{2}$.

5. Let R be the subset of the plane such that $(x, y) \in R$ provided that $y = x^4$.

6. Let $R = \{(x, y): x \geq 0, x^2 + y^2 = 26\}$. Find $R[0]$, $R[5]$, and $R[I]$, where $I = \{r: 0 \leq r \leq 1\}$; $R^{-1}[J]$, where $J = \{r: -1 \leq r \leq 1\}$.

7. Let $R = \{(x, y): x$ is real and $y = x(x - 1)(x - 2)\}$. Find $R[0]$, $R[1]$, $R[2]$, $R^{-1}[0]$, and $R[I]$, where $I = \{x: 0 \leq x \leq 2\}$.

8. Let R be a relation between sets X and Y, and suppose that A and B are subsets of X. In each of the following, tell whether the statement is necessarily true and give a justification of your answer: (a) $R[A \cap B] = R[A] \cap R[B]$, (b) $R[A \cap B] \subset R[A] \cap R[B]$, (c) $R[A \cap B] \supset R[A] \cap R[B]$.

9. Let $\mathbf{Z}$ be the set of all integers. For each m and n in $\mathbf{Z}$, let us write $m \, R \, n$ if and only if $m - n$ is an even integer. Thus, this relation R is the set $R = \{(m, n): m - n = 2k$ for some integer $k\}$. Find $R[1]$ and $R[2]$. How many distinct sets of the form $R[i]$ are there?

10. Let R be the relation defined as follows: For each ordered pair of integers (m, n), let $m \, R \, n$ if and only if $m - n$ is an integral multiple of 5 (including negative multiples of 5). Find $R[1]$, $R[2]$, and $R[6]$. How many distinct sets of the form $R[i]$ are there? Find $R^{-1}[1]$ and $R^{-1}[2]$. Is $R^{-1}[i] = R[i]$ for each i? For this relation R, if $i \, R \, j$ and $j \, R \, k$, does it follow that $i \, R \, k$?

12. SET INCLUSIONS FOR IMAGE AND INVERSE IMAGE SETS

There are important set inclusions and identities involving image sets and inverse image sets under a function $f: X \to Y$. Some of these can be established for both image sets and inverse image sets simultaneously because they are true for relations in general. There are other identities that are not true for relations in general but are true for the special case involving image sets or inverse image sets of functions.

12.1. Theorem. *Suppose that R is a relation between X and Y. Suppose $A \subset B \subset X$. Then $R[A] \subset R[B]$.*

PROOF. Suppose that $y \in R[A]$. Then there is an $x \in A$ such that $(x, y) \in R$. But since $A \subset B$, $x \in B$ and so $y \in R[B]$.

The reader should have discovered in Exercises 5(e) and 5(f), page 18 , that the statements found there are correct. The next theorem shows that these statements are true for relations in general.

12.2. Theorem. *Suppose that R is a relation between sets X and Y. Then if A and B are subsets of X, the following hold.*

12.2(a). $R[A \cup B] = R[A] \cup R[B]$.

12.2(b). $R[A \cap B] \subset R[A] \cap R[B]$.

PROOF. We establish first that $R[A \cup B] = R[A] \cup R[B]$. To see that $R[A \cup B] \subset R[A] \cup R[B]$, let $y \in R[A \cup B]$. Then there is an $x \in A \cup B$ such that $(x, y) \in R$. If $x \in A$, then $y \in R[A] \subset R[A] \cup R[B]$. On the other hand, if $x \in B$, then $y \in R[B] \subset R[A] \cup R[B]$. In either case, $y \in R[A] \cup R[B]$ and we have shown that $R[A \cup B] \subset R[A] \cup R[B]$. To show that $R[A] \cup R[B] \subset R[A \cup B]$, note from 12.1, that $R[A] \subset R[A \cup B]$ and $R[B] \subset R[A \cup B]$.

Hence, $R[A] \cup R[B] \subset R[A \cup B]$ (see Exercise 3, page 11). Thus we have established 12.2(a).

To prove that $R[A \cap B] \subset R[A] \cap R[B]$, note from 12.1 that $R[A \cap B] \subset R[A]$ and $R[A \cap B] \subset R[B]$. Hence, $R[A \cap B] \subset R[A] \cap R[B]$.

The inclusion $R[A] \cap R[B] \subset R[A \cap B]$ does not necessarily hold, even if a relation R is a function. The next example illustrates this.

12.3. *Example.* Refer back to Example 11.6 and note that R^{-1} is a function. Let $f = R^{-1}$, $I = \{x : 1 \leq x \leq 2\}$, $J = \{y : -\sqrt{2} \leq y \leq -1\}$, and $K = \{y : 1 \leq y \leq \sqrt{2}\}$. Note that $f[J \cap K] = f[\varnothing] = \varnothing$. On the other hand, $f[J] = f[K] = I$.

Example 12.3 shows that $R[A] \cap R[B] \subset R[A \cap B]$ does not necessarily hold even if R is a function. It turns out, however, that if R happens to be the inverse of a function, then the inclusion does hold.

12.4. **Theorem.** *Let $f : X \to Y$ be a function and suppose that C and D are subsets of Y. Then*

$$f^{-1}[C \cap D] = f^{-1}[C] \cap f^{-1}[D].$$

PROOF. Since f^{-1} is a relation, it follows from 12.2(b) that

$$f^{-1}[C \cap D] \subset f^{-1}[C] \cap f^{-1}[D].$$

To see that $f^{-1}[C] \cap f^{-1}[D] \subset f^{-1}[C \cap D]$, let $x \in f^{-1}[C] \cap f^{-1}[D]$. Then $f(x) \in C$ and $f(x) \in D$, so that $f(x) \in C \cap D$. Hence, $x \in f^{-1}[C \cap D]$. Thus, $f^{-1}[C] \cap f^{-1}[D] \subset f^{-1}[C \cap D]$.

The reader should note the role of the single-valuedness of f in the proof just completed.

The next theorem is an extension of Theorem 12.2 to collections of sets. The proof is left as an exercise.

12.5. **Theorem.** *Suppose that R is a relation between X and Y. Then if $\{A_\alpha : \alpha \in \Lambda\}$ is a nonempty collection of subsets of X, the following hold:*

12.5(a). $R[\bigcup \{A_\alpha : \alpha \in \Lambda\}] = \bigcup \{R[A_\alpha] : \alpha \in \Lambda\}$

12.5(b). $R[\bigcap \{A_\alpha : \alpha \in \Lambda\}] \subset \bigcap \{R[A_\alpha] : \alpha \in \Lambda\}$

The next theorem includes special cases of Theorem 12.5 and an extension of Theorem 12.4. The proof is left as an exercise.

12.6. **Theorem.** *Let $f : X \to Y$ be a function. Let $\{A_\delta : \delta \in \Delta\}$ and $\{B_\lambda : \lambda \in \Lambda\}$ be nonempty collection of subsets of X and Y respectively. Then*

12.6(a). $f[\bigcup \{A_\delta : \delta \in \Delta\}] = \bigcup \{f[A_\delta] : \delta \in \Delta\}$

12.6(b). $f[\bigcap \{A_\delta : \delta \in \Delta\}] \subset \bigcap \{f[A_\delta] : \delta \in \Delta\}$

12.6(c). $f^{-1}[\bigcup \{B_\lambda : \lambda \in \Lambda\}] = \bigcup \{f^{-1}[B_\lambda] : \lambda \in \Lambda\}$

$f^{-1}[\bigcap \{B_\lambda : \lambda \in \Lambda\}] = \bigcap \{f^{-1}[B_\lambda] : \lambda \in \Lambda\}$

In the next theorem we list a number of other set inclusions and identities of importance. Where a proof is not given, it is left as an exercise.

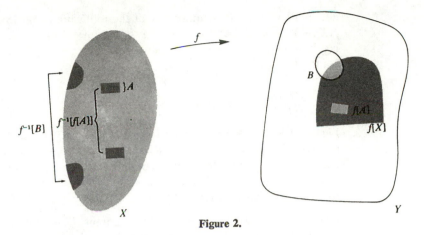

Figure 2.

12.7. Theorem. *Let $f: X \to Y$ be a function. Then each of the following holds.*

12.7(a). For each $x \in X$,

$$x \in f^{-1}[f[x]],$$

12.7(b). For each $A \subset X$,

$$A \subset f^{-1}[f[A]].$$

12.7(c). For each $y \in \text{Range } f$,

$$f[f^{-1}[y]] = \{y\}.$$

12.7(d). For each subset $B \subset Y$,

$$f[f^{-1}[B]] = B \cap \text{Range } f.$$

PROOF OF (d). Let $y \in f[f^{-1}[B]]$. Then there is an $x \in f^{-1}[B]$ such that $f(x) = y$. But then $f(x) \in B$ and obviously $f(x) \in \text{Range } f$, so that $f(x) = y \in B \cap \text{Range } f$. Hence, $f[f^{-1}[B]] \subset B \cap \text{Range } f$. Next choose a $y \in B \cap \text{Range } f$. Then there is an $x \in f^{-1}[B]$ such that $f(x) = y$. But since $x \in f^{-1}[B]$, $y = f(x) \in f[f^{-1}[B]]$. Hence, we have shown that $B \cap \text{Range } f \subset f[f^{-1}[B]]$. This completes the proof.

EXERCISES: SET INCLUSIONS FOR IMAGE AND INVERSE IMAGE SETS

1. Prove Theorem 12.5.

2. Prove Theorem 12.6.

3. Prove Theorem 12.7(a), (b), and (c).

4. Suppose that $f: X \to Y$ is a function and A and B are subsets of X. Suppose also that C and D are subsets of Y. For each of the following, determine whether the statement is necessarily true. In any case for which the statement is not necessarily true,

determine whether it is under any of the following conditions: $f:X \to Y$ is a surjection, $f:X \to Y$ is an injection, $f:X \to Y$ is a bijection.

(a) $f[A - B] = f[A] - f[B]$
(b) $f^{-1}[D - C] = f^{-1}[D] - f^{-1}[C]$
(c) $f^{-1}f[A] = A$
(d) $f[f^{-1}[C]] = C$

5. Let $M:\mathbf{R} \times \mathbf{R} \to \mathbf{R}$ be the map from $\mathbf{R} \times \mathbf{R}$ into $\mathbf{R}$ defined as follows: For each $(a, b) \in \mathbf{R} \times \mathbf{R}$, let $M((a, b)) = ab$. Is M a map from $\mathbf{R} \times \mathbf{R}$ onto $\mathbf{R}$? Representing $\mathbf{R} \times \mathbf{R}$ as a plane, draw a sketch of each of the following sets: $M^{-1}[0]$, $M^{-1}[1]$, $M^{-1}[I]$, where I is the closed interval $[0, 1]$.

6. Examine carefully the content of Theorem 12.6 and your answers to Exercise 4(a) and (b). Which seems to have a nicer behavior on collections of sets, f or f^{-1}?

13. THE RESTRICTION OF A FUNCTION

13.1. Definition. The restriction of a function. *Suppose that f is a function defined on a set X, and $A \subset X$. Suppose further that g is the function defined on A with the property that $g(x) = f(x)$ for each $x \in A$. Then g is called the restriction of f to A and is denoted by $f \mid A$. (Note that $f \mid A \subset f$.)*

The reader is already familiar with this notion from more elementary courses. For example, consider the sine function defined on all the reals. Often one is required to consider only the sine restricted to an interval of say 2π in length.

13.2. Theorem. *Suppose that $X = A \cup B$, and that $f:A \to Y$ and $g:B \to Y$ are maps such that $f \mid A \cap B = g \mid A \cap B$. Then $f \cup g$ is a function.*

Before proceeding with the proof we shall look at an example. Consider $g:[-1, 0] \to \mathbf{R}$ and $f:[0, 1] \to \mathbf{R}$ given by the following: $f(x) = x$ for $0 \leq x \leq 1$, and $g(x) = -x$ for $-1 \leq x \leq 0$. Note that the function h defined on $[-1, 1]$, given by $h(x) = x$ for $0 \leq x \leq 1$ and $h(x) = -x$ for $-1 \leq x \leq 0$, is precisely $f \cup g$. There is no ambiguity since f and g agree where their domains intersect.

PROOF. $f \cup g$ is a relation whose domain is $A \cup B$. To show that $f \cup g$ is a function, we need simply to show that $(x, y_1) \in f \cup g$ and $(x, y_2) \in f \cup g$ imply that $y_1 = y_2$. Notice that one of the following must be true: (i) $(x, y_1) \in f$ and $(x, y_2) \in f$, (ii) $(x, y_1) \in g$ and $(x, y_2) \in g$, (iii) $(x, y_1) \in f$ and $(x, y_2) \in g$, (iv) $(x, y_1) \in g$ and $(x, y_2) \in f$. Since f and g are functions, (i) and (ii) imply $y_1 = y_2$. If case (iii) holds, then $y_1 = f(x)$ and $y_2 = g(x)$. However, in that case $x \in A \cap B$, and from hypothesis, $y_1 = f(x) = g(x) = y_2$. In case (iv), the proof is similar.

EXERCISES: THE RESTRICTION OF A FUNCTION

1. Let $f:\mathbf{R} \to \mathbf{R}$ and $g:\mathbf{R} \to \mathbf{R}$ be mappings such that $f(x) = \sin x$ for each $x \in \mathbf{R}$ and $g(x) = \sqrt{1 - \cos^2 x}$ for each $x \in \mathbf{R}$. Find

the largest interval of real numbers, I, whose left hand endpoint is 0 and which satisfies $f \mid I = g \mid I$.

2. Let $f: \mathbf{R} \to \mathbf{R}$ be defined as follows: For each $x \in \mathbf{R}$, let $f(x) = |x - 1|$. Let $g: \mathbf{R} \to \mathbf{R}$ be defined by $g(x) = x - 1$ for each $x \in \mathbf{R}$. Find the largest set $S \subset \mathbf{R}$ for which $f \mid S = g \mid S$.

3. Let $f: \mathbf{R} \to \mathbf{R}$ and $g: \mathbf{R} \to \mathbf{R}$ be given by $g(x) = \cos x$ and $f(x) = \sqrt{1 - \sin^2 x}$ for $x \in \mathbf{R}$. Find the largest set $S \subset \mathbf{R}$ for which $f \mid S = g \mid S$.

14. COMPOSITION OF FUNCTIONS

Suppose one wanted to consider the function f defined on $\mathbf{R}$ and given by the formula $f(x) = \sin(\cos x)$. Actually this is a rather complicated function. However, since the sine and cosine are well-known functions, it is easier to study f by means of what is already known about the sine and cosine. This technique of studying complicated functions by representing them as the composition of simpler ones has proved to be an extremely useful technique at all levels of mathematical investigation. The reader who has studied elementary calculus will, for example, recall the technique of finding the derivative of a complicated function by means of the so-called chain rule for composite functions. As a matter of fact, the student is brought into contact with the notion of composition as early as in elementary high school algebra. For example, in finding the zeros of the function given by $F(x) = (x^2 - 1)^2 - 3(x^2 - 1) + 2$, one could introduce the functions given by $h(x) = x^2 - 1$ and $f(z) = z^2 - 3z + 2$ and represent $F(x)$ as $f(h(x))$. The problem then reduces itself to finding the zeros z_1 and z_2 of $f(z)$ and then, provided that z_1 and z_2 are in the range of h, solving the equations $h(x) = z_1$ and $h(x) = z_2$.

14.1. Definition. Composite function. *Suppose f and g are functions such that* Range $f \subset$ Dom g. *We then define the function $g \circ f$ (read the composition of f and g, or f followed by g) as the function defined on* Dom f *and given by $g \circ f(x) = g(f(x))$ for each $x \in$ Dom f.*

We make some simple observations implied by this definition.

$$\text{Dom } f = \text{Dom } g \circ f, \qquad \text{Range } g \circ f \subset \text{Range } g$$

The following schematic diagram can be of help in visualizing the various aspects of the definition.

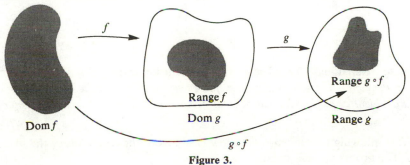

Figure 3.

14.2. *Example.* Let $f: \mathbf{R} \to \mathbf{R}$ and $g: \mathbf{R} \to \mathbf{R}$ be given by $f(x) = \sin x$ and $g(x) = x^3 + 1$. Then $g \circ f: \mathbf{R} \to \mathbf{R}$ and $f \circ g: \mathbf{R} \to \mathbf{R}$ are given by $g \circ f(x) = (\sin x)^3 + 1$ and $f \circ g(x) = \sin(x^3 + 1)$.

14.3. *Example.* Let $\mathbf{R}_+$ be defined as in 8.4. Suppose that $f: \mathbf{R}_+ \to \mathbf{R}_+$ is given by $f(x) = x^2$ and $g: \mathbf{R}_+ \to \mathbf{R}_+$ is given by $g(x) = \sqrt{x}$. Then $g \circ f(x) = \sqrt{x^2} = x$ for x in $\mathbf{R}_+$ and $f \circ g(x) = (\sqrt{x})^2 = x$ for x in $\mathbf{R}_+$. Note that in this example $g \circ f = f \circ g$ but that in 14.2 such is not the case.

14.4. **Theorem.** *Suppose that f, g, and h are functions such that* Range $f \subset$ Dom g *and* Range $g \subset$ Dom h. *Then* $h \circ (g \circ f) = (h \circ g) \circ f$.

PROOF. First note that the hypothesis implies that both $h \circ (g \circ f)$ and $(h \circ g) \circ f$ are well defined. Note also that Dom $f =$ Dom $h \circ (g \circ f) =$ Dom $(h \circ g) \circ f$. Also, for each $x \in$ Dom f, $((h \circ g) \circ f)(x) = h \circ g(f(x)) = h(g(f(x)))$, $(h \circ (g \circ f))(x) = h(g \circ f(x)) = h(g(f(x)))$. Thus, $(h \circ g) \circ f = h \circ (g \circ f)$.

We see from this that with the proper restrictions on their domains and ranges, composition of functions is associative. As in multiplication for real numbers, since the operation is associative we agree to write both $(h \circ g) \circ f$ and $h \circ (g \circ f)$ as $h \circ g \circ f$. (There are some texts, especially algebra texts, that write "$(x)f$" for "$f(x)$." In such books, "$f \circ g$" would be used instead of "$g \circ f$." This reversal of order would be natural, for then $(x)f \circ g = ((x)f)g$. On the other hand, in this text we would have $g \circ f(x) = g(f(x))$.)

14.5. **Function diagrams.** Suppose that $f: X \to Y$, $g: Y \to Z$, and $h: X \to Z$ are mappings such that $g \circ f = h$. A diagram such as the following will help the reader to visualize the situation.

In such a diagram it is customary to say that the diagram is commutative, to indicate that the same image point may be obtained by any sequence of mappings obtained by following the arrows. For example, the following commutative diagram would indicate that the mappings

$$f: X \to V, g: Y \to W, h: X \to Y, \quad \text{and} \quad k: V \to W \quad \text{satisfy} \quad g \circ h = k \circ f.$$

14.6. **Definitions. The identity and the inclusion maps.** *Suppose that $X \subset Y$ and i is the function given by $i(x) = x$ for each $x \in X$. Then the map $i: X \to X$ is called the identity map on X, and the map $i: X \to Y$ is called the inclusion map of X into Y.*

The following two statements follow easily from the definition of the identity map. The proofs are left as exercises.

14.7.   **Theorem.**   Suppose that $i: X \to X$ is the identity map on X and $f: X \to Y$ is a map. *Then $i \circ i = i$ and $f \circ i = f$.*

In view of the previous theorem, we have the following commutative diagrams:

14.8.   **Theorem.**   *Suppose that $f: X \to Y$ is a map and $i: Y \to Y$ is the identity map on Y. Then $i \circ f = f$.*

Theorem 14.8 gives the following commutative diagram:

EXERCISES: COMPOSITION OF FUNCTIONS

1.   Let $f: \mathbf{R} \to \mathbf{R}$ and $g: \mathbf{R} \to \mathbf{R}$ be given by $f(x) = x/(x^2 + 1)$ and $g(x) = x^2$. Give explicit formulas for $g \circ f(x)$ and $f \circ g(x)$. Determine the ranges of $g \circ f$ and $f \circ g$.

2.   Let $f: \mathbf{R} \to \mathbf{R}$ and $g: \mathbf{R} \to \mathbf{R}$ be defined as follows:
$$f(x) = x^2 \qquad \text{for } x \geq 0$$
$$= 2 \qquad \text{for } x < 0$$
$$g(x) = \sqrt{x} \qquad \text{for } x \geq 0$$
$$= x \qquad \text{for } x < 0$$
   (a)   Sketch the graph of $g \circ f$.
   (b)   Sketch the graph of $f \circ g$.
   (c)   Find $(f \circ g)^{-1}[x]$ for each $x \in \mathbf{R}$.

3.   Let $f: \mathbf{R} \to \mathbf{R}$ and $g: \mathbf{R} \to \mathbf{R}$ be given by $f(x) = \sin x$ and $g(x) = |x|$. Write explicit expressions for $g \circ f(x)$ and $f \circ g(x)$ and find the range of each.

4.   Let $f: \mathbf{R} \to \mathbf{R}$ and $g: \mathbf{R} \to \mathbf{R}$ be the maps given by $f(x) = x^2 + 2$ and $g(x) = x - 1$. Find expressions for $(f \circ g)(x)$ and $(g \circ f)(x)$ and note that $f \circ g \neq g \circ f$.

5.   Let $f: \mathbf{R} \to \mathbf{R}$, $g: \mathbf{R} \to \mathbf{R}$ and $h: \mathbf{R} \to \mathbf{R}$ be given by: $f(x) = x^2 + x$, $g(x) = (x - 1)^2$, and $h(x) = x + 1$ for each $x \in \mathbf{R}$. Find an expression for $h \circ g \circ f(x)$ for $x \in \mathbf{R}$.

6.   Suppose $f: X \to Y$ is a bijection. Show that $f^{-1} \circ f = i$ where $i: X \to X$ is the identity map on X and $f \circ f^{-1} = j$ where $j: Y \to Y$ is the identity map on Y.
     Solution to first part. Let $x \in X$. Then, $(x, f(x)) \in f$. Hence, $(f(x), x) \in f^{-1}$. Thus, $f^{-1}(f(x)) = x$ and $(f^{-1} \circ f)(x) = x$ for each $x \in X$. Hence, $f^{-1} \circ f = i$, the identity map on X.

7. Let $f:X \to Y$ and $g:Y \to X$ be surjections. Suppose $g \circ f = i$, where $i:X \to X$ is the identity map on X. Show: (a) f is one-to-one, (b) g is one-to-one, (c) $f \circ g = j$ where $j:Y \to Y$ is the identity map from Y onto Y, (d) $f = g^{-1}$, and (e) $g = f^{-1}$.

8. Recall that $\mathbf{R}_+ = \{x: x \in \mathbf{R}$ and $x > 0\}$. Recall also that the natural logarithm function, ln, is defined on $\mathbf{R}_+$ with range $\mathbf{R}$; the exponential function, $\{(x, e^x): x \in \mathbf{R}\}$, is the inverse of the ln function; $\cosh x = \dfrac{e^x + e^{-x}}{2}$ for each $x \in \mathbf{R}$.

   Sketch the cosh function and note that although cosh is not one-to-one, the restriction, $\cosh |\mathbf{R}_+$ is one-to-one and hence, its inverse is a function. By using results of Exercise 7, prove that $\cosh^{-1}(x) = \ln(x + \sqrt{x^2 - 1})$ for $x \geq 1$.

9. Consider the map $\alpha: \mathbf{R}^2 \to \mathbf{R}^2$ such that for each $(x, y) \in \mathbf{R}^2$ $\alpha((x, y))$ is the element $(u, v) \in \mathbf{R}^2$ given by $u = 2x - y$, $v = 5x + y$. Recall that $\mathbf{R}^2$ denotes $\mathbf{R} \times \mathbf{R}$.
   (a) Does $\alpha[\mathbf{R}^2] = \mathbf{R}^2$? (b) Is α one-to-one? (c) If α is one-to-one, find a rule for α^{-1} analogous to the rule given for α.

10. Consider the map $P: \mathbf{R}^2 \to \mathbf{R}$ such that for each $(x, y) \in \mathbf{R}^2$, $P((x, y)) = x$. Note that P is not one-to-one. Find a subset S of $\mathbf{R}^2$ such that $P \mid S: S \to \mathbf{R}$ is a one-to-one map from S onto $\mathbf{R}$.

11. Suppose $f:A \to B$ is a one-to-one map from A into B and $g:B \to C$ is a one-to-one map from B into C. Prove that $g \circ f: A \to C$ is a one-to-one map from A into C.

12. Let $f:\mathbf{R} \to \mathbf{R}$ be a map such that for each pair of numbers x and y, $f(x + y) = f(x) + f(y)$.
    (a) Show that $f(0) = 0$.
    (b) Show that $f(x) = -f(x)$ for all $x \in \mathbf{R}$.
    (c) Show that $f(mx) = mf(x)$ for each integer m and $x \in \mathbf{R}$.
    (d) Show that $f(rx) = rf(x)$ for each rational number r and $x \in \mathbf{R}$.

15. SEQUENCES

Recall that $\mathbf{P}$ denotes the set of all positive integers and $\mathbf{P}_n$ denotes the first n positive integers. Functions that have $\mathbf{P}$ or $\mathbf{P}_n$ as domains occur frequently, and such functions are given special names.

15.1. Definition. Sequence or infinite sequence. *A function defined on the set* $\mathbf{P}$ *of positive integers is called a sequence or an infinite sequence.*

15.2. Definition. Finite-sequence or n-tuple. *A function defined on the set* $\mathbf{P}_n$ *of the first n positive integers is known as a finite-sequence or an ordered n-tuple.*

We use the hyphen between the "finite" and "sequence" to indicate that "finite-sequence" is one word. Since "sequence" in mathematics is usually used synonymously with "infinite sequence," we do not wish to use "finite" as an adjective modifying sequence.

If f is an infinite sequence or a finite-sequence, then for a fixed n, we call $f(n)$ the nth *term* of the sequence. It is also customary to write $f(n)$ as f_n. Although, strictly speaking, an infinite sequence f is a function $\{(i, f(i)) : i \in \mathbf{P}\}$, one often encounters the notation $(f_i)_{i=1}^{\infty}$ or (f_i) for the sequence.

If $(a_i)_{i=1}^{\infty}$ is an infinite sequence, we shall sometimes write it as $(a_1, a_2, \ldots, a_n, \ldots)$. Likewise, we shall usually write $(a_1, a_2, \ldots, a_n)$ for the finite-sequence $\{(i, a_i) : i \in \mathbf{P}_n\}$. Observe that the notation $(a_1, a_2, \ldots, a_n)$ exhibits the range explicitly. The term in the ith position is the value of the sequence corresponding to i.

Recall that we built up the concept of function from the concept of ordered pair. At this point in our discussion, we now have a certain kind of function called a finite-sequence. For the special case where the finite-sequence is defined on $\mathbf{P}_2 = \{1, 2\}$, it would take the form $\{(1, a_1), (2, a_2)\}$ which we have agreed to abbreviate as (a_1, a_2). But (a_1, a_2) has also been used as the symbol for an ordered pair. This should lead to no confusion. As a matter of fact, it should be clear that two finite-sequences $(a_1, a_2, \ldots, a_n)$ and $(b_1, b_2, \ldots, b_n)$ are equal if and only if $a_i = b_i$ for $i \in \mathbf{P}_n$. Recall that this was a crucial property possessed by ordered pairs. Thus, ordered pairs and finite-sequences on $\mathbf{P}_2$ "act alike."

There are certain types of sequences that occur with sufficient frequency to warrant our having special names for them.

15.3. Definition. Increasing and decreasing sequences of numbers. *A sequence (a_i) of real numbers is said to be increasing (decreasing) provided that*

$$a_i \leq a_{i+1} (a_{i+1} \leq a_i) \quad \text{for} \quad i \in \mathbf{P}.$$

Moreover, the sequence (a_i) is said to be a strictly increasing (decreasing) sequence if

$$a_i < a_{i+1} (a_{i+1} < a_i) \quad \text{for} \quad i \in \mathbf{P}.$$

15.4. Definition. Increasing and decreasing sequences of sets. *A sequence of sets (K_i) is said to be increasing (decreasing) provided that*

$$K_i \subset K_{i+1} (K_{i+1} \subset K_i).$$

Moreover, the sequence of sets (K_i) is said to be strictly increasing (decreasing) provided that for each $i \in \mathbf{P}$,

$$K_i \neq K_{i+1} \quad \text{and} \quad K_i \subset K_{i+1} (K_{i+1} \subset K_i).$$

Sometimes the term *monotone* is used to refer to sequences of numbers or sets that are either increasing or decreasing.

In the next definition, we use the notions of *finite-sequence* and *infinite sequence* to extend the concept of Cartesian product.

15.5. Definition. Cartesian product of nonempty countable collections of sets. *Let n be a positive integer and let $\{A_i : i \in \mathbf{P}_n\}$ be a collection of sets indexed by $\mathbf{P}_n$. The Cartesian product of $\{A_i : i \in \mathbf{P}_n\}$, denoted by $\times \{A_i : i \in \mathbf{P}_n\}$, is defined to be the set of all finite sequences $(a_1, a_2, \ldots, a_n)$ such that $a_i \in A_i$ for $i \in \mathbf{P}_n$.*

Similarly, the Cartesian product of a collection of sets $\{A_i : i \in P\}$, denoted by $X\{A_i : i \in P\}$, is defined to be the collection of all infinite sequences (a_i) such that $a_i \in A_i$ for $i \in P$.

Often $X\{A_i : i \in P_n\}$ is denoted by $A_1 \times A_2 \times A_3 \times \ldots \times A_n$.

EXERCISES: SEQUENCES

1. In each of the following find a formula for the nth term a_n of an infinite sequence whose first five terms are given.
   (a) $a_1 = 1$, $a_2 = \frac{1}{2}$, $a_3 = \frac{1}{4}$, $a_4 = \frac{1}{8}$, $a_5 = \frac{1}{16}$.
   (b) $a_1 = 1$, $a_2 = 0$, $a_3 = 1$, $a_4 = 0$, $a_5 = 1$.
   (c) $a_1 = 1$, $a_2 = 0$, $a_3 = -1$, $a_4 = 0$, $a_5 = 1$.
   (d) $a_1 = 1$, $a_2 = 3$, $a_3 = 6$, $a_4 = 10$, $a_5 = 15$.

2. Let $f = (f_i)_{i=1}^{\infty}$, be the sequence defined as follows: Let $g(x) = \sin x$. For each positive integer i, let $f_i = g^{(i)}(0)$, where $g^{(i)}$ is the ith derivative of g. Write the terms of f sufficiently far to see the pattern followed.

3. For each $n \in P$, let $a_n = \sum_{j=1}^{n} j^2$. Try to discover a formula for a_n.

4. Suppose that a is a sequence such that $a_1 = 1$, $a_2 = 3$, $a_3 = a_1 + a_2$, and for $j \geq 3$, $a_j = a_{j-1} + a_{j-2}$. Find a_4, a_5, a_6, and a_7.

5. Suppose a sequence a is given by $a_n = 2^n$ for each positive integer n. For which values of n is it true that $a_n \geq 10,000$?

6. Let a be the sequence given by $a_n = n/(n+1)$. Find the smallest integer N such that for $n \geq N$, $a_n > \frac{9}{10}$.

7. Let a be the sequence given by $a_n = \sqrt{n+1} - \sqrt{n}$. Find an integer N such that for $n \geq N$, $a_{n+1} < a_n$. Find an integer M such that for $n \geq M$, $a_n \leq \frac{1}{10}$.

8. For each $(x, y) \in R^2$, let $f(x, y) = (x^2 - y^2, 2xy)$. Let the point p_1 in the plane be given by $f(\frac{1}{2}, \frac{1}{2})$, $p_2 = f(p_1)$, $p_3 = f(p_2)$, $p_4 = f(p_3)$, $p_5 = f(p_4)$. Calculate and plot the points p_1, $p_2, \ldots, p_5$.

9. Let $A_1 = \{1, 2, 3\}$, $A_2 = \{1, 2\}$, $A_3 = \{a, b\}$. Write out the elements of $A_1 \times A_2 \times A_3$. Write out the elements of $(A_1 \times A_2) \times A_3$. Show that there is a "natural" one-to-one correspondence between the elements of these two sets.

10. Suppose that for each $i \in P$, $A_i = \{0, 1\}$. Describe in words the set $X\{A_i : i \in P\}$.

11. Let A_1 be the set of all real numbers R. For each $i \in P$ such that $i \geq 2$, let $A_i = \{0\}$. Describe in words the set $X\{A_i : i \in P\}$. Show that there exists a bijection from $X\{A_i : i \in P\}$ onto R.

16. SUBSEQUENCES

We wish to give a formal definition of a subsequence of a sequence. Before doing so, however, we give an example of the concept we wish to define.

16.1. *Example.* Consider the sequence $f = \{(i, 3i) : i \in \mathbf{P}\}$. Consider the sequence h given by $h(1) = f(2)$, $h(2) = f(4)$, $h(3) = f(6)$, and in general $h(n) = f(2n) = 6n$. Notice that the original sequence f can be written $(3, 6, 9, \ldots, 3n, \ldots)$ and that h can be written as $(6, 12, \ldots, 6n, \ldots)$. Roughly, we can obtain $h = (6, 12, \ldots, 6n, \ldots)$ from $f = (3, 6, 9, \ldots, 3n \ldots)$ by leaving out some of the terms of f, keeping an infinite number of terms, and keeping the order the same. The sequence h given above is a subsequence of the sequence f. The concept can be made precise by appealing to the concept of composite function as indicated next.

16.2. **Definition. Subsequence of a sequence.** *Let $f : \mathbf{P} \to X$ be a sequence with functional values in a set X. Let $N : \mathbf{P} \to \mathbf{P}$ be a strictly increasing sequence from $\mathbf{P}$ into $\mathbf{P}$ (i.e., if $i > j$, then $N(i) > N(j)$). Then the composite function $h = f \circ N$ is said to be a subsequence of f.*

Notice that in 16.1, if map $N : \mathbf{P} \to \mathbf{P}$ is given by $N(i) = 2i$, then $h(i) = f \circ N(i) = f(2i) = 6i$.

It is often the case in mathematics that if $(a_i)_{i=1}^{\infty}$ is a sequence and $N : \mathbf{P} \to \mathbf{P}$ is a strictly increasing sequence, then the ith term of the subsequence $(a(N(i)))$ is written as a_{N_i}. Notice this is consistent with the notation introduced before, since $a(N(i)) = a_{N(i)} = a_{N_i}$.

EXERCISES: SEQUENCES AND SUBSEQUENCES

1. Consider the sequence $S = (1, \frac{1}{2}, \frac{1}{3}, \ldots, 1/n, \ldots)$. Find a map $N : \mathbf{P} \to \mathbf{P}$ such that $S \circ N$ is the sequence $(\frac{1}{3}, \frac{1}{6}, \ldots, 1/3n \ldots)$.

2. Consider the sequence S such that $S(n) = (-1)^n \, 1/2^n$. Find the nth term of the subsequence of S whose terms consist of all the positive terms of S and none of the negative terms of S.

3. For each positive integer n, let h_n be the function given by $h_n = \{(x, x^{n+1}) : 0 \le x \le 1\}$. Suppose that k is the sequence such that for each $n \in \mathbf{P}$, $k_n = \int_0^1 h_n(t) \, dt$. Find the nth term of k.

4. In each of the following determine whether or not the sequence is a strictly increasing sequence.

    (a) $f = \left(\dfrac{n}{n + 1} \right)_{n=1}^{\infty}$

    (b) $((n - \frac{1}{2})^2)_{n=1}^{\infty}$

    (c) $f = (50n - n^2)_{n=1}^{\infty}$

    (d) $g \circ f$, where f is a strictly increasing sequence of positive integers and g is a strictly increasing sequence of real numbers.

5. Suppose that h is a subsequence of a sequence k and f is a subsequence of h. Is f a subsequence of k?

17.   FINITE INDUCTION AND WELL-ORDERING FOR POSITIVE INTEGERS

The reader will recall that in algebra courses certain statements about natural numbers or positive integers were proved with the aid of the *principle of finite induction*. In any logical development of the natural number system, some form of this principle is given as an axiom or some axiom is given that implies it. We next give a formal statement of one form of the principle which we accept for the system of natural numbers. We prove it equivalent to the so-called well-ordering principle for the natural numbers. In this text we shall take the system of natural numbers to be the same as the set **P** of all positive integers.

17.1.   The principle of finite induction.   *Suppose that M is a subset of* **P** *such that $1 \in M$, and $h \in M$ implies that $h + 1 \in M$. Then $M =$* **P**.

17.2.   The principle of well-ordering for the set P of all positive integers.   *Let K be a nonempty set of positive integers. Then there is a first (smallest) element in K.*

17.3.   Theorem.   *The principle of finite induction and the well-ordering principle for positive integers are equivalent.*

PROOF.   We first assume the truth of the principle of finite induction and prove the well-ordering principle for positive integers. (Later we will show the converse.) Let **P** be the set of all positive integers and suppose that K is a nonempty subset of **P**. Assume that, contrary to the well-ordering principle, K has no first element. Let

$$M = \{n : n \in \mathbf{P} \quad \text{and} \quad \{1, 2, 3, \dots, n\} \subset \mathbf{P} - K\}.$$

Note that $1 \in M$, for otherwise 1 would be the first element in K. Next assume that $h \in M$. Then $h + 1 \in M$, for otherwise $h + 1$ would be the first element in K. Hence, by induction, $M = \mathbf{P}$. From the definition of M, it now follows that $K = \varnothing$ and we have a contradiction.

Next we assume the well-ordering principle for **P** and prove the principle for finite induction. Let M be a set of positive integers such that $1 \in M$, and $h \in M$ implies that $h + 1 \in M$. We wish to show that $M = \mathbf{P}$. If $M \neq \mathbf{P}$, then $\mathbf{P} - M$ is a nonempty set of positive integers. By the well-ordering principle, there is a smallest integer k in $\mathbf{P} - M$, and thus $k - 1 \notin \mathbf{P} - M$. Since $k \neq 1$, $k - 1 \in \mathbf{P}$ and, thus, $k - 1 \in M$. However, by our assumption for the set M, it now follows that $k \in M$ and we have a contradiction.

EXERCISES: FINITE INDUCTION AND WELL-ORDERING FOR POSITIVE INTEGERS

1. Prove that the following statement is equivalent to 17.1. Suppose that h is an integer. Suppose further that $S(n)$ is

a statement for each integer $n \geq h$, $S(h)$ is true, and $S(n)$ implies $S(n+1)$ for each integer $n \geq h$. Then $S(n)$ is true for each integer $n \geq h$.

2. Prove that the sum of the first n positive integers is $\frac{1}{2}n(n+1)$.

3. Prove that $1^2 + 2^2 + 3^2 + \cdots + n^2 = \frac{1}{6}n(n+1)(2n+1)$ for each $n \in \mathbf{P}$.

4. Prove or disprove the following statement: For each $n \in \mathbf{P}$,
$$1^3 + 2^3 + 3^3 + \cdots n^3 = \tfrac{1}{4}n^2(n+1)^2$$

5. Is the following statement true? Justify your answer. For each positive integer n, $2^n - 1 \geq n$.

6. Either prove or disprove the following statement: For each positive integer n,
$$1 \cdot 2 + 2 \cdot 3 + 3 \cdot 4 + \cdots + n(n+1)$$
$$= \tfrac{1}{3}[n(n+1)(n+2) + 3].$$

7. Either prove or disprove the following statement: For each $n \in \mathbf{P}$, $7^n - 3^n$ is divisible by 4.

8. Suppose that K is a nonempty collection of negative integers. Prove that there is a largest element in K.

9. Is $3n^2 + n$ an even integer for each positive integer n? Justify your answer.

10. Try to discover a formula for the number of subsets (including the empty set) of a set of n objects. Then prove by induction that your conjecture is correct.

11. Is $n(n+1)(n+2)$ divisible by 3 for each positive integer n? Justify your answer.

12. Is $[n(n+1)(n+2)(n+3)]/24$ an integer for each positive integer n? Justify your answer.

18. SEQUENCES DEFINED INDUCTIVELY

It often happens that a sequence is defined inductively. In this section we give an example illustrating the process and state a theorem that covers such an example. As a preliminary to the discussion we first state and prove a theorem that is basic to the process of defining a sequence inductively.

18.1. Union of Functions Theorem. *Suppose that $\{X_i : i \in \mathbf{P}\}$ is a collection of sets such that for each $i \in \mathbf{P}$, $X_{i+1} \supset X_i$ and X is a set such that $X = \bigcup \{X_i : i \in \mathbf{P}\}$. Suppose further that for each positive integer i, $g_i : X_i \to Y$ is a map from X_i into the set Y and $g_{i+1} | X_i = g_i$. Then, $\bigcup \{g_i : i \in \mathbf{P}\}$ is a function defined on X.*

PROOF. $g = \bigcup \{g_i : i \in \mathbf{P}\}$ is a relation whose domain is X. Suppose that g is not a function. Then for some $x \in X$ and positive integers i and h,

18.1(a). $x \in X_i$ and $g_i(x) \neq g_{i+h}(x)$.

Let k be the smallest positive integer for which $g_i(x) \neq g_{i+k}(x)$, where x and i are the same as in 18.1(a). Then $g_{i+k-1}(x) = g_i(x)$. But this is a contradiction, since $g_{i+k} | X_{i+k-1} = g_{i+k-1}$. This completes the proof.

18.2. *Remark.* Note that it follows from the proof of the preceding theorem that $g(x) = g_n(x)$ for each n for which $x \in X_n$. It will be important to recall this fact when applying Theorem 18.1.

18.3. *Example.* Let $f(1) = 2$, $f(2) = 7$. Furthermore, let it be given that for each positive integer $n \geq 3$, $f(n) = \frac{1}{2}[f(n-1) + f(n-2)]$. At this point it would be instructive for the reader to calculate a few terms of f. It is intuitively clear that there should exist a unique function satisfying the above properties. It furthermore seems reasonable that we should be able to prove by induction that such a function exists. This we can do, but not in as straightforward a manner as we might guess. To establish the existence of a function f satisfying the required properties, we proceed as follows: Let $f_2 = \{(1, 2), (2, 7)\}$, and for $n \geq 3$, let $S(n)$ be the following statement.

18.3(a). There exists a map $f_n : \mathbf{P}_n \to \mathbf{R}$ such that $f_n(1) = 2$ and $f_n(2) = 7$, and for $i \in \{3, 4, \ldots, n\}$,

$$f_n(i) = \tfrac{1}{2}[f_n(i-1) + f_n(i-2)].$$

We see that $S(3)$ is true by considering the function $f_3 = \{(1, 2), (2, 7), (3, \frac{9}{2})\}$. Let $h \geq 3$ and assume that $S(h)$ is true. We show that $S(h+1)$ is true. To see this let

$$f_{h+1} = f_h \cup \{(h+1, \tfrac{1}{2}(f_h(h) + f_h(h-1))\}.$$

It is easy to show that f_{h+1} satisfies the properties required of it so that $S(h+1)$ is true. Hence, by induction, $S(n)$ is true for each integer $n \geq 3$. Thus, there exists a collection of functions $\{f_n : n \geq 3\}$, each of which satisfies the conditions stated in 18.3(a).

We next prove that for each f_n in the collection, $f_{n+1} | \mathbf{P}_n = f_n$. To see this, suppose that for some fixed integer $n \geq 3$, $f_{n+1} | \mathbf{P}_n \neq f_n$. Let j be the first positive integer for which $f_{n+1}(j) \neq f_n(j)$. We observe that $3 \leq j \leq n$. Then

$$f_{n+1}(j-1) = f_n(j-1) \text{ and } f_{n+1}(j-2) = f_n(j-2).$$

From this, we obtain

$$f_{n+1}(j) = \tfrac{1}{2}[f_{n+1}(j-1) + f_{n+1}(j-2)] = \tfrac{1}{2}[f_n(j-1) + f_n(j-2)] = f_n(j)$$

and we have arrived at a contradiction.

For each $n \in \mathbf{P}$, let f_n be a function satisfying 18.3(a). We may apply 18.1 to $\{f_n : n \in \mathbf{P}\}$ and conclude that $f = \bigcup \{f_i : i \geq 3\}$ is a function. Next recall the remark in 18.2, about the union of functions theorem, which states that $f(x) = f_n(x)$ for all x for which $x \in \text{Dom } f_n$. Thus, $f(1) = f_3(1) = 2, f(2) = f_3(2) = 7$, and $f(3) = f_3(3) = \frac{1}{2}(f_3(2) + f_3(1))$. Also, for $i \geq 3$,

$$f(i) = f_i(i) = \tfrac{1}{2}(f_i(i-1) + f_i(i-2)) = \tfrac{1}{2}(f(i-1) + f(i-2)).$$

That f is a unique function that satisfies the given conditions is seen as follows: Suppose that there is another function f^* that satisfies the required conditions. Then the set $K = \{j : f^*(j) \neq f(j)\}$ would be a nonempty set of positive integers. Let m be the first element in K. Obviously $m \geq 4$. But then $f(i) = f^*(i)$ for $i < m$. However, $f(m) = \frac{1}{2}(f(m-1) + f(m-2)) = \frac{1}{2}(f^*(m-1) + f^*(m-2)) = f^*(m)$, and we have a contradiction.

Obviously we would not want to repeat this procedure for every example. It is convenient, therefore, to formulate a theorem that would include as special cases examples such as the one just discussed.

18.4. Theorem (Inductive definition). *Let X be a set, h a positive integer, and $a_i \in X$ for $i = 1, 2, \ldots, h$. Suppose further that $G : S \to X$ is a map defined on the set S of all finite-sequences with ranges in X. Then there is a unique sequence f with range in X such that $f(i) = a_i$ for $i = 1, 2, \ldots, h$ and $f(i) = G((f(1), f(2), \ldots, f(i-1)))$ for $i \geq h + 1$.*

Notice how this theorem covers our example. In the example $h = 2$, $a_1 = 2$, $a_2 = 7$, $f(1) = a_1$, $f(2) = a_2$, $G((s_1, s_2, s_3, \ldots, s_n)) = \frac{1}{2}(s_n + s_{n-1})$ for each finite sequence $(s_1, s_2, \ldots, s_n)$ with range in X. (We need not be concerned with the rule G for sequences with less than two terms.)

So we see that we can "inductively" define a sequence if we specify how to get it started and have a rule that tells us how to use the values already determined to get the next value.

The proof of the theorem can be patterned after the discussion in 18.3 and is left as an exercise. (See Exercise 1, following.)

EXERCISES: SEQUENCES DEFINED INDUCTIVELY

1. Prove Theorem 18.4, using the discussion in 18.3 as a hint.

2. Using Theorem 18.4, prove that there exists a unique function f on the set of all nonnegative integers that satisfies the following conditions.

   $f(0) = 1$ and $f(n) = nf(n-1)$ for each positive integer n.

   (Recall that common notation for $f(j)$ as defined inductively in this exercise is $j!$, read "j factorial.")

   Following are some exercises concerning the factorial function that are useful in various branches of mathematics.

3. For each positive integer n and each nonnegative integer r such that $r \leq n$ define $\binom{n}{r} = \dfrac{n!}{r!(n-r)!}$. Verify each of the following:

   (a) $\binom{n}{0} = 1$ for each positive integer n.

   (b) $\binom{n}{n} = 1$ for each positive integer n.

(c) For each positive integer h and for each positive integer $j \leq h$, $\binom{h}{j} + \binom{h}{j-1} = \binom{h+1}{j}$.

We can make use of Exercise 3 to prove the binomial expansion theorem in the next exercise.

4. Prove that for each positive integer n, $(a + b)^n = \binom{n}{0} a^n b^0 + \binom{n}{1} a^{n-1} b^1 + \binom{n}{2} a^{n-2} b^2 + \cdots + \binom{n}{n-1} a^1 b^{n-1} + \binom{n}{n} a^0 b^n$. Note that a short form for writing this, using summation notation, is

$$\sum_{i=0}^{n} \binom{n}{i} a^{n-i} b^i.$$

19. SOME IMPORTANT PROPERTIES OF RELATIONS

In Section 11 a *relation between sets X and Y* was defined as a subset of $X \times Y$. This gave us a generalization of the notion of function in which the single-valuedness property was dropped. So far most of our attention to the study of relations has been in connection with the study of functions and their inverses. In the next several sections we turn our attention to relations between a set X and itself. By imposing special conditions on such relations, we shall be able to study useful generalizations of relations of "ordering" such as $\leq$ and $<$ for the reals $\mathbf{R}$. Also there will be included a study of *equivalence relation* which is defined in this section.

19.1. Definition. Relations in a set. *Let S be a set, and suppose that $R \subset S \times S$. The relation R is then said to be a relation in S or the relation R is said to be defined in S.*

Note that if a relation R is defined in S, then Dom $R \subset S$ and Range $R \subset S$. It is *not* required that Dom $R = S$. Also, recall in what follows that $a R b$ means that $(a, b) \in R$.

Suppose that R is a relation in a set S so that $R \subset S \times S$. If $A \subset S$, then $R^* = R \cap (A \times A)$ is a relation *in A*. We shall refer to R^* as the *restriction* of R to A. Where there is no chance of confusion, we will use the same symbol for a relation R and a restriction of R. For example, if we consider $\geq$ in the reals $\mathbf{R}$ and we wish to restrict $\geq$ to a closed interval $[a, b]$, we still use the symbol "$\geq$."

19.2. Definition. Properties for relations. *Suppose that R is a relation in a set S. Then,*

R is transitive in S if and only if for all x, y, and z in S,

$$x R y \quad and \quad y R z \text{ imply that } x R z.$$

R is reflexive in S if and only if for all x in S,

$$x R x.$$

R is antireflexive in S if and only if for all x in S,

$$x \not\mathrel{R} x \; ((x, x) \notin R).$$

R is symmetric in S if and only if for all x and y in S,

$$x \mathrel{R} y \quad \text{implies} \quad y \mathrel{R} x.$$

R is antisymmetric in S if and only if for all x and y in S,

$$x \mathrel{R} y \quad \text{and} \quad y \mathrel{R} x \text{ imply that } x = y.$$

19.3. Definition. Equivalence relation. *A relation E in a set S is said to be an equivalence relation in S if and only if for all x, y, and z in S,*
$x \mathrel{E} x$ *(i.e., E is reflexive in S),*
$x \mathrel{E} y$ *implies* $y \mathrel{E} x$ *(i.e., E is symmetric in S),*
$x \mathrel{E} y$ *and* $y \mathrel{E} z$ *imply* $x \mathrel{E} z$ *(i.e., E is transitive in S).*

19.4. *Examples.* In the set of all real numbers **R**,
the relation $=$ is an equivalence relation,
the relations $\geq$ and $\leq$ are transitive, reflexive, and antisymmetric,
the relations $<$ and $>$ are transitive, antisymmetric, and antireflexive.

19.5. *Examples.* Let X be a set and let $\mathscr{P}(X)$ denote the power set of X. The relations $\subset$ and $\supset$ are transitive, reflexive, and antisymmetric in $\mathscr{P}(X)$.

EXERCISES: SOME IMPORTANT PROPERTIES OF RELATIONS

1. In each of the following, classify the relation as to which of the properties discussed in Section 19 it possesses.

    (a) Let S be the set of all triangles in the plane. Let R be the relation in S defined as follows: for all a and b in S,

    $$a \mathrel{R} b \text{ if and only if } a \text{ is congruent to } b.$$

    (b) Let **R** be the set of all real numbers. Let $S = \{(x, y): (x, y) \in \mathbf{R} \times \mathbf{R} \text{ and } y \neq 0\}$.
    For all (a, b) and (c, d) in S, let

    $$(a, b) \mathrel{R} (c, d) \text{ provided that } ad = bc.$$

    (c) Suppose j is a fixed positive integer. For each a and b in **Z**, let

    $$a \mathrel{R} b \text{ if and only if } a - b = jk \text{ for some integer } k.$$

    (See Exercises 9 and 10, page 23.)

2. Suppose that R is a relation that is transitive in a set S. Let us define a new relation in S as follows: For each a and b in S, let $a \mathrel{R^*} b$ if and only if $a = b$ or $a \mathrel{R} b$. Is R^* transitive in S? Is R^* reflexive in S? Illustrate with an R that is not reflexive.

3. Suppose that R is a relation in a set S. We define a new relation R^* as follows: For each a and b in S, let $a \mathrel{R^*} b$ if and only if $a \mathrel{R} b$ is true and $b \mathrel{R} a$ is false. Suppose that R is transitive. Is R^* also transitive? Is R^* necessarily antisymmetric?

4. Suppose that a relation R in a set S is transitive and antire-flexive. Is it necessarily antisymmetric?

5. Are the following propositions true?
    (a) Suppose that R is a relation in a set S. Then R is symmetric if and only if $R \subset R^{-1}$.
    (b) Suppose that R is a relation in a set S. Then R is symmetric if and only if $R = R^{-1}$.

6. Suppose that R is a relation defined in S. Is $R \cap R^{-1}$ a symmetric relation in S?

7. Suppose that R is a transitive relation in S. Is R^{-1} transitive in S?

8. Suppose that R is symmetric in S. Is R^{-1} symmetric in S?

9. Suppose that R is reflexive in S. Is R^{-1} reflexive in S?

10. Does there exist a nonempty set S and a relation R in S such that R is both symmetric and antisymmetric in S?

20. DECOMPOSITION OF A SET

20.1. Definition. Decomposition of a set. *Let $\mathscr{K}$ be a collection of sets. $\mathscr{K}$ is said to be pairwise disjoint provided that if $A \in \mathscr{K}$, $B \in \mathscr{K}$, and $A \neq B$, then $A \cap B = \varnothing$. Suppose that S is a set and $\mathscr{K}$ is a collection of nonempty pairwise disjoint subsets of S such that $\bigcup \mathscr{K} = S$. Then $\mathscr{K}$ is said to be a decomposition (or partition) of S.*

20.2. *Example.* Let **P** be the set of all positive integers. Let O be the set of all positive odd integers and let E be the set of all positive even integers. Then $\mathscr{K} = \{O, E\}$ is a decomposition of **P**.

20.3. *Example.* Let **Z** be the collection of all integers and let R be the relation defined as follows: For each ordered pair of integers (m, n), let $m \, R \, n$ if and only if $m - n$ is a multiple of 5. (See Exercise 10, page 23.) It can be shown that R is an equivalence relation in **Z**. The reader who did Exercise 10, page 23, should have determined that there are five distinct sets of the form $R[i]$. Now note that $\{R[i] : i = 0, 1, 2, 3, 4\}$ is a decomposition of **Z**.

EXERCISES: DECOMPOSITION OF A SET

1. For each real number r, let $F_r = \{(r, y) : y \in \mathbf{R}\}$. Is $\{F_r : r \in \mathbf{R}\}$ a partition of $\mathbf{R} \times \mathbf{R}$?

2. Let $A_0 = \{x : -1 \leq x \leq 1\}$. For each $x \in \mathbf{R} - A_0$, let $A_x = \{x\}$. Is the following collection $\mathscr{K}$ a decomposition of the real line $\mathbf{R}$?
    $$\mathscr{K} = \{A_x : x = 0 \quad \text{or} \quad |x| > 1\}$$

3. Suppose X is a nonempty set and $f : X \to Y$ is a surjection. Is $\{f^{-1}[y] : y \in Y\}$ a decomposition of X?

21. EQUIVALENCE CLASSES

Let S be a nonempty set and suppose that E is an equivalence relation in S (See 19.3). In terms of the notation introduced in Section 11, for each $a \in S$

$$E[a] = \{x : a \, E \, x\}.$$

We shall show in this section that $\{E[a] : a \in S\}$ is a decomposition of S.

21.1. Theorem. *Suppose that S is a nonempty set and E is an equivalence relation in S. Then:*

21.1. (a). *For all $a \in S$*

$$a \in E[a].$$

21.1. (b). *For all a and b in S,*

$$\text{if } b \in E[a], \quad \text{then} \quad E[a] = E[b].$$

21.1. (c). *For all a and b in S,*

$$E[a] \cap E[b] = \varnothing \quad \text{or} \quad E[a] = E[b]$$

and hence $\{E[a] : a \in S\}$ is a decomposition of S.

PROOF. Part (a) follows from the fact that E is a reflexive relation in S.

To prove Part (b), we first prove

21.1. (d). If $x \in E[y]$, then $E[x] \subset E[y]$.
To see this, assume that $x \in E[y]$ and choose a $w \in E[x]$. (We will show that $w \in E[y]$.) Since $w \in E[x]$ and since E is symmetric, $w \, E \, x$. Similarly, since $x \in E[y]$, $x \, E \, y$. Then the transitivity of E gives $w \, E \, y$. Consequently $w \in E[y]$ and we have shown that $E[x] \subset E[y]$.

Part (b) now follows easily. For suppose that $b \in E[a]$. Then since E is symmetric, we also have $a \in E[b]$. By using 21.1 (d) twice, we then get $E[b] \subset E[a]$ and $E[a] \subset E[b]$.

Part (c) follows from (b) by the following argument: Suppose that $z \in E[a] \cap E[b]$. Then from (b), $E[z] = E[a]$ and $E[z] = E[b]$. Hence, $E[a] = E[b]$. This completes the proof.

Each of the sets $E[a]$ in the previous discussion is called an *E-equivalence class*. Note that for each $a \in S$, $E[a]$ represents the set of all $y \in S$ to which a is E-related, or since E is symmetric, $E[a]$ is the set of all $y \in S$, each of which is related to a. Notice also that because of 21.1 (a) and (b), an equivalence class $E[a]$ can be written as $E[z]$ where z is any member of $E[a]$. Furthermore, the fact that $\{E[a] : a \in S\}$ is a decomposition of S shows that what happened in Example 20.3 happens in general. That is, if E is an equivalence relation in S, then E effects a decomposition of S. Moreover, it is of interest to note that if $\mathscr{K}$ is a partition of a set S, then there exists an equivalence relation E in S such that the E-equivalence classes are precisely the elements of $\mathscr{K}$. To see this, simply define E as follows: For each a and b in S, let $a \, E \, b$ if and only if a and b are in the same element of the partition $\mathscr{K}$. It is easy to see that E is indeed an equivalence relation in S.

EXERCISES: EQUIVALENCE CLASSES

1. For each ordered pair of real numbers (a, b) such that $a \neq 0$ and $b \neq 0$, let $E(a, b)$ be the equation $ax + by = 0$. Let $\xi = \{E(a, b): a \neq 0 \text{ and } b \neq 0\}$. For $E(a, b)$ and $E(c, d)$ in ξ, let us define a relation $\simeq$ as follows. $E(a, b) \simeq E(c, d)$ if and only if every solution (x, y) of $ax + by = 0$ is a solution of $cx + dy = 0$, and every solution of $cx + dy = 0$ is a solution of $ax + by = 0$. Note that $\simeq$ is an equivalence relation. Is it true that $E(2, 3) \simeq E(4, 6)$? Try to discover an equation relating a, b, c, and d so that $E(a, b) \simeq E(c, d)$ provided that a, b, c, and d satisfy your equation. Justify your conjecture. Write an equation that is in the same equivalence class as is $E(3, 2)$ but which is not the same as $E(3, 2)$.

2. Let $\mathbf{R}_+$ be the collection of all positive real numbers. For $a \in \mathbf{R}_+$ and $b \in \mathbf{R}_+$, let $a \, R \, b$ if and only if $a \div b$ is a rational number. Is R an equivalence relation on $\mathbf{R}_+$? Justify your answer. What is the form of all the numbers b such that $b \in R[\sqrt{2}]$? If a is an irrational positive number and $b \, R \, a$, is b necessarily an irrational number?

3. Let $\mathbf{R}$ be the set of all real numbers and let $\mathbf{R}^2 = \mathbf{R} \times \mathbf{R}$. Let m be a fixed real number. For each (x_1, y_1) and (x_2, y_2) in $\mathbf{R}^2$ let $(x_1, y_1) \, R \, (x_2, y_2)$ provided that $y_1 - mx_1 = y_2 - mx_2$. Is R an equivalence relation? Let $m = 3$. Sketch $R[(1, 2)]$.

4. Let $\mathcal{D}$ be the set of all real-valued functions which are defined and have derivatives on the open interval (a, b). For $f \in \mathcal{D}$ and $g \in \mathcal{D}$, let $f \, R \, g$ provided that $f' = g'$. Is R an equivalence relation in $\mathcal{D}$? Let $f(x) = x^2$ for $x \in (a, b)$. Find $R[f]$.

22. PARTIALLY ORDERED AND TOTALLY ORDERED SETS

In Examples 19.4 and 19.5 we saw examples of well-known relations, some of which were *transitive*, *reflexive*, and *antisymmetric*. Such relations are called *partial orders*. In the forthcoming discussion when we are speaking of a particular relation such as $\subset$ or $\supset$, we shall, of course, use one of the standard symbols for that relation. However, in a general discussion about partial orders, we shall use the symbol "$\leq$". This should not cause any confusion, since it will be known from the context that we are not necessarily referring to the usual "$\leq$" for real numbers.

22.1. Definitions. Partially ordered sets. *Suppose that S is a set and a relation $\leq$ in S is*

> *transitive:*   *for all a, b, and c in S, $a \leq b$ and $b \leq c$, imply $a \leq c$;*
>
> *reflexive:*   *for all a in S, $a \leq a$;*

and

> *antisymmetric: for all a and b in S, $a \leq b$ and $b \leq a$ imply $a = b$.*

Then, $\leq$ is said to be a partial order for S and the pair $(S, \leq)$ is said to be a partially ordered set or a partially ordered system.

It should be noted that if $\leq$ is a partial order for a set S, then for each subset $A \subset S$, the relation $\leq$ restricted to A is also a partial order for A.

If $(S, \leq)$ is a partially ordered set, we shall sometimes use the following terminology: For a and b in S, if $a \leq b$ we shall say that *a precedes b* or that *b follows a*.

22.2. *Example.* Let $\mathscr{K}$ be a collection of sets. Then $(\mathscr{K}, \supset)$ and $(\mathscr{K}, \subset)$ are partially ordered sets.

22.3. *Example.* Let S be a set of real numbers and let $\leq$ and $\geq$ have their usual meanings. Then $(S, \geq)$ and $(S, \leq)$ are partially ordered sets.

22.4. *Example.* Let $\mathscr{F}$ be the collection of all real-valued functions defined on $\mathbf{R}$. For all f and g in $\mathbf{R}$, let $f \leq g$ if and only if $f(x) \leq g(x)$ for all $x \in \mathbf{R}$. Then $(\mathscr{F}, \leq)$ is a partially ordered set.

There may be points a and b in a partially ordered set $(S, \leq)$ for which it is true that neither $a \leq b$ nor $b \leq a$. The partially ordered set $(\mathscr{K}, \subset)$ in Example 22.2 may be of this type, depending on the particular collection of sets $\mathscr{K}$. However, notice that in the example $(S, \leq)$ of 22.3, for each pair of numbers a and b in S, either $a \leq b$ or $b \leq a$. This is an example of a *totally ordered* set, which will be defined next.

22.5. **Definition. Totally ordered set.** *Suppose that S is a set and $\leq$ is a partial order for S that satisfies the following additional property:*

For each x and y in S, $x \leq y$ or $y \leq x$.

The relation $\leq$ is then said to be a total (or linear) order for S, and $(S, \leq)$ is said to be a totally (or linearly) ordered set.

The following special kind of set that is linearly ordered occurs so frequently that a special name has been given to it.

22.6. **Definition. A nested collection of sets.** *Let $\mathscr{K}$ be a collection of sets such that for each $A \in \mathscr{K}$ and $B \in \mathscr{K}$, either $A \subset B$ or $B \subset A$. Then $\mathscr{K}$ is said to be a nested collection of sets.*

22.7. *Example.* If (K_i) is an increasing or a decreasing sequence of sets (see 15.4), then the collection $\{K_i : i \in \mathbf{P}\}$ is nested. In particular, the collection $\{A_m : m \in \mathbf{P}\}$ in Exercise 2, page 13, is a nested collection.

22.8. *Example.* For each real number r, let $K_r = \{x : x \leq r\}$. The collection $\{K_r : r \in \mathbf{R}\}$ is a nested collection of subsets of $\mathbf{R}$.

23. PROPERTIES OF BOUNDEDNESS FOR PARTIALLY ORDERED SETS

The reader is probably familiar with the concepts of *upper bound*, *lower bound*, *least upper bound*, and *greatest lower bound* for subsets of real numbers. In this section, we shall define these and related concepts in the more general framework of partially ordered sets.

23.1. Definitions. Upper and lower bounds for partially ordered sets. *Suppose that $(S, \leq)$ is a partially ordered set. If $A \subset S$, $u \in S$, and*

$$a \leq u \quad \text{for all} \quad a \in A,$$

then u is said to be an upper bound of A (or for A). Similarly, if $l \in S$ and

$$l \leq a \quad \text{for all} \quad a \in A,$$

then l is said to be a lower bound of A.

23.2. Definition. Bounded subsets. *Let $(S, \leq)$ be a partially ordered set. If $A \subset S$ and A has both an upper and a lower bound, then A is said to be a bounded subset of S.*

23.3. Example. Let $[a, b]$ be a closed interval in **R**. Let $\leq$ be given the usual meaning of *equal to or less than*. Then for any $x \in \mathbf{R}$, if $x \leq a$, x is a lower bound for $[a, b]$. Similarly if $b \leq x$, x is an upper bound for $[a, b]$.

It should be pointed out that the choice of words *lower* and *upper* for partially ordered systems $(S, \leq)$ is purely arbitrary. Their usage is motivated by their usage in $(\mathbf{R}, \leq)$. If we consider $(\mathbf{R}, \geq)$ then *upper* and *lower* bounds as defined in 23.1 exchange roles with respect to their usual meanings. Thus, to be consistent with the usual language, it is preferable to deal with $(\mathbf{R}, \leq)$ rather than $(\mathbf{R}, \geq)$.

23.4. Example. Let X be a set and let $\mathscr{P}(X)$ be the power set of X. Consider the partially ordered system $(\mathscr{P}(X), \subset)$. Let $\mathscr{K} \subset \mathscr{P}(X)$. Then X is an upper bound for $\mathscr{K}$. Also $\bigcup \mathscr{K}$ is an upper bound for $\mathscr{K}$. Likewise, $\varnothing$ and $\bigcap \mathscr{K}$ are lower bounds for $\mathscr{K}$.

Observe that if A is a subset of a partially ordered set $(S, \leq)$, then at most one upper bound of A can be an element of A. This follows at once from the definition of upper bound and from the fact that the partial order $\leq$ is antisymmetric. Thus, in the following definition we use the term *the* greatest element in a set, and similarly, *the* least element.

23.5. Definitions. Least and greatest elements. *Suppose that $(S, \leq)$ is a partially ordered set. Suppose that $A \subset S$. Then l is said to be the least (or smallest or first or minimum) element of S provided that*

$$l \in A \quad \text{and} \quad l \leq x \quad \text{for all} \quad x \in A.$$

Similarly, g is said to be the greatest (largest or last or maximum) element of S provided that

$$g \in A \quad \text{and} \quad x \leq g \quad \text{for all} \quad x \in A.$$

23.6. Definition. Least upper bound and greatest lower bound. *Let $(S, \leq)$ be a partially ordered set. Suppose that A is a subset of S and A has an upper bound in S. If the set of upper bounds of S has a least element l, then l is called the least upper bound of A (l.u.b. (A)). If A has a lower bound and the set of lower bounds of S has a greatest element g, then g is called the greatest lower bound of A (g.l.b. (A)).*

23.7. Example. In Example 23.4, $\bigcap \mathscr{K} = $ g.l.b. $(\mathscr{K})$ and $\bigcup \mathscr{K} = $ l.u.b. $(\mathscr{K})$. Suppose in particular that $X = \{1, 2, 3, 4\}$ and $\mathscr{K} = \{\{1, 2\}, \{1, 3\}\}$. In this case, neither l.u.b. $(\mathscr{K})$ nor g.l.b. $(\mathscr{K})$ belongs to $\mathscr{K}$.

23.8. *Example.* Consider $(\mathbf{R}, \leq)$. Let U be the open interval $(0, 1)$ and let F be the closed interval $[0, 1]$. We observe that

$$\text{g.l.b.}\ (U) = \text{g.l.b.}\ (F) = 0$$

and

$$\text{l.u.b.}\ (U) = \text{l.u.b.}\ (F) = 1.$$

23.9. *Example.* Consider $(\mathbf{Q}, \leq)$, recalling that $\mathbf{Q}$ is the set of all rational numbers in $\mathbf{R}$. Let $S = \{x : x \in \mathbf{Q} \text{ and } x^2 < 2\}$. Note that S has many lower bounds and many upper bounds. However, S has neither a least upper bound nor a greatest lower bound in the system $(\mathbf{Q}, \leq)$.

Often the word *supremum* is used for least upper bound, and *infimum* is used for greatest lower bound. The corresponding abbreviations are: *supremum of* $A = sup\ (A)$ and *infimum of* $A = inf\ (A)$.

23.10. *Example.* Let a, b, and c be different objects and let $\mathscr{K} = \{\{a\}, \{b\}, \{c\}, \{b, c\}, \{a, c\}\}$. Consider the partially ordered set $(\mathscr{K}, \subset)$. Note that $\mathscr{K}$ has neither a maximum element nor a minimum element in $\mathscr{K}$. However, neither $\{a\}$ nor $\{b\}$ is preceded by any other element in $\mathscr{K}$. Likewise, neither $\{b, c\}$ nor $\{a, c\}$ is followed by any other element in $\mathscr{K}$. The properties illustrated by $\{a\}$ and $\{b\}$ on the one hand and by $\{b, c\}$ on the other are defined next.

23.11. **Definition. Maximal and minimal elements.** *Let $(S, \leq)$ be a partially ordered set. An element m in S is said to be a maximal element in S, if m is not followed by any other element in S, that is, if for all $s \in S$,*

$$m \leq s \text{ implies that } m = s.$$

Similarly, m is a minimal element in S provided that m is not preceded by any other element in S, that is, if for all $s \in S$

$$s \leq m \text{ implies that } s = m.$$

We see that in Example 23.10, $\{a\}$ and $\{b\}$ are *minimal* elements while $\{b, c\}$ and $\{a, c\}$ are *maximal* elements in $\mathscr{K}$.

23.12. **Definition. Well-ordered set.** *If $(S, \leq)$ is a partially ordered set such that every nonempty subset of S has a first element, then $\leq$ is said to be a well-ordering for S. Also, in such a case $(S, \leq)$ is said to be a well-ordered set.*

It should be clear to the reader that if $\leq$ is a well-ordering for S, then it is also a *linear ordering*. (See Exercise 6 in the next set of exercises.) It is to be noted that the set $\mathbf{P}$ of positive integers is well-ordered by $\leq$, where $\leq$ is given the usual meaning (See 17.2).

EXERCISES: PARTIALLY ORDERED AND TOTALLY ORDERED SETS

1. Consider the system $(\mathbf{R}, \leq)$ of the real line $\mathbf{R}$ together with the usual ordering $\leq$.
    (a) Give an example of subset A of $\mathbf{R}$ that is bounded below but not above. Similarly, give an example of a subset B of $\mathbf{R}$ that is bounded above but not below.
    (b) Give an example of a subset S of $\mathbf{R}$ that has a least upper bound but whose least upper bound does not belong to S.

2. Give an example of a collection of sets $\mathcal{K}$ such that the partially ordered set $(\mathcal{K}, \subset)$ satisfies the following two conditions:
    (a) $\mathcal{K}$ is not linearly ordered.
    (b) Every linearly ordered subset of $\mathcal{K}$ has an upper bound in $\mathcal{K}$. In your example, does $\mathcal{K}$ have a maximal element?

3. Suppose that S is a set and R is a relation in S that is transitive and antireflexive (19.2). Define $\underset{=}{R}$ in S as follows: For all x and y in S,
    $$x \underset{=}{R} y \text{ if and only if } x R y \text{ or } x = y.$$
    Is $\underset{=}{R}$ a partial ordering in S?

4. Let (A, R) and (B, Γ) be partially ordered sets. Define a relation $\leq$ in $A \times B$,
    $$(a_1, b_1) \leq (a_2, b_2) \text{ if and only if } a_1 R a_2 \text{ and } b_1 \Gamma b_2.$$
    (a) Show that $\leq$ is a partial ordering for $A \times B$.
    (b) Suppose that R and Γ are total orderings for A and B, respectively.
    Is $\leq$ necessarily a total ordering for $A \times B$?

5. Let $\mathcal{L}$ be the relation defined in $\mathbf{R} \times \mathbf{R}$ as follows. For all (a_1, a_2) and (b_1, b_2) in $\mathbf{R} \times \mathbf{R}$, let $(a_1, a_2) \mathcal{L} (b_1, b_2)$ if and only if
    $$a_1 \leq b_1, \quad \text{and if } a_1 = b_1, \quad \text{then } a_2 \leq b_2.$$
    For obvious reasons this relation, $\mathcal{L}$, is called a *dictionary* or *lexicographical* order for $\mathbf{R} \times \mathbf{R}$.
    (a) Is $\mathcal{L}$ a partial ordering for $\mathbf{R} \times \mathbf{R}$?
    (b) If the answer to (a) is yes, is $\mathcal{L}$ a total ordering for $\mathbf{R} \times \mathbf{R}$?

6. Prove that if $(S, \leq)$ is a well-ordered set, then it is a linearly ordered set.

24. AXIOM OF CHOICE AND ZORN'S LEMMA

Some form of the axiom of choice is usually included in any axiomatic treatment of set theory. The axiom of choice asserts that if $\mathcal{K}$ is a nonempty collection of nonempty sets, then there is a set that can be formed by choosing one element from each set in the collection $\mathcal{K}$. The statement known as Zorn's lemma is equivalent to the axiom of choice. We shall accept from set theory and make use of Zorn's lemma. The reader who is interested in reading a proof of the equivalence of the axiom of choice and Zorn's lemma is referred to, for example, [1], [20], or [26]. We now state a form of the axiom of choice in terms of a so-called choice function.

24.1. Axiom of Choice. *Suppose that $\mathcal{K}$ is a nonempty collection of nonempty sets. Then there is a function s defined on $\mathcal{K}$ such that $s(K) \in K$ for each $K \in \mathcal{K}$. (This function s selects or chooses an element out of each $K \in \mathcal{K}$.)*

In this statement, it is seen that if $\mathcal{K}$ is a nonempty collection of nonempty sets and s is a choice function for $\mathcal{K}$, then $s[\mathcal{K}]$ is a set formed by choosing one element from each $K \in \mathcal{K}$. Furthermore, if the elements of $\mathcal{K}$ are pairwise disjoint, we see that a choice function for $\mathcal{K}$ is a one-to-one function.

The following example illustrates a use of the axiom of choice.

24.2. *Example.* Let $f : X \to Y$ be a surjection. We show that there is an $X^* \subset X$ such that $f \,|\, X^* : X^* \to Y$ is a bijection. To see this let s be a choice function for the collection of sets $\{f^{-1}[y] : y \in Y\}$. Then $X^* = s[\{f^{-1}[y] : y \in Y\}]$ has the properties we are seeking. What we have done, using language which sounds more intuitive, is to choose one x from each $f^{-1}[y]$. The set X^* is the collection of all x's so chosen.

24.3. **Zorn's Lemma.** *Suppose that $(S, \leqq)$ is a nonempty partially ordered set such that every linearly ordered subset in S has an upper (lower) bound in S. Then S has a maximal (minimal) element in S.*

Suppose that X is a set and there exists a subset of X with some property P. Sometimes it may be of interest to know if there exists a *maximal* subset of X having that property (i.e., a subset S of X that has property P and is not contained in any other subset of X that has property P). If there is an affirmative answer to such a question, that fact is often proved by making use of Zorn's lemma. We illustrate this in the next proof.

24.4. **Hausdorff Maximality Principle.** *Let $(S, \leqq)$ be a partially ordered set. Then S contains a maximal linearly ordered subset.*

PROOF. Let $\mathcal{K}$ be the collection of all subsets of S that are linearly ordered with respect to $\leqq$. Then $(\mathcal{K}, \subset)$ is a partially ordered system. We shall show that $(\mathcal{K}, \subset)$ satisfies the condition needed to apply Zorn's lemma. Let $\mathcal{K}^*$ be a subset of $\mathcal{K}$ that is linearly ordered with respect to $\subset$. (Thus, if $A \in \mathcal{K}^*$ and $B \in \mathcal{K}^*$, then A and B are subsets of S that are linearly ordered with respect to $\leqq$, and $A \subset B$ or $B \subset A$.) Now $\bigcup \mathcal{K}^*$ is a subset of S. Furthermore, $\bigcup \mathcal{K}^*$ is linearly ordered by $\leqq$. For suppose that $a \in \bigcup \mathcal{K}^*$ and $b \in \bigcup \mathcal{K}^*$; then $a \in A$ and $b \in B$ for some $A \in \mathcal{K}^*$ and $B \in \mathcal{K}^*$. However, $A \subset B$ or $B \subset A$. Suppose, for example, that $A \subset B$. Then $a \in B$ and $b \in B$, and since B is linearly ordered with respect to $\leqq$, it follows that $a \leqq b$ or $b \leqq a$. Hence, we have shown that $\bigcup \mathcal{K}^*$ is linearly ordered by $\leqq$ and thus $\bigcup \mathcal{K}^* \in \mathcal{K}$. Moreover, for each $A \in \mathcal{K}^*$, $A \subset \bigcup \mathcal{K}^*$ so that $\bigcup \mathcal{K}^*$ is an upper bound for $\mathcal{K}^*$ with respect to the relation $\subset$. Hence, by Zorn's lemma, $\mathcal{K}$ has an element M that is a maximal element with respect to $\subset$. This means that there exists a subset M of S that is maximal with respect to the property of being a linearly ordered subset of S.

Note. We used Zorn's lemma to prove the Hausdorff maximality principle. Actually, the Hausdorff maximality principle is equivalent to Zorn's lemma and, hence, to the axiom of choice. Another statement that is equivalent to the axiom of choice is the *well-ordering principle*. Recall that the relation *equal to or less than* well-orders the positive integers (see 17.2). The *well-ordering principle* states that for each set S, there exists a partial ordering w for S such that with respect to this relation (S, w) is a well-ordered set (see 23.12). Of course, this does not mean that for a given set we will necessarily know how to find such an ordering explicitly.

EXERCISES: AXIOM OF CHOICE AND ZORN'S LEMMA

1. Show that the two forms of Zorn's lemma given in 24.3 are equivalent.

2. Prove the following variation of 24.4. Let $(S, \leqq)$ be a partially ordered set. If A is a linearly ordered subset of S, then there exists a maximal linearly ordered subset M of S such that $A \subset M$.

3. Let $\mathscr{K}$ be a collection of sets. Prove that there exists a maximal nested subcollection of $\mathscr{K}$.

25. CARDINALITY OF SETS (INTRODUCTION)

Suppose that two piles of pennies are set before us. We can determine whether one pile has the same number of pennies as the second pile by pairing off the pennies. If the piles do not have the same number, we can determine which pile has more pennies. Here we are considering so-called finite sets. Can such a comparison also be made between infinite sets? This and the next several sections deal with certain concepts suggested by this question.

In the previous paragraph we used the terms finite set and infinite set. At this point we give precise definitions of these terms.

25.1. Definitions. Finite and infinite sets. *Suppose that S is a set and for some positive integer n, there exists a one-to-one map from $\mathbf{P}_n = \{1, 2, \ldots, n\}$ onto S. Then S is said to be a finite set and n is said to be its cardinal number. If a set is empty it is also called a finite set, and it is said to have 0 as its cardinal number. If a set is not a finite set it is said to be an infinite set.*

Notice that if we count the number of elements in a finite collection and determine that it has n elements, we are essentially putting the set into one-to-one correspondence with a "standard set," $\mathbf{P}_n$. (Recall that a one-to-one correspondence between two sets is a one-to-one map from one set onto the other.) However, one could compare two sets without actually having the positive numbers available. Suppose, for example, that each of two children has a collection of marbles. Let us imagine that the children do not know how to count with natural numbers but do have an aptitude for counting. They could pair off each marble in the first collection with a marble in the second collection. If both collections are exhausted by the pairing, then the collections are the same "size." However, if the first collection is exhausted but the second collection still has elements, we would agree, in the case of the marble collection, that the second collection is larger. This approach is suggestive of what could be done in the infinite case, but it must be modified because, as illustrated in the next paragraph, something happens in the infinite case that does not happen in the finite case.

Consider the set $\mathbf{P}$ of all positive integers and the set $\mathbf{N}$ of negative integers. Consider the following pairing (or function): $\{(1, -1), (2, -2), (3, -3), \ldots, (n, -n), \ldots, \}$. We see that we have defined a one-to-one function whose domain is $\mathbf{P}$ and whose range is $\mathbf{N}$. Thus, we will want to consider $\mathbf{P}$ and $\mathbf{N}$ as having the same number of elements. On the other hand, consider the map $f: \mathbf{P} \to \mathbf{N}$ given by $f(n) = -2n$ for each $n \in \mathbf{P}$. Here we have a one-to-one function from $\mathbf{P}$ into a proper subset of $\mathbf{N}$ so that $\mathbf{N}$ is not exhausted. When this situation occurred in our consideration of finite collections, we would have been willing to say that the second set had more elements in it than did the first set. However, here we see a situation in which, from the standpoint of one map, we might be tempted to say that one set has more elements than does the other, but from the standpoint of another map we would want to say that the two sets have the same number of elements. As we shall see, the situation can be handled by making appropriate definitions.

We now define a concept that will generalize the notion of two sets having the same number of points.

25.2. Definition. Equivalent sets. *Two sets A and B are said to be equivalent (with respect to cardinality) provided that there is a one-to-one map from A onto B. We shall use the notation $A \sim B$ to indicate this.*

It is to be emphasized that for A and B to be equivalent there must exist at least one map which is one-to-one and which takes A onto B. In the finite case, if there exists a one-to-one mapping from A onto B, there cannot exist a one-to-one mapping from A onto a proper subset of B. This is not so in the infinite case. It can be shown that if S is an infinite set, then there exists a one-to-one mapping from S onto a proper subset of itself (see Exercise 5, page 53). Thus, just because there is a one-to-one correspondence between A and a proper subset of B, we would not wish to think of A as having fewer points than does B. However, there is a famous theorem, called the Schröder-Bernstein theorem, which asserts that if A is equivalent to a subset of B and B is equivalent to a subset of A, then A and B are equivalent. Thus, we would want to think of A as having fewer points than B if and only if A is equivalent to a proper subset of B but is not equivalent to B.

25.3. Theorem. *Let $\mathscr{K}$ be a nonempty collection of sets. Then $\sim$ as defined in 25.2 defines an equivalence relation in $\mathscr{K}$.*

The proof is straightforward and is left as an exercise.

EXERCISES: CARDINALITY OF SETS (INTRODUCTION)

1. Prove Theorem 25.3.

2. Show that the set $\mathbf{P}$ of all positive integers is equivalent to the set of all positive even integers.

3. Show that the set $\mathbf{P}$ of all positive integers is equivalent to the set $\mathbf{P} - \{1\}$.

4. Recall that $\mathbf{P}_n = \{1, 2, \ldots, n\}$. Show that $\mathbf{P} \sim (\mathbf{P} - \mathbf{P}_n)$.

26. COUNTABLE SETS

We observe that the notions of finite set and infinite set are disjoint alternatives. It is useful to introduce other classifications of sets with respect to cardinality. For this section we shall consider finite sets and also sets which are equivalent to the set **P** of positive integers. In the next section we shall show that there are infinite sets that have so many elements that they cannot be put into one-to-one correspondence with the set **P**.

26.1. **Definitions.** **Countable set, countably infinite set.** *A set C that is either finite or equivalent to the set* **P** *of all positive integers is said to be a countable set. If C is countable and infinite, it is called a countably infinite set.*

The following theorem is easy to see and we omit a formal proof.

26.2. **Theorem.** *Let* $\{K_1, K_2, \ldots, K_n\}$ *be a finite collection of finite sets* K_i. *Then* $\bigcup \{K_1, K_2, \ldots, K_n\}$ *is a finite set.*

26.3. **Theorem.** *If A is a countably infinite set and B is an infinite subset of A, then B is countably infinite.*

PROOF. We shall show that there is a one-to-one sequence β that maps **P** onto B. Since A is countably infinite, there is a one-to-one sequence ρ that maps **P** onto A. Now let $N(1)$ be the first positive integer such that $\rho(N(1)) \in B$. Let $N(2)$ be the first positive integer such that $\rho(N(2)) \in B - \{\rho(N(1))\}$. After $N(1)$, $N(2), \ldots, N(j)$ have been chosen, let $N(j + 1)$ be the first positive integer such that $\rho(N(j + 1)) \in B - \{\rho(N(i)) : i \in \mathbf{P}_j\}$. Since B is an infinite set, $B - \{\rho(N(i)) : i \in \mathbf{P}_j\}$ is an infinite set, so that $N(j + 1)$ exists. By 18.4, the sequence $(N(i))$ is well defined. Next, for each $i \in \mathbf{P}$, let $\beta(i) = \rho(N(i))$. It is clear that β is one-to-one since N and ρ are each one-to-one. To see that β maps **P** onto B, let $x \in B$. There exists a $j \in \mathbf{P}$ such that $\rho(j) = x$. Notice that if $j \notin \{N(1), N(2), \ldots, N(j - 1)\}$, then $x \in B - \{\rho(N(i)) : i \in \mathbf{P}_{j-1}\}$. However, in that case, $N(j) = j$ and $\beta(j) = \rho(N(j)) = x$. This completes the proof.

The following is an immediate consequence of the last two theorems.

26.4. **Theorem.** *Every subset of a countable set is a countable set.*

The following theorem is easy to prove and is left as an exercise.

26.5. **Theorem.** *If X is a set that is equivalent to a countable set, then it is also a countable set.*

26.6. **Theorem.** *If* $f : X \to Y$ *is a surjection and X is countable, then Y is countable.*

PROOF. Let $\mathbf{P}^*$ be $\mathbf{P}_n$ or $\mathbf{P}$ according to whether X is finite with cardinality $n \geq 1$ or is infinite. Then there exists a bijection $\alpha : \mathbf{P}^* \to X$. Using the well-ordering principle for positive integers, define $\beta : Y \to \mathbf{P}^*$ by $\beta(y) = $ least element of $\alpha^{-1}[f^{-1}[y]]$. The mapping $\beta : Y \to \mathbf{P}^*$ is an injection (one-to-one) and $\beta[Y]$ is a countable set by 26.4. The fact that $\beta : Y \to \beta[Y]$ is a bijection shows, by 26.5, that Y is countable.

26.7. **Theorem.** $\mathbf{P} \times \mathbf{P}$ *is countably infinite.*

PROOF. For each $(m, n) \in \mathbf{P} \times \mathbf{P}$, let $f((m, n)) = 2^m \cdot 3^n$. This defines a map f from $\mathbf{P} \times \mathbf{P}$ onto a subset of **P**. Further, f is a one-to-one map. Then f^{-1}: $f[\mathbf{P} \times \mathbf{P}] \to \mathbf{P} \times \mathbf{P}$ is a one-to-one map from $f[\mathbf{P} \times \mathbf{P}]$ onto $\mathbf{P} \times \mathbf{P}$. $f[\mathbf{P} \times \mathbf{P}]$ is a subset of the countable set **P** so that it is countable. Hence, $\mathbf{P} \times \mathbf{P}$ is countable. Obviously $\mathbf{P} \times \mathbf{P}$ is an infinite set, so that it is countably infinite.

26.8. **Theorem.** *The union of a countable collection of countable sets is countable.*

PROOF. Let $\mathcal{K}$ be a countable collection of countable sets. Without loss of generality, we may assume that $\mathcal{K}$ is nonempty and that none of the elements of $\mathcal{K}$ is empty. We may write $\mathcal{K} = \{K_i : i \in \mathbf{P}^*\}$, where $\mathbf{P}^* = \mathbf{P}_j$ or $\mathbf{P}$ according to whether $\mathcal{K}$ is finite with j elements or is countably infinite. Since for each $i \in \mathbf{P}^*$, K_i is a countable set, there is a one-to-one map $\alpha^{(i)}$ from $P^{(i)}$ onto K_i, where $P^{(i)}$ is $\mathbf{P}$ if K_i is infinite, or $P^{(i)}$ is the set $\{1, 2, \ldots, n(i)\}$ where $n(i)$ is the number of points in K_i if K_i is finite. Next let $X = \{(m, n) : m \in \mathbf{P}^* \text{ and } n \in P^{(m)}\}$. (For example, if K_1 had exactly six elements, then $(1, 1), (1, 2), \ldots, (1, 6)$ would each be a member of X, but $(1, 7)$ would not be. More generally, if $(m, n) \in X$, then $K_m \in \mathcal{K}$, and there is an nth element in K_m.) Notice that X is a subset of $\mathbf{P} \times \mathbf{P}$, and hence it is countable.

We will next define a map ψ from X onto $\bigcup \mathcal{K}$. By 26.6 it will then follow that $\bigcup \mathcal{K}$ is countable and the proof will be complete.

Define $\psi : X \to \bigcup \mathcal{K}$ as follows. For each $(m, n) \in X$, let $\psi((m, n)) = \alpha^{(m)}(n)$. (Thus, $\psi((m, n))$ is simply the "nth element" in the set K_m.) This completes the proof.

Note that in the proof of 26.8, we did not know that ψ was one-to-one, since we did not know if the K_i's were pairwise disjoint.

EXERCISES: **COUNTABLE SETS**

1. Prove that the set of all rational real numbers is a countably infinite set.

2. Prove that if X is a set that is equivalent to a countable set, then X is also a countable set.

3. Suppose that $f : X \to Y$ is a map from a set X onto a countable set Y. Suppose that for each $y \in Y$, $f^{-1}[y]$ is a countable set. Is X necessarily a countable set?

4. Prove that if A and B are countable sets, then so is $A \times B$.

5. Let $f : X \to Y$ be a surjection. Show that there is a subset of X that is equivalent to Y.

27. UNCOUNTABLE SETS

In the last section, we defined a set as being a countably infinite set if the set $\mathbf{P}$ of all positive integers could be put into one-to-one correspondence with it. We show next, by an example, that there are infinite sets that have "so many points" in them that they are not countably infinite.

27.1. *Example.* Let $\mathcal{S}$ be the collection of all infinite sequences s such that $s(i) = 0$ or 1 for each $i \in \mathbf{P}$. It is easy to see that $\mathcal{S}$ is an infinite set by considering the sequences $(1, 0, 0, 0, \ldots), (0, 1, 0, 0, \ldots), (0, 0, 1, 0, 0, 0, \ldots), \ldots$. We prove that $\mathcal{A}$ is not countably infinite by contradiction. Suppose that there existed a one-to-one map f from **P** onto $\mathcal{A}$. Notice that for each i, $f(i)$ is a sequence whose range is contained in the set $\{0, 1\}$. Designate the jth term of $f(i)$ as $f(i)_j$. We shall obtain a contradiction by exhibiting a sequence α that is an element of $\mathcal{A}$ and at the same time could not be an element of $\mathcal{A}$. Toward this end, let

$$\alpha(j) = 1 - f(j)_j, \quad j \in \mathbf{P}.$$

Note that the range of α is contained in $\{0, 1\}$, so that $\alpha \in \mathcal{A}$. However, since $\alpha \in \mathcal{A}$, and f maps **P** onto $\mathcal{A}$, there must be an integer k such that $f(k) = \alpha$. Then $f(k)_k = \alpha_k$. But $\alpha_k = 1 - f(k)_k$. Hence, $f(k)_k = 1 - f(k)_k$, from which we obtain a contradiction. Thus, $\alpha \in \mathcal{A}$ implies a contradiction and so the map f cannot exist.

The proof given in 27.1 is known as a "Cantor diagonal proof." The use of the term diagonal will become apparent in the following, less formal version of the same proof. If the set $\mathcal{A}$ in Example 27.1 were countable, we could think of being able to "list" its elements $s(i)$ as follows:

$$s(1) = (s(1)_1, s(1)_2, s(1)_3, \ldots, s(1)_n, \ldots)$$
$$s(2) = (s(2)_1, s(2)_2, s(2)_3, \ldots, s(2)_n, \ldots)$$
$$\cdot$$
$$\cdot$$
$$\cdot$$
$$s(n) = (s(n)_1, s(n)_2, s(n)_3, \ldots, s(n)_n, \ldots)$$
$$\cdot$$
$$\cdot$$
$$\cdot$$

Now define the sequence α as follows:

$$\alpha(i) = 1 - s(i)_i$$

Notice that every term of α is different from the corresponding term down the main diagonal of the array shown. For example, $\alpha(1) \neq s(1)_1$, $\alpha(2) \neq s(2)_2$, etc. Next, note that $\alpha \in \mathcal{A}$, so that it must appear in the listing. Suppose that it were $s(j)$. This could not be so, since $s(j)_j \neq \alpha(j)$.

27.2. Definition. Uncountable set. *A set that is not countable is said to be an uncountable set.*

It is clear that if a set is uncountable, then it is an infinite set. Thus, the following diagram schematically summarizes the classification of sets discussed in the last several sections.

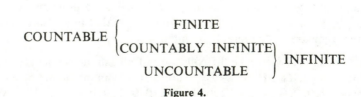

Figure 4.

EXERCISES: UNCOUNTABLE SETS

1. It is known that every real number between 0 and 1 inclusive has a (binary) representation in the form $. a_1 a_2 a_3 \cdots a_n \cdots$ where each a_i is either 0 or 1. However, as in the decimal system, the representation is not unique. For example, .01100 (remaining terms 0) represents the same number as .0101111 (remaining terms 1). But each real number between 0 and 1 has at least one and no more than two representations. Use this information to prove that the reals are uncountable. Point out why this implies that the set of all irrational numbers is uncountable.

2. Suppose that A is an uncountable set and C is a countable subset of A. Show that $A - C$ is an uncountable set.

3. Suppose that A, B, C, and D are sets such that $A \cap C = B \cap D = \varnothing$. Suppose further that $A \sim B$ and $C \sim D$. Is $(A \cup C) \sim (B \cup D)$?

4. Prove that every infinite set contains a countably infinite subset.

5. Prove that if S is an infinite set and $x \in S$, then $S - \{x\} \sim S$.

6. Suppose S is an uncountable set and C is a countable set. Show that $S - C \sim S$.

28. NONEQUIVALENT SETS

We wish to formalize the notion of one set having "fewer points" than another. To do this we introduce the following notation: If A is a set and B is a set, then we shall say $A < B$ (A is less than B with respect to cardinality) provided A is equivalent to a proper subset of B and A is not equivalent to B. The following fact about this relation between sets is of interest. If A and B are sets, then exactly one of the following holds: (i) $A < B$ (ii) $B < A$ (iii) $A \sim B$. The proof of this will not be taken up.

In Exercise 1, page 53, it was pointed out that $\mathbf{Q} < \mathbf{R}$. Does there exist a set M such that $\mathbf{R} < M$? The answer is yes. As a matter of fact it can be shown that for any set S, there is a set S^* such that $S < S^*$. To prove this fact, we shall consider the so-called power set of a set. Recall that if S is a set, by the power set $\mathscr{P}(S)$ of S we mean the collection of all subsets of S. For example, if $S = \{1, 2, 3\}$, then $\mathscr{P}(S) = \{\varnothing, \{1\}, \{2\}, \{3\}, \{1, 2\}, \{1, 3\}, \{2, 3\}, \{1, 2, 3\}\}$. Note that in this case $S < \mathscr{P}(S)$. What we propose to do is show that, in general, if S is a set then $S < \mathscr{P}(S)$. Before doing so, we shall discuss the special case for which S is a finite nonempty set. Notice that in the example in which $S = \{1, 2, 3\}$, we listed the elements of $\mathscr{P}(S)$. There were eight elements. It would be instructive at this point for the reader to list the elements in $\mathscr{P}(S)$ if $S = \{1, 2, 3, 4\}$. After doing this and observing that the number of elements in $\mathscr{P}(S)$ is 16, make a guess as to the cardinal number of $\mathscr{P}(S)$ if the cardinal number of S is n. (See Exercise 10, page 35.)

28.1.   Theorem.   *Let S be a finite set whose cardinal number is n. Then the cardinal number of its power set is 2^n.*

PROOF.   We see by inspection that the statement is true when $n = 0$ (i.e., when $S = \varnothing$). Next we assume that h is a nonnegative integer and that the statement is true for h. Suppose that S is a set with cardinal number $h + 1$. Let $x \in S$ and let $M = S - \{x\}$. From the inductive hypothesis, M has exactly 2^h subsets. Notice that a set is a subset of S if and only if it is a subset of M or if it is a set of the form $L \cup \{x\}$ where L is a subset of M. Furthermore, there is a one-to-one correspondence between the subsets of M and those of the form $L \cup \{x\}$, $L \subset M$. Thus, we can set up a two-to-one map from $\mathscr{P}(S)$ onto $\mathscr{P}(M)$. Hence, $\mathscr{P}(S)$ has twice as many elements as $\mathscr{P}(M)$, so that the cardinal number of $\mathscr{P}(S)$ is $2(2^h) = 2^{h+1}$. By induction, the proof is complete.

28.2.   Theorem.   *Let S be a set. Then $S < \mathscr{P}(S)$, where $\mathscr{P}(S)$ is the power set of S.*

PROOF.   The statement is clearly true if $S = \varnothing$. We prove the statement for the case $S \neq \varnothing$ by showing that S is equivalent to a subset of $\mathscr{P}(S)$ and that the assumption $S \sim \mathscr{P}(S)$ leads to a contradiction. It will then follow that $S < \mathscr{P}(S)$ from the definition of $<$ (less than with respect to cardinality).

Let $S^* = \{\{x\} : x \in S\}$. Each $\{x\} \in \mathscr{P}(S)$, so that $S^* \subset \mathscr{P}(S)$, and further, it is clear that $S \sim S^*$.

Next assume that there is a one-to-one map ψ from S onto $\mathscr{P}(S)$. Let $Q = \{x : x \in S \text{ and } x \notin \psi(x)\}$. Now $Q \in \mathscr{P}(S)$. So there must be a $q \in S$ such that $\psi(q) = Q$. This, however, leads to an interesting contradiction. Note first that $q \in Q$ or $q \notin Q$. Suppose that $q \in Q$. Then $q \in \psi(q)$. But then from the definition of Q, $q \notin Q$. Thus, it is false that $q \in Q$, and we have therefore shown that $q \notin Q$. But then $q \notin \psi(q) = Q$, and it follows from the definition of Q that $q \in Q$. We have arrived at a contradiction and the proof is complete.

29.   REVIEW EXERCISES

I.   Let p and q, r, and s be statements. Consider the following compound statement: If (p and q), then (r or s).

Choose the statement or statements below which would be the correct way to state the contrapositive of the given compound statement.

1.   If (r is false and s is false), then (p is false or q is false).

2.   If r is false or s is false, then p is false and q is false.

3.   If (p and q) is a false statement, then (r or s) is a false statement.

4.   If (r or s), then (p and q).

5.   None of the previous choices is correct, but a correct one

is_____.

II. Let A and B and C be nonempty sets. Either prove the following statement or give a counterexample.

$$(A \times B) \cap (C \times D) = (A \cap C) \times (B \cap D).$$

III. **1.** Give the names of the three properties that a relation must possess in order for it to be called an equivalence relation.

2. Suppose that R is a relation defined on the set of real numbers as follows: $x \, R \, y$ if and only if there exists an integer k such that $x - y = k$. Is R an equivalence relation? Justify your answer.

3. Suppose that R is an equivalence relation defined in a set A. Which of the following is necessarily true? Justify your answers.
  (a) $R = A \times A$
  (b) $R \subset A \times A$
  (c) $\{(x, x) : x \in A\} \subset R$
  (d) If a and b are distinct elements in A, then $R[a] \cap R[b] = \varnothing$
  (e) If $R[a] \cap R[b] \neq \varnothing$, then $R[a] = R[b]$
  (f) $R = R^{-1}$

IV. Let R be the relation defined as follows:

$$R = \{(x, y) : x \text{ is real}, y \text{ is real, and } |x - y| = 5\}$$

1. Is R a symmetric relation?

2. Is R a transitive relation?

3. Determine $R[2]$.

4. Is R a function?

5. Find the domain of R^{-1}.

V. Let $f : X \to Y$ be a bijection. Let A and B be subsets of X.

1. Prove that $f[A \cap B] = f[A] \cap f[B]$.

2. Give a counterexample to show that the preceding is not true if f is not a one-to-one function but simply a function.

3. Explain what step in your proof of the first part breaks down if the hypothesis does not say "one-to-one."

VI. Consider the map $f : \mathbf{R} \to \mathbf{R}$ given by the following:

$$f(x) = x(x - 2) \text{ for each } x \text{ in } \mathbf{R}$$

1. Find the range of f.

2. Determine the set $f^{-1}[\{y : -1 \leq y \leq 0\}]$.

     3.   Find the largest number z such that f restricted to the set $\{x : 0 \leq x \leq z\}$ is a one-to-one function.

VII. **1.**   Give a precise statement of the principle of finite induction.

     2.   Give a precise statement of the well-ordering principle for integers.

     3.   Prove that the well-ordering principle implies the principle of finite induction.

     4.   Prove that $9^n - 8n - 1$ is divisible by 64, if n is any positive integer.

VIII. **1.**   Define what is meant by an infinite sequence.

     2.   Define what is meant by a subsequence of an infinite sequence.

     3.   Define what is meant by a decomposition of a set.

     4.   Define what is meant by a function that is one-to-one.

     5.   Suppose that $f : X \to Y$ is a one-to-one function, and $g : Y \to Z$ is also a one-to-one function. Prove that the composition $g \circ f$ is a one-to-one function.

IX. Is the following statement necessarily true? Justify your answer.

     If R is a relation and $R^{-1} \subset R$, then R is a symmetric relation.

X. **1.**   Define what is meant by a partially ordered set.

     2.   Define what is meant by a totally ordered set.

     3.   Give an example of a partially ordered set that is not totally ordered.

     4.   Give an example of a partially ordered set that has a maximal element but no greatest element.

     5.   Prove that a set has at most one greatest lower bound.

XI. **1.**   Use the axiom of choice to give an alternate proof of Theorem 26.6.

     2.   Let $\mathscr{S}$ be the collection of all finite-sequences of integers. Is $\mathscr{S}$ a countable set?

XII. Let $\mathscr{F}$ be the set of all functions that map the closed interval $[0, 1]$ into $[0, 1]$. Prove that $[0, 1] < \mathscr{F}$. (Hint: Imitate somewhat the proof of 27.1.)

XIII. **1.**   Give an example of a set A and a relation R in A that is symmetric and transitive in A but is not reflexive.

2. Tell what is wrong with the following argument, which claims to show that if a relation R in a set A is transitive and symmetric, then it is also reflexive.

Let $a \in A$. Choose an element $b \in A$ such that $a\,R\,b$. Since R is symmetric, it then follows that $b\,R\,a$. Since R is transitive, $a\,R\,b$ and $b\,R\,a$ imply that $a\,R\,a$. Hence, we have shown that R is reflexive.

XIV. Let $(S, \leq)$ be a nonempty partially ordered set such that every linearly ordered subset has an upper bound. Show that if $a \in A$, then there is a maximal element m in S such that $a \leq m$.

2

Structure of **R** *and* **R**ⁿ

Thus far our study has been largely restricted to that of abstract sets and functions, or relations defined on abstract sets. The study of sets generally becomes more useful when a structure of one kind or another is imposed on the sets to be studied. For example, sometimes the set studied is endowed with an algebraic structure and sometimes with a geometric structure. Often the sets studied have both an algebraic and a geometric structure. When the domain and range of a function are endowed with enough geometric structure so that the concept of distance between points or some generalization of the notion is available, then the notion of a continuous function is also available. For, roughly, the idea of a function f being continuous at x_0 is that $f(x)$ can be made as close as we wish to $f(x_0)$ provided x is sufficiently close to x_0. Similarly, if the domain and range of a function are endowed with an algebraic structure, it is often useful to study certain algebraic properties that the function might possess.

The real number system **R** has both a geometric and an algebraic structure. The reader is already familiar with the distance formula for points in the plane. Recall that the distance between points (x_1, x_2) and (y_1, y_2) in the plane is given by $[(x_1 - y_1)^2 + (x_2 - y_2)^2]^{\frac{1}{2}}$. Similarly in three-dimensional space, the distance between points (x_1, x_2, x_3) and (y_1, y_2, y_3) is $[(x_1 - y_1)^2 + (x_2 - y_2)^2 + (x_3 - y_3)^2]^{\frac{1}{2}}$. In this chapter, we shall extend this distance formula to **R**ⁿ, the collection of all n-tuples of real numbers. There is also a useful algebraic structure that can be imposed on **R**ⁿ. We shall give detailed consideration to the geometric (metric) structure of **R** and **R**ⁿ. However, since the geometric structure we will consider is related to certain algebraic considerations, it will be useful for us to also consider, at least to some small extent, the algebraic structure of those spaces.

By making use of the notion of distance in **R**ⁿ, we will define the important concepts of open set, closed set, limit point, and convergent sequence. Some important theorems in classical analysis are related to these concepts. Among those covered in this chapter are the Bolzano-Weierstrass, Heine-Borel, and Lindelöf theorems.

30. ALGEBRAIC STRUCTURE OF R

For reference we list in this section those properties of the real number system that give it algebraic structure.

30.1. Properties of addition. *To each ordered pair of numbers (a, b), there corresponds a unique number $a + b$. Addition satisfies the following:*

30.1(a). *Addition is commutative.*
> *For each pair of numbers a and b in* **R**,

$$a + b = b + a.$$

30.1(b). *Addition is associative.*
> *For all numbers a, b, and c*

$$a + (b + c) = (a + b) + c.$$

30.1(c). *The zero element.*
> *There is a number* 0 *that has the property that*

$$a + 0 = a \text{ for each } a \in \mathbf{R}.$$

30.1(d). *Existence of additive inverse.*
> *To each $a \in$* **R** *there corresponds an element $-a \in$* **R** *such that*

$$a + (-a) = 0.$$

30.2. Properties of multiplication. *To each pair of numbers a and b in* **R** *there corresponds a unique number $a \cdot b$ called the product of a and b.*

The properties of product (or binary operation of multiplication) are:

30.2(a). *Multiplication is commutative.*
> *For each a and b in* **R**,

$$a \cdot b = b \cdot a.$$

30.2(b). *Multiplication is associative.*
> *For a, b, and c in* **R**,

$$a \cdot (b \cdot c) = (a \cdot b) \cdot c.$$

30.2(c). *Existence of a unit or multiplicative identity.*
> *There is a number* 1 *such that*

$$1 \cdot a = a \text{ for each } a \in \mathbf{R}.$$

30.2(d). *Multiplicative inverse for nonzero numbers.*
> *To each $a \in$* **R** *such that $a \neq 0$, there exists a number a^{-1} such that*

$$a \cdot a^{-1} = 1.$$

The following property relates to both addition and multiplication.
30.2(e). *Distributive law. For all a, b, and c in* **R**,

$$a \cdot (b + c) = a \cdot b + a \cdot c.$$

(The reader familiar with the notion of *group* and *field* should notice that the properties listed in 30.1 tell us that $(\mathbf{R}, +)$ is a *commutative group*. From the properties listed in 30.2(a) through (d), we see that $(\mathbf{R} - \{0\}, \cdot)$ is also a commutative group. Furthermore, the properties listed in 30.1 and 30.2 make $(\mathbf{R}, +, \cdot)$ a *field*.)

30.3. Order relation in R. *The set of real numbers* $\mathbf{R}$ *possesses an order* $(<)$ *structure satisfying the following:*

30.3(a). *For all a and b in* $\mathbf{R}$, *exactly one of the following holds.*

$$a = b, \quad a < b, \quad b < a$$

For all a, b, and c in $\mathbf{R}$,

30.3(b). *if* $a < b$, *then* $a + c < b + c$,

30.3(c). *if* $a < b$ *and* $0 < c$, *then* $a \cdot c < b \cdot c$,

30.3(d). *if* $a < b$ *and* $b < c$, *then* $a < c$; *that is, the relation is transitive.*

Thus $\{\mathbf{R}_+, \mathbf{R}_-, \{0\}\}$ *is a decomposition of* $\mathbf{R}$, *where*

$$\mathbf{R}_+ = \{x : 0 < x\}, \text{ the set of positive real numbers}$$

and

$$\mathbf{R}_- = \{x : x < 0\}, \text{ the set of negative real numbers.}$$

(If $a < b$, we may alternatively write $b > a$.)

30.4. Archimedean property. *If* x *and* y *are real numbers and* $y > 0$, *then there is a positive integer* n *such that* $x < n \cdot y$.

By making use of the Archimedean property, we can show that the set $\mathbf{Q}$ of all rational numbers is *dense* in the real number system. By this, we mean that if x and y are real numbers, then there is a rational number between x and y. Equivalently, for every open interval $(a, b) \subset \mathbf{R}$, $(a, b) \cap \mathbf{Q} \neq \varnothing$.

30.5. Theorem. Density of rationals in the reals. *Let* x *and* y *be two real numbers such that* $x < y$. *Then there is a rational number* r *such that* $x < r < y$.

PROOF.   We first consider the case $0 < x < y$. (Once we prove the theorem for this case, the general case will follow easily.) Since $y - x > 0$, it follows from the Archimedean property that we may choose an integer n such that $1 < n(y - x)$. Hence, $\dfrac{1}{n} < y - x$. Again, by the Archimedean property, there is a positive integer M such that $x < M\left(\dfrac{1}{n}\right)$. We choose the first positive integer m such that $x < \dfrac{m}{n}$. We claim that $x < \dfrac{m}{n} < y$ and justify this assertion as follows: If it is false that $0 < x < \dfrac{m}{n} < y$, then $y \leq \dfrac{m}{n}$. But from the way in which m was chosen, it would then have to follow that

$$\frac{m-1}{n} \leq x < y \leq \frac{m}{n}.$$

So

$$y - x \leq \frac{m}{n} - \frac{m-1}{n} = \frac{1}{n},$$

a contradiction. Hence, the rational number $\dfrac{m}{n}$ satisfies $0 < x < \dfrac{m}{n} < y$.

We next consider the case $x < y$ and $x \leq 0$. Let N be a positive integer such that $0 < N + x$. Then $0 < N + x < N + y$. From the proof of the first part, there is a rational number r such that $N + x < r < N + y$. But then $r - N$ is a rational number and $x < r - N < y$.

In terms of some of the language used in the previous chapter, the relation $\leq$ is transitive, reflexive, and antisymmetric in **R** (Sections 22 and 23) and, hence, (**R**, $\leq$) is a partially ordered system. Moreover, because of 30.3(a), (**R**, $\leq$) is a linearly ordered set. Recall that we have already used this fact in some of our examples in Section 23. Notions of *upper bound, lower bound, bounded set, least and greatest elements, least upper bound* and *greatest lower bound* have already been defined in that section. To apply these terms in the usual way to the real number system **R**, we simply consider the partially ordered system (**R**, $\leq$) where $\leq$ has the usual meaning. Thus, a subset $A \subset$ **R** *is bounded above* (*below*) by b provided that $a \leq b$ ($b \leq a$) for all $a \in A$. The number b is then called an upper (*lower*) *bound* for A. A subset of **R** is a bounded set if it has both an upper and a lower bound. The number b is the *greatest* (*least*) element of A provided it belongs to the set A and is greater (less) than all other elements of A. A number l is the *least upper bound* of A, (l.u.b. (A)) provided that A is bounded above and l is the least element of the set of all upper bounds of A. More explicitly, $l =$ l.u.b. (A) if and only if

$$a \leq l \text{ for all } a \in A$$
$$l \leq b \text{ for all upper bounds } b \text{ of } A$$

Also, a number g is the greatest lower bound of A (g.l.b. (A)) if and only if

$$g \leq a \text{ for all } a \in A$$
$$b \leq g \text{ for all lower bounds } b \text{ of } A$$

Recall that for a partially ordered set (S, $\leq$), a nonempty subset might have an upper bound but not necessarily a least upper bound in S (see Example 23.9). This cannot happen in the real number system.

30.6.   The least upper bound axiom for the real number system R. *If S is a nonempty subset of* **R** *and S has an upper bound then S has a least upper bound in* **R**.

The algebraic and order properties stated in 30.1, 30.2, and 30.3 are still not sufficient to make **R** "act geometrically like a line." For example, the set of rationals **Q** satisfies all these properties. However, the least upper bound axiom makes **R** "connected," as will be shown in a later chapter.

30.7.   Definition.   Maximum value and minimum value of a real-valued function on a set. *Suppose that f is a real-valued function and $S \subset$ Dom f. Then M is said to be the maximum value of f on S provided M is the maximum or greatest element in the set $f[S]$. Similarly, m is said to be the minimum value of f on S provided m is the minimum or least element of the set $f[S]$.*

We see that if M is the maximum value of f on S, then

$$M = \text{l.u.b. } (f[S])$$

and

$$M \in f[S].$$

30.8. *Example.* Consider the function $f: \mathbf{R} \to \mathbf{R}$ given by $f(x) = x^2$. $f[\mathbf{R}]$ is not bounded above, so that f does not have a maximum value on $\mathbf{R}$. However, f has a minimum value of 0 on $\mathbf{R}$. Also, f has a minimum value of 0 on $[0, 2]$ and a maximum value of 4 on $[0, 2]$. On the open interval $(0, 2)$, f has neither a minimum value nor a maximum value. Note that l.u.b. $(f[(0, 2)]) = 4$ but $4 \notin f[(0, 2)]$.

At this point it would be helpful to work with these concepts, and the following exercises are provided for that purpose. In doing these and subsequent exercises, the reader should use his knowledge of elementary algebra freely rather than show that every step can ultimately be justified on the basis of the properties stated in 30.1, 30.2, and 30.3.

EXERCISES: ALGEBRAIC STRUCTURE OF R

All sets are understood to be subsets of **R**.

1. Explain by example why it is that the system of all rational numbers does not satisfy the least upper bound axiom.

2. Prove that the least upper bound property implies the following: If S is a nonempty subset of real numbers that has a lower bound, then S has a greatest lower bound.

3. Let $S = \left\{ x : x = 1 - \dfrac{1}{n}, n \in \mathbf{P} \right\}$. Find l.u.b. (S) and g.l.b. (S) if they exist.

4. Let $f: \mathbf{R} \to \mathbf{R}$ be given by $f(x) = x^3$. Find the l.u.b. $(f[\{x : 0 < x < 1\}])$. Find l.u.b. $(f[\{x : 0 \le x \le 1\}])$.

5. Suppose that $f : \{x : 0 < x\} \to \mathbf{R}$ is given by $f(x) = \dfrac{1}{x}$ for $0 < x$. Does $f[\{x : 0 < x\}]$ have a l.u.b.? Does it have a g.l.b.?

6. Give an example of a function f defined on a closed interval S such that l.u.b. $(f[S])$ exists but f does not attain a maximum value on S.

7. Prove the following statement: If $a = $ l.u.b. (A), then for each $\varepsilon > 0$, there is an $x \in A$ such that $a - \varepsilon < x \le a$. State and prove an analogous proposition for g.l.b. (A).

8. Prove that if A is a nonempty bounded set of real numbers then g.l.b. $(A) \le$ l.u.b. (A). For what kind of set A is g.l.b. $(A) = $ l.u.b. (A)?

9. Prove that if A and B are nonempty subsets of $\mathbf{R}$ and $A \subset B$, then g.l.b. $(B) \le$ g.l.b. $(A) \le$ l.u.b. $(A) \le$ l.u.b. (B).

31. DISTANCE BETWEEN TWO POINTS IN R

Recall that by the absolute value $|x|$ of a number x we mean x if $x \geq 0$ and $-x$ if $x < 0$. Using the absolute value function, we can introduce the notion of the distance, $d(x, y)$, between points x and y in **R** as follows.

31.1. Definition. Distance formula for R. *For* $x \in$ **R** *and* $y \in$ **R**, *by the distance between* x *and* y *we shall mean the number* $d(x, y)$ *given by the following formula*:

$$d(x, y) = |x - y|.$$

The distance function as defined in 31.1 satisfies the following. For all a, b, and c in **R**,

31.2(a). $d(a, b) = 0$ if and only if $a = b$

31.2(b). $d(a, b) = d(b, a)$

31.2(c). $d(a, b) + d(b, c) \geq d(a, c)$ (Triangle inequality).

31.3. Definition. ε-Neighborhood of a point in R. *Let* $p \in$ **R** *and* $\varepsilon > 0$. *By an ε-neighborhood of* p, $N(p; \varepsilon)$, *we shall mean the following subset of* **R**:

$$N(p; \varepsilon) = \{q : d(q, p) < \varepsilon\} = \{q : |q - p| < \varepsilon\}.$$

Notice that $N(p; \varepsilon)$ *is the open interval* $(p - \varepsilon, p + \varepsilon)$.

We shall sometimes refer to an ε-neighborhood simply as a neighborhood.

EXERCISES: DISTANCE BETWEEN TWO POINTS IN R

1. Verify the properties stated in 31.2.

2. Let $p \in$ **R**. Give an example of a collection $\mathscr{K}$ of neighborhoods of p such that $\bigcap \mathscr{K}$ is not a neighborhood. Show that if $\mathscr{K}$ is a nonempty finite collection of neighborhoods of p, then $\bigcup \mathscr{K}$ and $\bigcap \mathscr{K}$ are neighborhoods of p.

32. LIMIT OF A SEQUENCE IN R

The following concept should be familiar to the reader from his study of calculus.

32.1. Definition. Limit of a sequence in R. *Let* (a_n) *be a sequence in* **R**. *Then a number* A *is said to be the limit of* (a_n) *provided that for each* $\varepsilon > 0$, *there is an integer* N *such that*

$$if \ n \geq N \ then \ |a_n - A| < \varepsilon.$$

In that case, we write $\lim_{n \to \infty} a_n = A$, *or* $\lim (a_n) = A$.

Notice that if N satisfies the requirement just stated, then so does any integer $M \geq N$.

If $\lim (a_n) = A$, we shall say that the sequence (a_n) *converges* to A. It should be noted that if a sequence converges, its limit is unique. (See Exercise 6 in the following set of exercises.)

EXERCISES: LIMIT OF A SEQUENCE IN R

Assume that all sequences are in **R**. Prove each of the following.

1. Suppose that (a_n) is a sequence such that $a_n \leq a_{n+1}$, $(a_{n+1} \leq a_n)$ for each positive integer n. Suppose further that the sequence (a_n) is bounded above (below). Then $\lim (a_n)$ exists.

2. Let the sequence (c_n) in **R** be given by $c_n = a_n + b_n$, where $\lim (a_n) = A$ and $\lim (b_n) = B$. Then, $\lim (c_n) = A + B$.

3. Let the sequence (c_n) in **R** be given by $c_n = ka_n$ where k is a constant and $\lim (a_n) = A$. Then $\lim (c_n) = kA$.

4. Let the sequence (c_n) in **R** be given by $c_n = a_n \cdot b_n$ where $\lim (a_n) = A$ and $\lim (b_n) = B$. Then $\lim (c_n) = A \cdot B$.

5. Let the sequence $(c_n) = (a_n/b_n)$ where $b_n \neq 0$, $\lim (a_n) = A$, $\lim (b_n) = B \neq 0$. Then $\lim (c_n) = A/B$.

6. If a sequence (a_i) has a limit, it is unique.

33. THE NESTED INTERVAL THEOREM FOR R

In Exercise 3, page 13 , it should have been noted that it is possible to have a decreasing sequence $\{K_i\}$ of nonempty sets such that $\bigcap \{K_i\} = \varnothing$. By making use of the least upper bound and greatest lower bound properties of the real number system, one can prove that if the K_i have certain restrictions, then $\bigcap \{K_i\} \neq \varnothing$. We shall prove this fact in this section for closed intervals (i.e., sets of the form $\{x : a \leq x \leq b\}$, allowing the possibility that $a = b$) in **R**. At various points in our development, we shall extend this important result to certain more general types of sets. The following theorem is of considerable importance.

33.1. Nested interval theorem. *For each positive integer i, let K_i be the closed interval $[a_i, b_i]$. Suppose that $K_i \supset K_{i+1}$, $i \in \mathbf{P}$. Then $\bigcap_{i=1}^{\infty} K_i \neq \varnothing$. Furthermore, if $\lim (|b_i - a_i|) = 0$, then $\bigcap_{i=1}^{\infty} K_i$ has exactly one element.*

An outline of the proof follows. The details are left as an exercise (see Exercise 1, page 6 5).

(i) $a_1 \leq a_i \leq a_{i+1} \leq b_1$ and $a_1 \leq b_{i+1} \leq b_i \leq b_1$.

(ii) Show that $\lim (a_i)$ exists; call it A. Show that $\lim (b_i)$ exists; call it B. Further, $A \leq B$.

(iii) The closed interval $[A, B] \subset K_n$ for each $n \in \mathbf{P}$ and the conclusion that $\bigcap\limits_{i=1}^{\infty} K_i \neq \phi$ follows.

(iv) If x and y are two different points in $\bigcap\limits_{i=1}^{\infty} K_i$, then $|b_i - a_i| \geq |x - y| > 0$ for all $i \in \mathbf{P}$. Hence, $\lim (|b_i - a_i|) \neq 0$.

In the previous theorem we used the notion of closed interval $[a, b] = \{x : a \leq x \leq b\}$ where we allow the possibility of $a = b$. If $a = b$, then we call the interval *degenerate*. Previously we have also used the notion of open intervals of the form $\{x : a < x < b\}$. Next, we define a more general type of interval in **R**.

33.2. Definition. Interval in R. *A set S in* **R** *is called an interval provided that it satisfies the following condition. If $a \in S$ and $b \in S$ with $a \leq b$, then the closed interval $[a, b] \subset S$.*

EXERCISES: THE NESTED INTERVAL THEOREM FOR **R**

1. Give the details of the proof of Theorem 33.1.

2. Give an example of nonempty intervals I_i (see 33.2) such that
$$I_1 \supset I_2 \supset, \ldots, I_n \supset \cdots \text{ and } \bigcap_{i=1}^{\infty} I_i = \varnothing.$$

3. Is the following statement true? Given an interval I and a point $p \in I$, there exists a countable collection of closed intervals $\{I_i\}$ such that $p \in I_1 \subset I_2 \subset I_3 \cdots \subset I_n \subset \cdots$ (possibly only a finite number needed) and such that $I = \bigcup\limits_{i=1}^{\infty} \{I_i\}$.

4. Suppose that $\mathscr{K}$ is a collection of intervals such that $\bigcap \mathscr{K} \neq \varnothing$. Is $\bigcup K$ necessarily an interval?

34. ALGEBRAIC STRUCTURE FOR R^n

We shall use the symbol **R**^{n} to represent the Cartesian product **R** $\times$ **R** $\times \cdots \times$ **R** of **R** taken n times. Thus, **R**^{n} is the collection of n-tuples of real numbers. The reader is probably familiar with the vector space in which elements are equivalence classes of directed line segments with vector addition, scalar multiplication, and inner product defined. That is but one model of a so-called 3-dimensional vector space with inner product. The algebraic structure that we shall discuss for **R**^{n} represents a model "algebraically equivalent" to the above model when $n = 3$. (Also, **R**^{1} = **R**.)

34.1. Definitions

34.1(a). *Vector addition and multiplication by a scalar. Let $x = (x_1, x_2, \ldots, x_n)$ and $y = (y_1, y_2, \ldots, y_n)$. By $x + y$, the vector sum of x and y, we mean the element of* **R**^{n} *given by $(x_1 + y_1, x_2 + y_2, \ldots, x_n + y_n)$. For each scalar (real number) α and $x = (x_1, x_2, \ldots, x_n) \in$ **R**^{n}, αx, the scalar multiple of x by α, is defined to be the element in* **R**^{n} *given by $(\alpha x_1, \alpha x_2, \ldots, \alpha x_n)$.*

34.1(b). *The zero element for* $\mathbf{R}^n$. *The element in* $\mathbf{R}^n$, *each of whose coordinates is* 0, *is called the zero vector for* $\mathbf{R}^n$ *and will be designated by* θ.

34.1(c). *The negative of an element in* $\mathbf{R}^n$. *For each* $x = (x_1, x_2, \ldots, x_n)$ *we shall mean by* $-x$, *the element in* $\mathbf{R}^n$ *given by* $(-x_1, -x_2, \ldots, -x_n)$.

34.2. Properties of vector addition and scalar multiplication

34.2(a). *For vector addition: For all* x, y, *and* z *in* $\mathbf{R}^n$,

(i) $x + y = y + x$
(ii) $x + (y + z) = (x + y) + z$
(iii) $\theta + x = x$
(iv) $x - x = \theta$ *(by* $x - y$ *we shall mean* $x + (-y)$)

34.2(b). *For scalar multiplication: For scalars* α, β *and vectors* x *and* y,

(i) $\alpha(\beta x) = \alpha\beta(x)$
(ii) $(\alpha + \beta)x = \alpha x + \beta x$
(iii) $\alpha(x + y) = \alpha x + \alpha y$
(iv) $1x = x$

34.3. Definition. Inner (or dot) product. *For* $x = (x_1, x_2, \ldots, x_n)$ *and* $y = (y_1, y_2, \ldots, y_n)$, *the inner product of* x *and* y *is denoted by* $x \cdot y$ *and is defined by the real number* $x_1y_1 + x_2y_2 + \cdots + x_ny_n$.

34.4. Properties of the inner product. *For* x, y, *and* z *in* $\mathbf{R}^n$ *and scalars* α *and* β,

(i) $x \cdot y = y \cdot x$
(ii) $x \cdot (\alpha y + \beta z) = \alpha(x \cdot y) + \beta(x \cdot z)$
(iii) $(\alpha x + \beta y) \cdot z = \alpha(x \cdot z) + \beta(y \cdot z)$
(iv) $x \cdot x > 0$ *if* $x \neq \theta$ *and* $\theta \cdot \theta = 0$

34.5. Definition. The magnitude of a vector $x \in \mathbf{R}^n$. *By* $|x|$, *the magnitude of* $x \in \mathbf{R}^n$, *we shall mean the nonnegative real number* $(x \cdot x)^{\frac{1}{2}}$ *(i.e.,* $(x_1^2 + x_2^2 + \cdots + x_n^2)^{\frac{1}{2}})$. *If* $|x| = 1$, x *is called a unit vector.*

The following exercises should help the reader to see the geometric motivation for the previous definitions and should serve to motivate several forthcoming notions.

EXERCISES: ALGEBRAIC STRUCTURE FOR $\mathbf{R}^n$

1. Verify the properties stated in 34.2 and 34.4.

2. Show that for points in $\mathbf{R}^2$, $|x - y|$ gives the usual distance formula with which the reader is familiar from analytic geometry.

3. Let x, y, and z be distinct points in $\mathbf{R}^2$. Let L_1 be the line segment with endpoints x and z. Let L_2 be the line segment with endpoints y and z. Let α be the smaller angle (or one of the angles if equal) formed by L_1 and L_2 at z. Show that

$$\cos \alpha = \frac{(x - z) \cdot (y - z)}{|x - z|\,|y - z|}$$

4. Let x and y be two elements in **R**^{2}. Show that $|x \cdot y| \leq |x| |y|$ and $|x + y| \leq |x| + |y|$.

5. The results of this exercise will be needed in the next section. Consider the function f given by $f(x) = Ax^2 + 2Bx + C$, where $A > 0$, and which further satisfies $f(x) \geq 0$ for all real x. Prove that $B^2 - AC \leq 0$.

35. THE CAUCHY-SCHWARZ INEQUALITY

In Exercise 4, this page, it was pointed out that for vectors x and y in **R**^{2}, $|x \cdot y| \leq |x| |y|$. This inequality is true for vectors in **R**^{n} and is of considerable importance in analysis. For us, it will be used in proving the triangle inequality for **R**^{n}. The "natural" way to try to prove the inequality would be to write the inequality in terms of coordinates. This can be done, and it would be instructive for the reader to try this. However, in proving general relations for vectors in **R**^{n}, one can often use to good advantage the general properties of vector addition, scalar multiplication, and inner product. Such is the case in the proof of the inequality presented next.

35.1. Theorem. Cauchy-Schwarz inequality for R^n. *Let x and y be elements in* **R**^{n}. *Then* $|x \cdot y| \leq |x| |y|$.

PROOF. For each real number α, $(\alpha x + y) \cdot (\alpha x + y) \geq 0$, by 34.4 (iv). Then by making use of various properties listed in 34.2 and 34.4,

$$(\alpha x + y) \cdot (\alpha x + y) = (\alpha x + y) \cdot (\alpha x) + (\alpha x + y) \cdot y$$
$$= \alpha^2 (x \cdot x) + \alpha(y \cdot x) + \alpha(x \cdot y) + (y \cdot y)$$
$$= (x \cdot x)\alpha^2 + 2(x \cdot y)\alpha + (y \cdot y) \geq 0$$

Notice that the function here is a quadradic in α that is nonnegative for all real α. Hence, its discriminant must be nonpositive (see Exercise 5, this page). Hence, $(x \cdot y)^2 - (x \cdot x)(y \cdot y) \leq 0$. From this it follows that $(x \cdot y)^2 \leq |x|^2 |y|^2$ or $|x \cdot y| \leq |x| |y|$. This completes the proof.

By means of this last inequality we can now prove the important triangle inequality for **R**^{n}.

35.2. Theorem (triangle inequality). *Let x and y be elements in* **R**^{n}. *Then*

$$|x + y| \leq |x| + |y|$$

PROOF. Each of the following first four statements is clearly equivalent to the next. The last statement is true because of 35.1 and, hence, the first statement is correct.

$$|x + y| \leq |x| + |y|$$
$$|x + y|^2 \leq (|x| + |y|)^2.$$
$$(x + y) \cdot (x + y) \leq (x \cdot x) + 2|x| |y| + (y \cdot y)$$
$$x \cdot x + 2(x \cdot y) + y \cdot y \leq (x \cdot x) + 2|x| |y| + (y \cdot y)$$
$$x \cdot y \leq |x| |y|.$$

35.3.   Theorem.   *For all x, y, and z in $\mathbf{R}^n$ and real numbers α,*

35.3(a). $|x| > 0$ if $x \neq 0$ and $|0| = 0$

35.3(b). $|\alpha x| = |\alpha|\,|x|$

35.3(c). $|x| = |-x|$

35.3(d). $|x - y| + |y - z| \geq |x - z|$

The proofs are left as exercises.

35.4.   Definition.   Orthogonal vectors.   *Vectors x and y in $\mathbf{R}^n$ are said to be orthogonal provided that $x \cdot y = 0$.   (See Exercise 3, page 66 for motivation.)*

35.5.   Definitions.   Line segments and lines in $\mathbf{R}^n$.   *Let a and b be points in $\mathbf{R}^n$. Then the set $\{x : x \in \mathbf{R}^n$ and $x = (1 - t)a + tb$ for $0 \leq t \leq 1\}$ is called the line segment $L(a, b)$ with endpoints a and b. If $a = b$, the line segment is said to be degenerate. If $a \neq b$ and $0 < s < 1$, then the point $x = (1 - s)a + sb$ is said to be between a and b.*

Let $a \neq b$.   The set $\{x : x = (1 - t)a + tb,\ t$ real$\}$ is called the line in $\mathbf{R}^n$ determined by a and b.

The reader should convince himself that these definitions are consistent with the notion of line and line segment in $\mathbf{R}^2$ and $\mathbf{R}^3$, with which he is familiar.

35.6.   Definition.   Polygon in $\mathbf{R}^n$.   *We shall call S a polygon in $\mathbf{R}^n$ provided that there exist points $x_0, x_1, x_2, \ldots, x_m$, not necessarily distinct, such that S is the union of the collection of line segments $\{L(x_i, x_{i+1}) : i = 0, 1, 2, \ldots, m - 1\}$. In that case we shall say that S joins x_0 to x_m.*

EXERCISES: THE CAUCHY-SCHWARZ INEQUALITY

　　1.   Verify 35.3

　　2.   Suppose that $a \in \mathbf{R}^n$.   Consider the collection $S = \{\alpha a : -\infty < \alpha < \infty\}$. Show that S is a line.

36.   THE DISTANCE FORMULA IN $\mathbf{R}^n$

Recall that for $\mathbf{R}^n$, $|x|$ is defined as $(x \cdot x)^{\frac{1}{2}} = (x_1^2 + x_2^2 + \cdots + x_n^2)^{\frac{1}{2}}$ where $x = (x_1, x_2, \ldots, x_n)$.   Thus,

$$|x - y| = [(x_1 - y_1)^2 + (x_2 - y_2)^2 + \cdots + (x_n - y_n)^2]^{\frac{1}{2}}$$

We see that for $n = 1$ we have, in this expression, the distance formula already introduced for $\mathbf{R}$. For $n = 2$ and 3, we recognize the distance formula from analytic geometry.   Thus, we would expect that if we define the distance between x and y in $\mathbf{R}^n$ as $d(x, y) = |x - y|$, this function should satisfy the properties listed in 31.2. Such is the case, and we have the following.

36.1.   Theorem.   *Let $d(x, y) = |x - y|$ for x and y in $\mathbf{R}^n$.   Then for every x, y, and z in $\mathbf{R}^n$*

36.1(a). $d(x, y) \geq 0$

36.1(b). $d(x, y) = 0$ *if and only if* $x = y$

36.1(c). $d(x, y) = d(y, x)$

36.1(d). $d(x, y) + d(y, z) \geqq d(x, z)$.

The proof follows directly from the properties proved about the magnitude function.

    36.2. **Definition.** **Euclidean metric for R^n.** *By the Euclidean metric, or the Euclidean distance formula for* R^n, *we shall mean the function* $d: \mathbf{R}^n \times \mathbf{R}^n \to \mathbf{R}$ *given by*

$$d(x, y) = |x - y| = [(x_1 - y_1)^2 + (x_2 - y_2)^2 + \cdots + (x_n - y_n)^2]^{\frac{1}{2}}$$

    In the remainder of this chapter, when reference is made to the distance between points it shall be understood that distance is measured by the Euclidean distance formula. The reader should verify the following remark, which will be useful for some of our subsequent considerations.

    36.3. *Remark.* *Let* $x = (x_1, x_2, \ldots, x_n)$ *and* $y = (y_1, y_2, \ldots, y_n)$ *be points in* R^n. *Then for each* $i \in \mathbf{P}_n$,

$$|x_i - y_i| \leq d(x, y)$$

and

$$d(x, y) \leq \sqrt{n} \max \{|x_i - y_i| : i \in \mathbf{P}_n\}$$

    36.4. **Definition.** **ε-Neighborhood.** *Let* $\varepsilon > 0$. *By the ε-neighborhood of a point* $p \in \mathbf{R}^n$ *we shall mean the set of all points in* R^n *that are less than* ε *distance from* p. $N(p; \varepsilon)$ *will be the symbol used to designate that set. Thus,*

$$N(p; \varepsilon) = \{x : d(x, p) < \varepsilon\}.$$

    If we visualize R^3 as we did in solid analytic geometry, the set $\{x : d(x, p) \leqq r\}$, where $r > 0$, has the appearance of a solid ball. On the other hand, the set $\{x : d(x, p) = r\}$ is the spherical surface of the solid ball $\{x : d(x, p) \leqq r\}$. The corresponding sets in R^n take their names from these familiar objects in R^3 as follows.

    36.5. **Definitions.** **Closed balls and** $(n - 1)$**-spheres in R^n.** *Let* $r > 0$. *The set* $\{x : d(x, p) \leqq r\}$ *in* R^n *is called the closed ball with center* p *and radius* r. *The set* $\{x : d(x, p) = r\}$ *in* R^n *is known as the* $(n - 1)$*-sphere with center* p *and radius* r.

    Notice in particular that if $n = 1$, the closed ball with center $p \in \mathbf{R}$ and radius r is the closed interval $[p - r, p + r]$. Its corresponding spherical boundary is the two-point set $\{p - r, p + r\}$. In R^2 the closed ball with center $p \in \mathbf{R}^2$ and radius r is often known as a *disc*, and of course its spherical boundary is the *circle* with center p and radius r.

EXERCISES: THE DISTANCE FORMULA IN R^n

    1.   Prove Theorem 36.1.

    2.   Verify that if properties (b), (c), and (d) of 36.1 are assumed for a real-valued function defined on R^n X R^n, then (a) follows automatically.

    3.   Prove Remark 36.3.

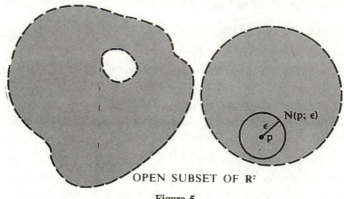

OPEN SUBSET OF $\mathbf{R}^2$

Figure 5

37. OPEN SUBSETS OF $\mathbf{R}^n$

The notion of an open subset of $\mathbf{R}^n$ will be of fundamental importance in our study.

37.1. Definition. Open set in $\mathbf{R}^n$. *A subset U of $\mathbf{R}^n$ is said to be open (in $\mathbf{R}^n$) provided that for each $p \in U$, there is an $\varepsilon > 0$ such that $N(p; \varepsilon) \subset U$.*

To get a feeling for this concept, the reader should verify what is called for in each of the following examples.

37.2. *Example.* $\varnothing$ and $\mathbf{R}^n$ are open subsets of $\mathbf{R}^n$.

37.3. *Example.* $\mathbf{R} \times \{0\}$ is not an open subset of $\mathbf{R}^2$.

37.4. *Example.* The following subset S of $\mathbf{R}^2$ is an open set in $\mathbf{R}^2$.

$$S = \{(x, y) : x + y < 2\}$$

37.5. *Example.* Let A and B be the following subsets of $\mathbf{R}^2$. $A = \{(x, y) : x^2 + y^2 < 1\}$, $B = \{(x, y) : (x - \frac{1}{2})^2 + (y - \frac{1}{2})^2 < 1\}$. A, B, $A \cap B$, $A \cup B$ are open subsets of $\mathbf{R}^2$.

EXERCISES: OPEN SUBSETS OF $\mathbf{R}^n$

1. Verify Examples 37.2, 37.3, 37.4, and 37.5. Let $p \in \mathbf{R}^n$ and let $\varepsilon > 0$. Prove that the set $N(p; \varepsilon)$ is an open subset of $\mathbf{R}^n$.

2. Prove that if A and B are open subsets of $\mathbf{R}^n$, then $A \cup B$ and $A \cap B$ are open subsets of $\mathbf{R}^n$.

3. Let $\mathscr{K}$ be a collection of open subsets of $\mathbf{R}^n$. Prove that $\bigcup \mathscr{K}$ is an open subset of $\mathbf{R}^n$. Prove that if $\mathscr{K}$ is a nonempty finite collection of open subsets of $\mathbf{R}^n$, then $\bigcap \mathscr{K}$ is open.

    The reader, by virtue of having proved the previous exercise has proved the following very important theorem. *The union of an arbitrary collection of open subsets of $\mathbf{R}^n$ is open. The intersection of a finite collection of open subsets of $\mathbf{R}^n$ is open.*

38. LIMIT POINTS IN R^n

38.1. Definition. Limit point of a subset of R^n. *A point $p \in$ **R**^{n} is said to be a limit point (or an accumulation point) of a subset S of **R**^{n} provided that each neighborhood $N(p; \varepsilon)$ of p intersects S in at least one point distinct from p; i.e., for each $\varepsilon > 0$,*

$$(N(p; \varepsilon) - \{p\}) \cap S \neq \varnothing.$$

In each of the subsets of **R**^{2} pictured next, p_1, p_2, and p_3 are limit points of the set and z is not a limit point of the set   (Figure 6).

38.2. *Example*. Each point x in the real line **R** is a limit point of the set **Q** of rationals. Likewise, each point $p \in$ **R** is a limit point of **R**.

38.3. *Example*. Let $r > 0$. Each point in the disc $\{x : d(x, p) \le r\} \subset$ **R**^{2} is a limit point of the r-neighborhood $N(p; r)$.

38.4. *Example*. Let $S = \left\{ \dfrac{1}{n} : n \in \mathbf{P} \right\} \subset$ **R**. The number 0 is the only limit point of S.

EXERCISES: LIMIT POINTS IN R^n

1. Let F be a finite subset of **R**^{n}. Can F have any limit points?

2. Give an example of a subset S of **R**^{2} such that every point of S is a limit point of S.

3. Give an example of a subset S of **R**^{2} that is infinite and has no limit points.

4. Suppose S is a nonempty open subset of **R**^{n}. Is every point of S a limit point of S? Give an example of an open nonempty subset of **R**^{2} that contains all of its limit points.

5. Suppose that $\mathscr{K}$ is a collection of subsets of **R**^{n}. Suppose p is a limit point of $\bigcup \mathscr{K}$. Is p necessarily a limit point of at least one $A \in \mathscr{K}$?

6. If your answer to Exercise 5 is no, prove the following: Suppose that $\mathscr{K}$ is a finite collection of subsets of **R**^{n}. If p is a limit point of $\bigcup \mathscr{K}$, then p is a limit point of at least one $A \in \mathscr{K}$.

7. Let $S \subset$ **R**^{n} and let z be a limit point of S. Show that for every $\varepsilon > 0$, $N(z; \varepsilon) \cap S$ is an infinite set.

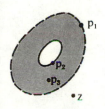
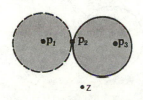
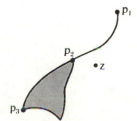

Figure 6

39. CLOSED SUBSETS OF R^n

Among the examples in Section 38, some sets contain all their limit points (see 38.3) and some sets do not (see 38.2). Those sets that contain all their limit points will play a significant role in what is to follow.

39.1. Definition. Closed subsets of R^n. *A subset S of R^n is said to be a closed subset of R^n provided that S contains all its limit points.*

The following theorem characterizes closed sets of R^n in terms of open sets, and its easy proof is left as an exercise.

39.2. Theorem. *Let $S \subset R^n$. Then S is closed if and only if its complement is open in R^n.*

Recall from Example 37.2 that $\varnothing$ and R^n are open sets. Also, the union of an arbitrary collection of open sets is open and the intersection of a nonempty finite collection of open sets is open (see comment in Exercise 3, page 70). By making use of these facts, 39.2, and De Morgan's Laws (Exercise 5, page 13), we can prove the following. It is equivalent to the corresponding facts just quoted for open sets, with intersection and union replacing each other.

39.3. Theorem. (a) *The empty set and R^n are closed subsets of R^n.* (b) *The intersection of an arbitrary nonempty collection of closed sets is closed.* (c) *The union of a finite collection of closed sets is closed.*

39.4. Definition. The closure of a set in R^n. *Let S be a subset of R^n. The closure of S is defined to be the union of S and the set of all limit points of S. We shall use the notation* cl (S) *to denote the closure of S.*

39.5. Theorem. *For each subset of R^n,* cl (S) *is a closed subset of R^n.*

PROOF. Let x be a limit point of cl (S). We wish to show that $x \in$ cl (S). If $x \notin$ cl (S), then $x \notin S$ and x is not a limit point of S. Hence, there is an $\varepsilon > 0$ such that $N(x; \varepsilon) \subset R^n - S$. However, since x is a limit point of cl (S), there is a point $z \in$ cl (S) such that $0 < d(x, z) < \varepsilon$. Then since $z \notin S$, z is a limit point of

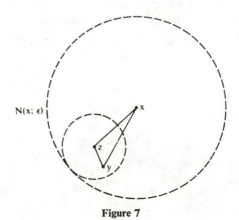

Figure 7

S. Since $\varepsilon - d(x, z) > 0$ and z is a limit point of S, there is a point $y \in S$ such that $d(y, z) < \varepsilon - d(x, z)$. However, $d(x, y) \leq d(y, z) + d(z, x) < \varepsilon - d(x, z) + d(z, x) = \varepsilon$ (see accompanying figure). Hence, $y \in S \cap N(x; \varepsilon)$, contrary to the way in which ε was chosen.

EXERCISES: CLOSED SUBSETS IN R^n

1. Suppose that F is a finite subset of **R**^{n}. Is F necessarily closed ·in **R**^{n}?

2. Prove Theorems 39.2 and 39.3.

3. Is the set in Example 37.3 a closed set?

4. Let a and b be real numbers and let S be the following subset of **R**^{2}. $S = \{(x, y): ax + by \leq 1\}$. Is S a closed set?

5. Let S be a subset of **R**^{n}. Let $S' = \{x: x \in \mathbf{R}^n$ and x is a limit point of $S\}$. Is S' necessarily a closed set?

6. Give an example of a countable collection $\mathscr{K}$ of closed subsets of **R**^{2} such that $\bigcup \mathscr{K}$ is not closed.

7. Let S be a subset of **R**^{n}. Let F consist of all points p in **R**^{n} such that for each $\varepsilon > 0$, $N(p; \varepsilon) \cap S \neq \emptyset$ and $N(p; \varepsilon) \cap (\sim S) \neq \emptyset$. Is F necessarily a closed set?

8. Show that lines and polygons are closed subsets of **R**^{n} (see 35.5 and 35.6).

40. BOUNDED SUBSETS OF R^n

Recall that a subset of **R** is a bounded set if it has both an upper and a lower bound. Equivalently, we may say that S is a bounded subset of **R** provided the distance function is bounded on $S \times S$, that is, for some number M, $d(x, y) = |x - y| \leq M$ for all x and y in S. It is this form of the definition that we use in **R**^{n}.

40.1. Definitions. Bounded subsets of R^n; the diameter of bounded sets. *A subset S of* **R**^{n} *is said to be bounded if there exists a number $M > 0$ such that*

$$d(x, y) \leq M \quad \text{for all } x \text{ and } y \text{ in } S.$$

If S is a bounded set, then the diameter of S, written diam (S) *is defined as follows*:

$$\text{diam } (S) = \text{l.u.b. } \{d(x, y): x \in S, y \in S\}, \text{ if } S \neq \emptyset$$
$$\text{diam } (\emptyset) = 0.$$

Note that the diameter of a closed interval in the real line is the length of the interval. The notion of closed interval can be extended to **R**^{n}, and a formula can be given for its diameter as shown next.

40.2. Definition. Closed interval in $\mathbf{R}^n$. *Let* $\alpha_i \leq \beta_i$ *for* $i = 1, 2, \ldots, n$. *The set* $\{(x_1, x_2, \ldots, x_n): \text{ for each } i \in \mathbf{P}_n, \alpha_i \leq x_i \leq \beta_i\}$ *is known as a closed interval in* $\mathbf{R}^n$.

Notice that

$$\text{diam } \{(x_1, x_2, \ldots, x_n) : \alpha_i \leq x_i \leq \beta_i\} = \left[\sum_{i=1}^{n} (\beta_i - \alpha_i)^2 \right]^{\frac{1}{2}}.$$

Recall from Exercise 3, page 7 1, that there exist infinite subsets of $\mathbf{R}^n$ that have no limit points. However, for an infinite set to have no limit point it must be unbounded. Stated another way, every bounded infinite subset of $\mathbf{R}^n$ has a limit point in $\mathbf{R}^n$. In order to prove this important theorem, we shall first extend the nested interval theorem (33.1) to $\mathbf{R}^n$.

40.3. Theorem (Nested Interval Theorem). *Let* (K_j) *be a sequence of nonempty closed intervals of* $\mathbf{R}^n$ *such that* $K_j \supset K_{j+1}$ *for each* $j \in \mathbf{P}$. *Then* $\bigcap \{K_j : j \in \mathbf{P}\} \neq \varnothing$. *Further, if* $\lim (\text{diam } (K_i)) = 0$, *then there is only one point in the intersection.*

PROOF. (See accompanying figure.) For $i \in \mathbf{P}_n$ and $j \in \mathbf{P}$, let $\alpha_{j,i}$ and $\beta_{j,i}$ be such that $K_j = \{(y_1, y_2, \ldots, y_n): \alpha_{j,i} \leq y_i \leq \beta_{j,i} \text{ for } i \in \mathbf{P}_n\}$. For each $i \in \mathbf{P}_n$, the collection $\{[\alpha_{j,i}, \beta_{j,i}]: j \in \mathbf{P}\}$ satisfies the hypothesis for the nested interval theorem (33.1). So there is an $x_i \in \bigcap \{[\alpha_{j,i}, \beta_{j,i}]: j \in \mathbf{P}\}$. Choose one such x_i for each $i \in \mathbf{P}_n$ and let $x = (x_1, x_2, \ldots, x_n)$. It follows that $x \in K_j$ for each $j \in \mathbf{P}$ and, thus, $x \in \bigcap \{K_j : j \in \mathbf{P}\}$.

We next prove the contrapositive of the last part of the conclusion. Suppose there were two different points $x = (x_1, x_2, \ldots, x_n)$ and $y = (y_1, y_2, \ldots, y_n)$ in $\bigcap \{K_j : j \in \mathbf{P}\}$. Then $d(x, y) > 0$, and since $\{x, y\} \subset K_j$, diam $(K_j) \geq d(x, y)$ for each $j \in \mathbf{P}$. Thus, $\lim (\text{diam } (K_j)) \geq d(x, y) > 0$, a contradiction.

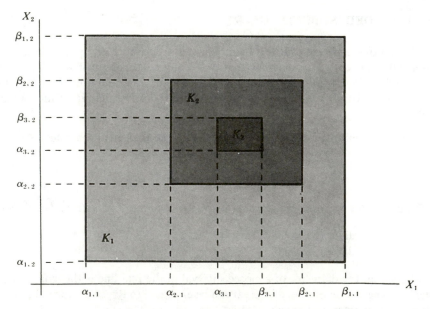

Figure 8

40.4. The Bolzano-Weierstrass Theorem. *Suppose that S is a bounded infinite subset of* **R**^{n}. *Then S has at least one limit point in* **R**^{n}.

PROOF. (It will be helpful if the reader will draw a "picture" of the various steps in the proof for the special case of **R**^{2}.) Suppose that S is a bounded infinite subset of **R**^{n}. Then there is a closed interval K_0 in **R**^{n} that contains S and that can be taken to have sides all equal in length. Thus, we may take K_0 as $[\alpha, \beta]_1 \times [\alpha, \beta]_2 \times \cdots \times [\alpha, \beta]_n$, where the subscript i will designate the ith side. Now for each i, the ith side can be subdivided into two equal subintervals. By choosing a subinterval J_i of $[\alpha, \beta]_i$ for each $i \in \mathbf{P}_n$, we can construct a closed interval $J_1 \times J_2 \times \cdots \times J_n$. This closed interval is a subset of K_0 and each of its sides has length $\frac{1}{2}(\beta - \alpha)$. But each J_i could have been chosen in one of two ways, so that there are 2^n such subintervals of K_0. Since K_0 is the union of these subintervals, at least one of the 2^n subintervals of K_0 must contain an infinite subset of S. Choose one such subinterval that has an infinite intersection with S and call it K_1. Now assume that for $i = 1, 2, \ldots, k$, K_i has been chosen as a closed subinterval of K_{i-1} such that

$K_i \cap S$ is an infinite set and such that the length of each side of K_i is $\dfrac{1}{2^i}(\beta - \alpha)$.

We can proceed to the $(k + 1)$th step in a manner similar to the way we took the first step. Thus, by induction we have a collection $\{K_i : i \in \mathbf{P}\}$ of closed intervals such that $K_i \subset K_{i-1}$, $K_i \cap S$ is infinite, and each K_i has all sides of length $\dfrac{1}{2^i}(\beta - \alpha)$. By the nested interval theorem, there is single point $p \in \bigcap \{K_i : i \in \mathbf{P}\}$.

Let us show that p is a limit point of S. Let $\varepsilon > 0$. Choose i so that $\dfrac{\beta - \alpha}{2^i} < n^{-\frac{1}{2}}\varepsilon$. Then, it is easy to verify that $K_i \subset N(p; \varepsilon)$. For if $x = (x_1, x_2, \ldots x_n) \in K_i$, then for each $j \in \mathbf{P}_n$, $|p_j - x_j| \leq \dfrac{\beta - \alpha}{2^i}$. Thus, $d(x, p) \leq \left[n\left(\dfrac{\beta - \alpha}{2^i}\right)^2 \right]^{\frac{1}{2}} < \varepsilon$. But $K_i \cap S$ is an infinite set so that $N(p; \varepsilon) \cap S$ certainly contains a point of S distinct from p. Hence, p is a limit point of S.

41. CONVERGENT SEQUENCES IN **R**^{n}

We have already dealt briefly with convergent sequences in **R**. We can take over the definition to **R**^{n} without change.

41.1. Definition. Convergent sequence. *If* (a_i) *is a sequence in* **R**^{n} *and* $l \in$ **R**^{n}, *we say that* (a_i) *converges to* l *provided that for each* $\varepsilon > 0$, *there is a positive integer* N *such that* $d(a_i, l) < \varepsilon$ *whenever* $i \geq N$ *(or equivalently* $|a_i - l| < \varepsilon$ *whenever* $i \geq N$). *In that case we write* $\lim (a_i) = l$ *or* $\lim_{i \to \infty} a_i = l$.

It should be observed that if a sequence in **R**^{n} has a limit, then that limit is unique.

One would guess, looking at a picture such as the accompanying figure, that if $(P_i) = ((x_i, y_i))$ is a sequence in **R**^{2}, then $\lim P_i = P_0 = (x_0, y_0)$ if and only if

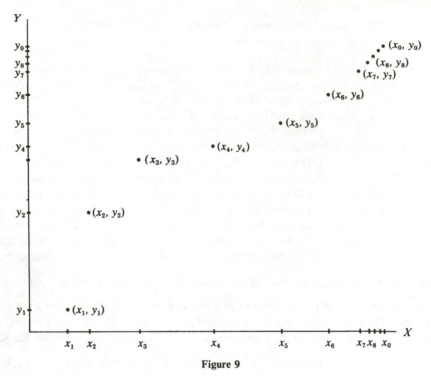

Figure 9

$\lim x_i = x_0$ and $\lim y_i = y_0$. Such is indeed the case, and the following theorem states this result for $\mathbf{R}^n$.

41.2. Theorem. *Suppose that $x_i = (x_{i,1},\ x_{i,2}, \ldots, x_{i,n})$ for each $i \in \mathbf{P}$ and $x_0 = (x_{0,1}, x_{0,2}, \ldots, x_{0,n})$. Then $\lim (x_i) = x_0$ if and only if $\lim_{i \to \infty} x_{i,k} = x_{0,k}$ for each $k \in \mathbf{P}_n$.*

The proof follows easily from 36.3 and is left as an important exercise for the reader.

We shall say that *a sequence in $\mathbf{R}^n$ is bounded provided that its range is bounded.* Note that a sequence with finite range is bounded. Likewise, note that if (a_i) converges to L, then $N(L; 1)$ contains all but k terms for some positive integer k. Thus, if $M = \max\{d(a_i, L) : i \in \mathbf{P}_k\}$, then the range of (a_i) is contained in $N(L; M + 1)$. Thus, we have verified the following statement.

41.3. Remark. If (a_i) is a convergent sequence, then (a_i) is a bounded sequence.

It is certainly not true that every bounded sequence in $\mathbf{R}^n$ is convergent. However, by making use of the Bolzano-Weierstrass theorem we can prove the following.

41.4. Theorem. *Let (a_i) be a bounded sequence in $\mathbf{R}^n$. Then there is a subsequence (a_{n_i}) of (a_i) that converges.*

PROOF. Suppose the range of (a_i) is a finite set. Then there is a strictly increasing sequence of positive integers (n_i) and a point $z \in \mathbf{R}^n$, such that $a_{n_i} = z$ for $i \in \mathbf{P}$. Hence, $\lim (a_{n_i}) = z$ and we are through. Next suppose the range S of (a_i) is an infinite set. By the Bolzano-Weierstrass theorem there is a point z that is a limit point of S. (We shall find a subsequence (a_{n_i}) of (a_i) by choosing for each $i \in \mathbf{P}$, an a_{n_i} such that $d(a_{n_i}, z) < \dfrac{1}{i}$. However, we must be careful to choose these in such a way that (n_i) is a strictly increasing sequence of positive integers. We shall do this inductively.) Since z is a limit point of S, there is a positive integer n_1 such that

$$0 < d(a_{n_1}, z) < 1.$$

Assume that for $1, 2, \ldots, h$, increasing integers $n_1, n_2, \ldots, n_h$ have been chosen such that

$$0 < d(a_{n_i}, z) < \frac{1}{i}.$$

Since $S \cap N\left(z; \dfrac{1}{i+1}\right)$ is an infinite set (see Exercise 7, page 71), there is an integer n_{h+1} such that $n_h < n_{h+1}$ and for which $d(a_{n_{h+1}}, z) < \dfrac{1}{h+1}$. Hence, by induction (see 18.4), a strictly increasing sequence of integers (n_i) has been chosen such that $d(a_{n_i}, z) < \dfrac{1}{i}$. It is clear from the way in which the integers (n_i) were chosen that (a_{n_i}) is a subsequence of (a_i) and that $\lim (a_{n_i}) = z$.

EXERCISES: CONVERGENT SEQUENCES IN $\mathbf{R}^n$

1. Prove that if S is a subset of $\mathbf{R}^n$ and $z \in \mathbf{R}^n$, then z is a limit point of S if and only if there exists a sequence (a_i) in S that converges to z and which is such that $a_i \neq a_j$ for $i \neq j$.

2. Suppose (a_i) is a sequence in $\mathbf{R}^n$ that has the following property: For each $\varepsilon > 0$, there is an integer N such that $d(a_m, a_n) < \varepsilon$ for $m \geq N$ and $n \geq N$. Show that (a_i) is a bounded sequence.

3. Prove that if (a_i) is a convergent sequence in $\mathbf{R}^n$, then for each subsequence (a_{N_i}) of (a_i), $\lim (a_i) = \lim (a_{N_i})$.

4. Give the details of the proof of 41.2.

5. Prove that a subset S of $\mathbf{R}^n$ is closed and bounded if and only if every sequence in S has a convergent subsequence whose limit is a point in S.

6. Suppose (S_i) is a sequence of nonempty bounded and closed subsets of $\mathbf{R}^n$ such that for each $i \in \mathbf{P}$, $S_{i+1} \subset S_i$. Prove that there is at least one point $p \in \bigcap \{S_i : i \in \mathbf{P}\}$. (Note that this is a generalization of the nested interval theorem. After proving the proposition, see if your proof depended on some consequence of the nested interval theorem.)

7. Suppose that (a_i) and (b_i) are sequences in $\mathbf{R}^n$. Suppose also that $c_i = a_i + b_i$ for $i \in \mathbf{P}$. Show that if two of the sequences (a_i), (b_i), (c_i) converge, then so does the remaining one and further $\lim (c_i) = \lim (a_i) + \lim (b_i)$.

8. Prove that $\lim (a_i) = a$ if and only if $\lim (d(a_i, a)) = 0$.

42. CAUCHY CRITERION FOR CONVERGENCE

To use the definition directly to determine if a sequence (x_i) converges to a point x, one generally first has to find a "candidate" x and then determine if indeed the terms are getting arbitrarily close to x. In fact, it may be impossible to determine what the limit is likely to be. If it could be determined that a sequence converges without first determining a "candidate," then it might be possible to determine how large i must be for x_i to be a sufficiently good approximation to the limit. The so-called Cauchy criterion for sequences gives us a criterion involving the sequence alone.

42.1. Cauchy criterion for convergence in $\mathbf{R}^n$. *Suppose that (a_i) is a sequence in $\mathbf{R}^n$. Then (a_i) converges if and only if the following condition is satisfied.*

42.1(a). *For each $\varepsilon > 0$, there is a positive integer M such that for $m \geq M$ and $n \geq M$, $d(a_m, a_n) < \varepsilon$.*

PROOF. We first prove that if (a_i) converges, then (a_i) satisfies the stated condition. To see this, let $\lim a_i = a$ and suppose that $\varepsilon > 0$. Then there is an integer M such that for $m \geq M$, $d(a_m, a) < \varepsilon/2$. Then if $m \geq M$ and $n \geq M$, $d(a_m, a_n) \leq d(a_m, a) + d(a_n, a) < \varepsilon$.

Next, suppose that (a_i) satisfies the given condition. In that case it is not hard to prove that (a_i) is a bounded sequence in $\mathbf{R}^n$ (see Exercise 2, page 77). By 41.4, since (a_i) is a bounded sequence in $\mathbf{R}^n$, there is a subsequence (a_{n_i}) of (a_i) that converges to a point z. We complete the proof by showing that because of the Cauchy condition, the sequence (a_i) itself converges to z. To see this let $\varepsilon > 0$. There is an integer N such that for $m \geq N$ and $n \geq N$, $d(a_m, a_n) < \frac{1}{2}\varepsilon$. Now there is an $M = n_k \geq N$ such that $d(a_M, z) < \frac{1}{2}\varepsilon$ since (a_{n_i}) converges to z. Then for $m > M$,

$$d(a_m, z) \leq d(a_m, a_M) + d(a_M, z)$$

$$< \frac{\varepsilon}{2} + \frac{\varepsilon}{2} = \varepsilon.$$

This completes the proof.

EXERCISES: CAUCHY CRITERION FOR CONVERGENCE

1. Show directly from the definition that the sequence $(1/i)$ satisfies the Cauchy criterion.

2. Define the sequence (x_n) in $\mathbf{R}$ inductively as follows: Let $x_1 = a$, $x_2 = b$, $x_n = \frac{1}{2}(x_{n-1} + x_{n-2})$, $n \geq 3$. Show that (x_n) converges.

3. Suppose that (x_i) is a sequence in $\mathbf{R}^n$. For each i, let $x_i = (x_{i,1}, x_{i,2}, \ldots, x_{i,n})$. Show that (x_i) satisfies the Cauchy condition for $\mathbf{R}^n$ stated in 42.1(a) if and only if for each $j \in \mathbf{P}_n$, the sequence $(x_{i,j})_{i=1}^{\infty}$ satisfies the condition for $\mathbf{R}$.

4. Suppose that (a_i) is a sequence in R^n that satisfies the following property. There exists a θ, where $0 \le \theta < 1$, such that for each $n \in P$, $d(a_{n+1}, a_{n+2}) < \theta \, d(a_n, a_{n+1})$. Show that (a_n) converges.

43. SOME ADDITIONAL PROPERTIES FOR R^n

In this section we state several important theorems. These theorems are not only important in their own right, but they will be useful in some of our considerations in subsequent chapters. The proof of each of these theorems is left as an exercise.

A subset S of R^n is said to be *dense* in R^n provided that each nonempty open subset of R^n has in it at least one point of S. This is equivalent to requiring that cl $(S) = R^n$. From 30.5, the set of rationals Q is dense in R. Thus, R contains a countable dense subset. Similarly, the following is true for R^n.

43.1. Theorem. *Let D be the collection of all points $(x_1, x_2, \ldots, x_n)$ in R^n such that for $i \in P_n$, x_i is rational. Then D is dense in R^n. Hence, R^n contains a countable dense subset.*

By making use of the previous theorem, we can prove the following.

43.2. Theorem. *There exists a countable collection $\mathscr{B} = \{B_i : i \in P\}$ of open subsets of R^n that have the following property:*
If U is open in R^n and $p \in U$, then there is a $B_i \in \mathscr{B}$ such that

$$p \in B_i \subset U.$$

Hence, each open set $U \subset R^n$ is the union of a countable subcollection of $\mathscr{B}$.

The previous theorem can be used to prove the following important theorem.

43.3. Lindelöf Theorem for R^n. *Let X be a subset of R^n. Suppose that $\mathscr{B}$ is a collection of open subsets of R^n such that $\bigcup \mathscr{B} \supset X$. Then for some countable subcollection $\mathscr{B}^* \subset \mathscr{B}$, $\bigcup \mathscr{B}^* \supset X$.*

Suppose that S is a subset of a space X and $\mathscr{B}$ is a collection of subsets of X such that $\bigcup \mathscr{B} \supset S$. Then $\mathscr{B}$ is called a *covering* of S and is said to *cover* S.

43.4. Heine-Borel Theorem. *Suppose that X is a bounded and closed subset of R^n. Let $\mathscr{K}$ be a collection of open subsets of R^n that covers X. Then some finite subcollection of $\mathscr{K}$ also covers X.*

(Hint: Use 43.3 to first find a countable subcollection $\mathscr{K}^*$ that covers X. Then assume that no finite subcollection of $\mathscr{K}^*$ covers X.)

EXERCISES: SOME ADDITIONAL PROPERTIES FOR R^n

1. Prove the theorems in this section.

2. Suppose $\mathscr{B}$ is a collection of pairwise disjoint open subsets of R^n. Show that $\mathscr{B}$ is a countable collection.

44. SOME FURTHER REMARKS ABOUT $\mathbf{R}^n$

We have discussed in this chapter some important properties possessed by $\mathbf{R}^n$. Other important properties possessed by $\mathbf{R}^n$ will be considered in subsequent chapters. We shall also discuss in detail the nature of continuous and uniformly continuous mappings from $\mathbf{R}^m$ into $\mathbf{R}^n$. However, much of our discussion will depend on the fact that $\mathbf{R}^n$ has a distance function defined on it and will hold equally well for the more general type of metric spaces to be studied next. On the other hand, there are properties possessed by $\mathbf{R}^n$ because of its particular structure that do not necessarily hold for the more general spaces. The Cauchy criterion for convergence (42.1) and the properties stated in the Bolzano-Weierstrass (40.4), the Heine-Borel (43.4), and the Lindelöf (43.3) theorems are examples. Such properties have motivated many useful definitions and concepts. These permit analogies to be made between theorems and proofs in $\mathbf{R}^n$ and in certain other more general spaces.

REVIEW EXERCISES

1. Suppose that K is an uncountable collection of real numbers. Prove that there exists at least one $z \in K$ such that z is a limit point of K. Prove that there is an uncountable collection of limit points of K in K.

2. Suppose that $S \subset \mathbf{R}$, $S \neq \varnothing$ and $S \neq \mathbf{R}$. Prove that S is not both open and closed.

3. Suppose that $\{K_i : i \in P\}$ is a collection of nonempty bounded and closed subsets of $\mathbf{R}^n$ such that $K_{i+1} \subset K_i$ for $i \in P$. Is $\bigcap \{K_i : i \in P\}$ necessarily nonempty?

4. Suppose that K is a nonempty closed and bounded subset of $\mathbf{R}^n$. Do there exist points x and y in K such that $d(x, y) = \operatorname{diam}(K)$?

5. Suppose that M is a countable subset of $\mathbf{R}^2$ and $p \in \mathbf{R}^2 - M$. Does there necessarily exist a line $L \subset \mathbf{R}^2$ such that $p \in L \subset \mathbf{R}^2 - M$?

6. Suppose that A and B are subsets of $\mathbf{R}^n$. Is $\operatorname{cl}(A \cup B) = \operatorname{cl}(A) \cup \operatorname{cl}(B)$ (see 39.4)? Is $\operatorname{cl}(A) \cap \operatorname{cl}(B) = \operatorname{cl}(A \cap B)$?

7. Determine whether the following proposition is correct.
    Suppose that X is a bounded subset of $\mathbf{R}^n$, and $\varepsilon > 0$. Then there exists a finite set $F \subset X$ such that

$$X \subset \bigcup \{N(x; \varepsilon) : x \in F\}.$$

3

Metric Spaces: Introduction

At this juncture it should be apparent to the reader that the notion of distance between points in $\mathbf{R}^n$ has been of fundamental importance in our discussion. There are numerous other types of sets, in addition to subsets of $\mathbf{R}^n$, that mathematicians have had to study. Often in studying some set of mathematical objects, it is possible to measure in some useful way the distance between each pair of these objects. The word "useful" should be emphasized, for, as we shall see, it is always possible to define a mapping d that satisfies the properties listed in 36.1. However, to make the notion useful, the distance function has to be introduced in a way that is compatible with the proposed application. This is not always possible, and one often has to generalize the notion of "closeness" in ways that do not depend on distance.

In this chapter we shall formalize the notion of a metric space and we shall give definitions of various concepts which have been introduced earlier only for $\mathbf{R}^n$. For example, the notions of *open set, closed set, limit point,* and *convergent sequence* will be defined and studied for metric spaces. The very important concepts of continuous and uniformly continuous mappings will be introduced and will continue to be studied in subsequent chapters.

45. DISTANCE FUNCTION AND METRIC SPACES

In 36 a distance function was defined and in 36.1 some important properties for this function were listed. These properties give motivation for the definition which follows.

45.1. Definitions. Distance function and metric space. *Let X be a set. Let d be a nonnegative real-valued function defined on $X \times X$ that satisfies the following: For all $x, y,$ and z in X,*

45.1(a). $d(x, y) = 0$ *if and only if* $x = y$

45.1(b). $d(x, y) = d(y, x)$

45.1(c). $d(x, y) + d(y, z) \geq d(x, z)$ *(triangle inequality)*.

The function d is called a distance function or a metric for X and (X, d) is called a metric space.

When there seems to be no chance for confusion, we shall sometimes speak of a "metric space X," with the notation for the metric being understood. It should be emphasized, however, that for a given set there are, in general, many possible metrics. The reader is familiar with the Euclidean metric for $\mathbf{R}^n$ (see 36). Next, we give other examples of metrics. In each case the reader should verify that the function is indeed a metric for the given set.

45.2. *Example.* Let $k > 0$ and $\rho : \mathbf{R}^n \times \mathbf{R}^n \to \mathbf{R}$ be given by

$$\rho(x, y) = k \, |x - y| \quad \text{for all } x \text{ and } y \text{ in } \mathbf{R}^n.$$

45.3. *Example.* The function $g : \mathbf{R}^2 \times \mathbf{R}^2 \to \mathbf{R}$ given by

$$g(x, y) = |x_1 - y_1| + |x_2 - y_2|$$

where $x = (x_1, x_2)$ and $y = (y_1, y_2)$.

45.4. *Example.* Let X be an arbitrary set. Define $m : X \times X \to \mathbf{R}$ as follows:

$$m(x, y) = 0 \quad \text{if} \quad x = y$$

$$m(x, y) = 1 \quad \text{if} \quad x \neq y.$$

45.5. *Example.* Let p be the real-valued function defined on $\mathbf{R}^2 \times \mathbf{R}^2$ by

$$p(x, y) = \max \{|x_i - y_i| : i = 1, 2\}$$

where $x = (x_1, x_2)$ and $y = (y_1, y_2)$. The function p is a metric for $\mathbf{R}^2$.

45.6. *Example.* Let the function h be defined as follows: For all x and y in $\mathbf{R}^2$, let

$$h(x, y) = \frac{|x - y|}{1 + |x - y|}.$$

The function h is a metric for $\mathbf{R}^2$.

45.7. *Example.* For all $x \in \mathbf{Q}$ and $y \in \mathbf{Q}$, where $\mathbf{Q}$ is the set of all rationals, let

$$d(x, y) = |x - y|.$$

(Note that $d : \mathbf{Q} \times \mathbf{Q} \to \mathbf{R}$ is the restriction to $\mathbf{Q} \times \mathbf{Q}$ of the Euclidean distance function for $\mathbf{R}$.) $(\mathbf{Q}, d)$ is a metric space.

EXERCISES: DISTANCE FUNCTION AND METRIC SPACES

1. Verify that the functions given in 45.2 through 45.6 are metrics.

2. Let (X, d) be a metric space. Define $d^* : X \times X \to \mathbf{R}$ as follows:

$$d^*(x, y) = \min \{1, d(x, y)\}$$

for all x and y in X. Show that d^* is a metric for X.

3. Let $d: X \times X \to \mathbf{R}$ be a function that satisfies the following: For all x, y, and z, in X

$$d(x, y) = 0 \quad \text{if and only if} \quad x = y$$

$$d(z, x) + d(z, y) \geq d(x, y).$$

Show that for all x and y in X, $d(x, y) \geq 0$ and $d(x, y) = d(y, x)$. (Hence, a function satisfying the two given properties is a metric for X.)

46. OPEN SETS AND CLOSED SETS

In 36.4 we defined for $\mathbf{R}^n$ the notion of ε-neighborhood with respect to the Euclidean distance formula given in 36. The notion of *limit point* was defined in terms of ε-neighborhood and, in turn, *closed set* was defined in terms of limit point. Thus, ultimately all these concepts rest on the idea of a metric. In this section we extend all these concepts to metric spaces. The Euclidean metric given in 36 will continue to come up in our discussions. If we use the phrase "the space $\mathbf{R}^n$" or more simply "$\mathbf{R}^n$" without specific mention of a metric, it will be understood to stand for $(\mathbf{R}^n, d)$, where d is the Euclidean metric defined in 36. In some books this space is referred to as E^n (Euclidean n-dimensional metric space).

In the following definitions (X, d) is a metric space.

46.1. Definitions. Open spheres and closed balls. *Let $\varepsilon > 0$ and $p \in X$. Then by the ε-neighborhood of p or the open sphere with center p and radius ε, we shall mean the set*

$$N(p; \varepsilon) = \{x : x \in X \quad \text{and} \quad d(x, p) < \varepsilon\}.$$

By the closed ball with p as center and radius ε, we shall mean the set

$$B(p; \varepsilon) = \{x : x \in X \quad \text{and} \quad d(x, p) \leq \varepsilon\}.$$

If more than one metric is used in a discussion, we shall sometimes use subscripts to denote which metric is involved. For example, $B_d(x; \varepsilon)$ would denote an ε-ball with respect to the metric d. We shall follow the same convention with other special types of sets involving a metric.

In order for one to get a feel for the concepts under discussion, it is helpful to draw pictures of open spheres and closed balls using various metrics for the set $\mathbf{R}^2$ that were given in Examples 45.2 through 45.6. We give some consideration to such pictures in the following examples, and the reader will be asked to draw further pictures in the next set of exercises.

46.2. *Example.* Let the metrics g and p be as in Example 45.3 and Example 45.5, respectively. Let d be the Euclidean metric for $\mathbf{R}^2$. The 1-neighborhoods of $0 = (0, 0)$ for these three metrics are shown in Figure 10.

46.3. Definition. Open set. *Let S be a subset of X. S is said to be an open subset of X provided that for each $p \in S$, there exists an $\varepsilon > 0$ such that $N(p; \varepsilon) \subset S$.*

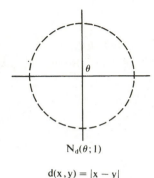

$N_d(\theta; 1)$

$d(x,y) = |x - y|$

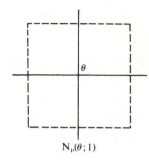

$N_p(\theta; 1)$

$p(x,y) = \max \{|x_1 - y_1|, |x_2 - y_2|\}$

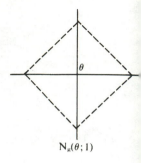
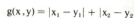

$N_g(\theta; 1)$

$g(x,y) = |x_1 - y_1| + |x_2 - y_2|$

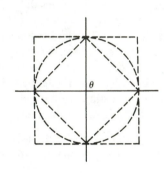

Figure 10

46.4. Definition. Limit point of a set. *Let $S \subset X$ and let $p \in X$. The point p is said to be a limit point of S provided that each neighborhood $N(p; \varepsilon)$ of p intersects S in at least one point different from p.*

46.5. Definition. Closed set. *A subset S of X is said to be closed provided S contains all its limit points.*

46.6. Definition. Closure of a set. *Let $S \subset X$. The set $S \cup S'$, where S' is the set of all limit points of S, is known as the closure of S, abbreviated* cl (S).

Just as in $\mathbf{R}^n$, closed subsets can be characterized in terms of open subsets as follows.

46.7. Theorem. *A subset S of a metric space X is closed if and only if its complement is open.*

In the definitions just given and in other definitions related to a space (X, d) the following should be kept in mind: The definitions are made in the setting of a given fixed metric space. It would probably be more accurate to say, for example, that S is closed with respect to (X, d) or S is a d-closed set. However, where there is no chance for confusion, we will generally not refer to the space.

Since thus far our experience with these concepts is related to the Euclidean space $\mathbf{R}^2$, we must be cautious about dealing with sets when other metrics are used. For example, for the Euclidean metric d for $\mathbf{R}^n$, cl $(N_d(p; \varepsilon)) = B_d(p; \varepsilon)$. Further, in the space $\mathbf{R}^n$, $N(p; \varepsilon)$ is a *proper* subset of $B(p; \varepsilon)$. Again, this need not be so in general. There will be examples in the next set of exercises that will bring out these facts.

EXERCISES: OPEN SETS AND CLOSED SETS

1. In this exercise, let d be the Euclidean metric for $\mathbf{R}^2$. Further, let g, m, p, and h be the metrics given in Examples 45.3 through 45.6. Let $\theta = (0, 0)$. Draw each of the following sets (some will simply be verifications of the pictures given in the text).

   (a) $N_d(\theta; 1)$, $B_d(\theta; 1)$
   (b) $N_g(\theta; 1)$, $B_g(\theta; 1)$
   (c) $N_m(\theta; \frac{1}{2})$, $N_m(\theta; 1)$, $B_m(\theta; 1)$
   (d) $N_p(\theta; 1)$, $B_p(\theta; 1)$
   (e) $N_h(\theta; 1)$, $B_h(\theta; 1)$

2. Using Exercise 1 or other examples, show that for some metrics it need not be so that cl $(N(p; \varepsilon)) = B(p; \varepsilon)$. Further, show that $N(p; \varepsilon)$ need not be a proper subset of $B(p; \varepsilon)$.

3. Prove that if (X, d) is a metric space, $p \in X$, and $\varepsilon > 0$, then $N(p; \varepsilon)$ is an open set. Is it possible in some cases that $N(p; \varepsilon)$ is also a closed set?

4. Are closed balls $B(p; \varepsilon)$ necessarily closed sets?

5. Show that if (X, d) is a metric space, then X can be expressed as the union of a countable collection of neighborhoods with a common center.

6. Let (X, d) be a metric space. Prove that if $A \subset B \subset X$, then cl $(A) \subset$ cl (B).

7. In each of the following either prove the statement or give a counterexample:

   (a) cl $(A \cap B) \subset$ cl $(A) \cap$ cl (B)
   (b) cl $(A \cap B) =$ cl $(A) \cap$ cl (B)
   (c) cl $(A \cup B) \subset$ cl $(A) \cup$ cl (B)
   (d) cl $(A \cup B) =$ cl $(A) \cup$ cl (B)

47. SOME BASIC THEOREMS CONCERNING OPEN AND CLOSED SETS

The theorems listed in this section include some already met by the reader in the context of $\mathbf{R}^n$ (see 39 and Exercises on page 7 0). The reader should provide a proof in each case in which a proof is not given in the text.

47.1. Theorem. (a) $\varnothing$ *and* X *are open subsets of* X. (b) *The intersection of any two open subsets of* X *is open.* (c) *The union of each arbitrary collection of open subsets is open.*

The dual of 47.1 is the following important theorem.

47.2. Theorem. (a) $\varnothing$ *and* X *are closed subsets of* X. (b) *The union of any two closed subsets of* X *is closed.* (c) *The intersection of each nonempty arbitrary collection of closed subsets is closed.*

The proof of Theorem 39.5 for the closure of sets in $\mathbf{R}^n$ carries over for metric spaces to give the following.

47.3. Theorem. *For each subset* S *of* X, cl (S) *is closed.*

47.4. Theorem. *For each subset* S *of* X, cl (S) *is the intersection of the collection of all closed subsets of* X *that contain* S, *that is,*

$$\text{cl }(S) = \bigcap \{F : S \subset F \text{ and } F \text{ is a closed subset of } X\}.$$

Proof. Let $K = \bigcap \{F : S \subset F$ and F is a closed subset of $X\}$. By 47.3, cl (S) is a closed set and since $S \subset$ cl (S), it follows that $K \subset$ cl (S). We complete the proof by showing that cl $(S) \subset K$. For each $H \in \{F : S \subset F$ and F is closed$\}$, $S \subset H$ and, hence, cl $(S) \subset$ cl $(H) = H$. Therefore, cl $(S) \subset \bigcap \{F : S \subset F$ and F is closed$\}$.

EXERCISE: SOME BASIC THEOREMS CONCERNING OPEN AND CLOSED SETS

      1. (a) Prove Theorem 47.1.
          (b) Prove Theorem 47.2.
          (c) Prove Theorem 47.3.

48. TOPOLOGY GENERATED BY A METRIC

Let (X, d) be a metric space. Let $\mathcal{T}(d)$ be the collection of all open subsets of X with respect to the metric d. Then $\mathcal{T}(d)$ is said to be the *topology* for X generated by d. As we proceed with our discussion we shall see that many of the concepts with which we deal can be completely defined or characterized in terms of the open subsets of X.

We shall also see that different metrics for a set X may generate the same collection of open sets. Hence, those properties that depend only on the open sets will be possessed by both (X, d_1) and (X, d_2) if d_1 and d_2 generate the same open sets. Furthermore, as we shall see later, concepts characterized by open sets lend themselves to useful generalizations.

48.1. Definition. Topology generated by a metric. *Let* (X, d) *be a metric space. The collection* $\mathcal{T}(d)$ *of all* d-*open subsets of* X *is called the topology for* X *generated by* d.

48.2. Definition. Equivalent metrics. *Let* d_1 *and* d_2 *be metrics for a set* X. *The metrics are said to be equivalent provided that the topologies* $\mathcal{T}(d_1)$ *and* $\mathcal{T}(d_2)$ *generated by* d_1 *and* d_2 *are the same (i.e.,* $\mathcal{T}(d_1) = \mathcal{T}(d_2)$).

48.3. Definition. Topological property for metric spaces. *If P is a property for metric spaces that can be characterized completely by the topology (i.e., the open sets) of the space, then P is said to be a topological property.*

The following theorem is a useful device for determining whether two metrics are equivalent. The easy proof is left as an exercise.

48.4. Theorem. *Let X be a set and let d_1 and d_2 be metrics for X. Then d_1 and d_2 are equivalent if and only if the following condition is satisfied. For each $p \in X$ and $\varepsilon > 0$, there is a $\delta_1 > 0$ and $\delta_2 > 0$ such that*

$$N_{d_1}(p; \delta_1) \subset N_{d_2}(p; \varepsilon)$$

and

$$N_{d_2}(p; \delta_2) \subset N_{d_1}(p; \varepsilon).$$

48.5. *Example.* Let d be the Euclidean metric for $\mathbf{R}^2$, p the metric for $\mathbf{R}^2$ as given in Example 45.5 and g the metric of Example 45.3. Then each of these three metrics is equivalent to the others. See the figure accompanying Example 46.2 (Figure 10).

EXERCISES: TOPOLOGY GENERATED BY A METRIC

1. Prove Theorem 48.4.

2. Verify Example 48.5.

3. Let (X, d) be a metric space. Prove that there is a metric d^* which is equivalent to d and which satisfies the inequality $d^*(x, y) \leq 1$ for each x and y in X. Hint: Let $d^*(x, y) = \min\{1, d(x, y)\}$.

4. Let (X, d) be a metric space. Let $k > 0$ and d^* be defined on $X \times X$ as follows:
$$d^*(x, y) = k\, d(x, y).$$
Show that d^* is a metric for X and that d^* is equivalent to d.

5. Let X be a set and m be as in Example 45.4. Show that every subset of X is an m-open set. Hence, $\mathscr{T}(m)$, the topology for X generated by m, is the largest topology for X that can be generated by a metric d; that is, $\mathscr{T}(d) \subset \mathscr{T}(m)$ for each metric d for X.

49. SUBSPACE OF A METRIC SPACE

Suppose we wish to have a measure of distance between points on a circle in $\mathbf{R}^2$. One obvious way to do this is simply to apply the Euclidean metric for $\mathbf{R}^2$ to the circle. Similarly the metric for $\mathbf{R}^3$ can be used to measure the distance between points on a 2-sphere in $\mathbf{R}^3$. More generally, if (X, d) is a metric space and $Y \subset X$, the restriction $d \mid Y \times Y$ is a metric for Y.

49.1. Definition. Subspace of a metric space. *Let (X, d) be a metric space and suppose $Y \subset X$. Then $(Y, d \mid Y \times Y)$ is called a subspace of (X, d).*

Generally we will use the shorter notation (Y, d) for $(Y, d \mid Y \times Y)$. Also, when there is little chance for confusion, we shall say that Y is a subspace of X, the reference to the metric d being understood.

49.2. *Example.* Consider the line $L = \{(x, 0) : x \in \mathbf{R}\} \subset \mathbf{R}^2$. If we restrict the Euclidean metric d for $\mathbf{R}^2$ to points in L, then we see that

$$d((x, 0), (y, 0)) = |x - y|$$

for all $(x, 0)$ and $(y, 0)$ in L.

It should be noted especially that just because a set S is open (closed) in a subspace (Y, d) of (X, d) it does not necessarily follow that S is open (closed) in (X, d). For example, in 49.2 any set of the form $\{(x, 0) : a < x < b\}$ is open in (L, d) but is not open in $(\mathbf{R}^2, d)$.

49.3. *Example.* Let d be the Euclidean metric for $\mathbf{R}$. Let $\mathbf{Q}$ be the set of all rational numbers in $\mathbf{R}$. We note that $\mathbf{Q}$ is both open and closed in $(\mathbf{Q}, d)$ but neither open nor closed in $(\mathbf{R}, d)$.

Note that if we call $d^* = d \mid Y \times Y$ and let $p \in Y$, then $N_d(p; \varepsilon) = \{y : y \in X$ and $d(y, p) < \varepsilon\}$ and $N_{d*}(p; \varepsilon) = \{y : y \in Y$ and $d^*(y, p) = d(y, p) < \varepsilon\}$. Thus, $N_{d*}(p; \varepsilon) = N_d(p; \varepsilon) \cap Y$.

This observation is useful in proving the following theorem.

49.4. Theorem. *Let (Y, d) be a subspace of (X, d). Then $S \subset Y$ is open (closed) in (Y, d) if and only if there is a set $Q \subset X$, open (closed) in X, such that $S = Q \cap Y$.*

PROOF. We shall prove the part for open sets and leave the part for closed sets as an exercise.

Let $d^* = d \mid Y \times Y$. First we assume that S is open in Y and that $p \in S$. Since S is open in Y, there is an $\varepsilon(p) > 0$ such that $N_{d*}(p; \varepsilon(p)) \subset S$. Choose one such $\varepsilon(p)$ for each $p \in S$. Recall that $N_{d*}(p; \varepsilon(p)) = N_d(p; \varepsilon(p)) \cap Y$. Let $Q = \bigcup \{N_d(p; \varepsilon(p)) : p \in S\}$. Note that Q is open in X and that $Q \cap Y = S$.

We next prove the converse. Suppose $S = Q \cap Y$ where Q is open in X. We show that S is open in Y. Let $p \in S = Q \cap Y$. Since Q is open in X, there is an $\varepsilon > 0$ such that $N_d(p; \varepsilon) \subset Q$. But then, $N_{d*}(p; \varepsilon) = N_d(p; \varepsilon) \cap Y \subset Q \cap Y$. Thus, S is open in Y and the proof for this part is complete.

EXERCISES: SUBSPACE OF A METRIC SPACE

1. Prove the part of Theorem 49.4 for closed sets.

2. Let (Y, d) be an open (a closed) subspace of (X, d). Then $W \subset Y$ is open (closed) in Y if and only if W is open (closed) in X.

3. Let $A = (a_1, a_2)$ and $B = (b_1, b_2)$ be distinct points in $\mathbf{R}^2$. Let S be the line $\{(1 - t) A + tB : t \in \mathbf{R}\}$ in $\mathbf{R}^2$. Notice that each point in S can be represented uniquely in the form

$$P(t) = (1 - t) A + tB.$$

Define $d^* : S \times S \to \mathbf{R}$ by

$$d^*(P(t_1), P(t_2)) = |t_1 - t_2|.$$

Is d^* a metric for S? If your answer is yes, is d^* equivalent to $d \mid S \times S$, where d is the Euclidean metric?

50. CONVERGENT SEQUENCES IN METRIC SPACES

We have already considered the notion of *convergent sequence* in the real line (32) and, more generally, in $\mathbf{R}^n$ (41). The definition carries over directly to metric spaces.

50.1. Definition. Convergent sequences in metric spaces. *Let (X, d) be a metric space. Let (x_n) be a sequence in X and let $x \in X$. We say that (x_n) converges to x or $\lim (x_n) = x$ provided that for each $\varepsilon > 0$ there is a positive integer N_ε such that*

$$\text{if } n \geq N_\varepsilon, \quad \text{then } d(x_n, x) < \varepsilon.$$

Just as in $\mathbf{R}^n$, it follows easily from the definition that a sequence in a metric space can converge to, at most, one point. Also, it should be observed that that convergence is defined with reference to a space. Suppose, for example, that (x_i) is a sequence in (X, d) such that each $x_i \in S \subset X$. If (x_i) converges in (X, d) to a point in $X - S$, then (x_i) converges in (X, d) but does not converge in the subspace (S, d).

50.2. *Example.* Consider the space $(\mathbf{R}, d)$ where d is the Euclidean metric and S is the open interval $(0, 1) \subset \mathbf{R}$. Note that the sequence $(1/n)$ converges in $(\mathbf{R}, d)$ but does not converge in (S, d).

The next theorem follows immediately from the definition of convergence in a metric space.

50.3. Theorem. *Let (X, d) be a metric space and let (S, d) be a subspace of (X, d). If (x_i) is a sequence in S, then (x_i) converges in (S, d) if and only if (x_i) converges in (X, d) to a point in S.*

The proof of the following statement is easy and is left as an exercise.

50.4. Theorem. *Suppose that (x_n) is a sequence in a metric space (X, d) and $x \in X$. The sequence (x_n) converges to x if and only if for each open set U in X that contains x, there is a positive integer N_U such that if $n \geq N_U$, then $x_n \in U$.*

Suppose that d_1 and d_2 are equivalent metrics for a set X. It follows easily from 50.4 that a sequence converges in (X, d_1) if and only if it converges in (X, d_2).

For metric spaces, limit points of sets and, hence, closed sets can be characterized in a very useful way by means of sequences.

50.5. Theorem. *Let $S \subset X$ and $l \in X$. Then,*

50.5(a). *l is a limit point of the set S if and only if there exists a sequence of points in $S - \{l\}$ that converges to l.*

50.5(b). *S is a closed subset of X if and only if every convergent sequence in S converges to a point in S.*

PROOF. We prove (a) and leave (b) as an exercise.

First suppose that l is a limit point of S. Then for each positive integer i, choose $x_i \in N(l; 1/i) \cap [S - \{l\}]$. Obviously $\lim (x_i) = l$.

Next suppose that (x_i) is a sequence of points in $S - \{l\}$ such that $\lim (x_i) = l$. Given a neighborhood $N(l; \varepsilon)$, there is an n such that $x_n \in N(l; \varepsilon)$. Upon noting that $x_n \neq l$, we conclude that l is a limit point of S.

EXERCISES: CONVERGENT SEQUENCES IN METRIC SPACES

1. Prove Theorem 50.4 and Theorem 50.5(b).

2. Suppose that (x_i) is a convergent sequence in a metric space (X, d). Show that for each $\varepsilon > 0$, there is a positive integer M such that if $m \geq M$ and $n \geq M$, then $d(x_m, x_n) < \varepsilon$.

3. Let $\mathbf{Q}$ be the set of all rationals and let d be the Euclidean metric for $\mathbf{R}$. Consider the subspace $(\mathbf{Q}, d)$ of $(\mathbf{R}, d)$. Show by an example that if a sequence in $\mathbf{Q}$ satisfies the condition in the conclusion of the previous exercise, the sequence need not converge to a point in $\mathbf{Q}$.

4. Consider the space of Example 45.4. Describe the nature of convergent sequences in that space.

5. Suppose that (x_i) and (y_i) are two convergent sequences in (X, d) that converge to the same point x. Let z_i be given by $z_{2i-1} = x_i$ and $z_{2i} = y_i$ so that $z = (x_1, y_1, x_2, y_2, x_3, y_3, \cdots)$. Show that $\lim (z_i) = x$.

6. Prove the following:
    Let (X, d) be a metric space. Let S be a subset of X. Then $x \in \mathrm{cl}\ (S)$ if and only if x is a limit of a sequence in S.

7. Suppose (X, d) is a metric space and (x_i) is a sequence in X and $l \in X$. Prove:
    (a) $\lim (x_i) = l$ if and only if $\lim (d(x_i, l)) = 0$.
    (b) If (y_i) is also a sequence in X and $\lim (x_i) = l$, then $\lim (y_i) = l$ if and only if $\lim (d(x_i, y_i)) = 0$.

51. CARTESIAN PRODUCT OF A FINITE NUMBER OF METRIC SPACES

Recall that in Chapter 2 the distance function d for $\mathbf{R}^n$ was given as

$$d(x, y) = \left[\sum_{i=1}^{n} |x_i - y_i|^2 \right]^{\frac{1}{2}}$$

Notice that in this formula $|x_i - y_i|$ represents the distance between the ith coordinates of x and y. This observation motivates what turns out to be a very useful metric for the Cartesian product set $\mathbf{X} \{X_i : i \in \mathbf{P}_n\}$, where $\{(X_i, d_i) : i \in \mathbf{P}_n\}$ is a finite collection of metric spaces.

51.1.   Theorem.   *Let $\{(X_i, d_i) : i \in \mathbf{P}_n\}$ be a finite collection of metric spaces. Let $X = \bigtimes \{X_i : i \in \mathbf{P}_n\}$. Then the function $d : X \times X \to \mathbf{R}$, given by the following formula, is a metric for X.*

$$d(x, y) = \left[\sum_{i=1}^{n} (d_i(x_i, y_i))^2 \right]^{\frac{1}{2}}$$

where $x = (x_1, x_2, \ldots, x_n)$ and $y = (y_1, y_2, \ldots, y_n)$.

Before beginning the proof of this proposition, let us recall that we proved the triangle inequality for the Euclidean metric for $\mathbf{R}^n$ by proving the triangle inequality for the magnitude function (see 35.2). We shall also make strategic use of the magnitude function in the proof of 51.1, by noticing, for example, that if we set

$$A = (d_1(x_1, y_1), d_2(x_2, y_2), \ldots, d_n(x_n, y_n)),$$

then $A \in \mathbf{R}^n$ and

$$|A| = d(x, y).$$

PROOF.   That $d(x, y) = 0$ if and only if $x = y$ and that $d(x, y) = d(y, x)$ are obvious from the definition. We prove next that for x, y, and z in X, $d(x, y) + d(y, z) \geqq d(x, z)$. Set

$$A = (d_1(x_1, y_1), d_2(x_2, y_2), \ldots, d_n(x_n, y_n))$$

and

$$B = (d_1(y_1, z_1), d_2(y_2, z_2), \ldots, d_n(y_n, z_n)).$$

Then,

$$d(x, y) + d(y, z)$$
$$= |A| + |B|$$
$$\geqq |A + B|$$
$$= |(d_1(x_1, y_1) + d_1(y_1, z_1), d_2(x_2, y_2) + d_2(y_2, z_2), \ldots, d_n(x_n, y_n) + d_n(y_n, z_n))|$$
$$= \left[\sum_{i=1}^{n} (d_i(x_i, y_i) + d_i(y_i, z_i))^2 \right]^{\frac{1}{2}}.$$

Then, since for each $i \in \mathbf{P}_n$

$$d_i(x_i, y_i) + d_i(y_i, z_i) \geqq d_i(x_i, z_i),$$

we get

$$d(x, y) + d(y, z) \geqq \left[\sum_{i=1}^{n} (d_i(x_i, z_i))^2 \right]^{\frac{1}{2}}$$
$$= d(x, z).$$

We shall refer to the metric just defined as the product metric and designate the corresponding metric space by $\bigtimes \{(X_i, d_i) : i \in \mathbf{P}_n\}$.

We have shown that, given a finite collection of metric spaces, we can define a product metric in a manner that generalizes the Euclidean metric for $\mathbf{R}^n$. Many of the concepts that we shall study depend entirely on the topology, that is, on the sets which are open with respect to the product metric. Thus, in dealing with the product space, it is often not necessary to make explicit use of the product metric itself. For this reason it is useful to have a characterization of the open sets in terms of the open sets for the coordinate spaces. We accomplish this by means of the following important theorem.

51.2.   Theorem.   *Let $\{(X_i, d_i) : i \in \mathbf{P}_n\}$ be a finite collection of metric spaces and let (X, d) be the corresponding product space. Then U is open in (X, d) if and only if U is the union of sets of the form*

51.2(a).
$$\underset{}{\mathsf{X}} \{U_i : i \in \mathbf{P}_n\} \text{ where each } U_i \text{ is open in } X_i.$$

PROOF.   We first show that a set of the form 51.2(a) is open in (X, d) and, hence, a union of such sets is open. Let $(p_1, p_2, \ldots, p_n) \in \underset{}{\mathsf{X}} \{U_i : i \in \mathbf{P}_n\}$, where each U_i is open in X_i and $p_i \in U_i$. Since each U_i is open, for each $i \in \mathbf{P}_n$, there is a $\delta_i > 0$ such that $N_{d_i}(p_i; \delta_i) \subset U_i$. Now let $\delta = \min \{\delta_i : i \in \mathbf{P}_n\}$. If $x \in N(p; \delta)$, then $d_i(p_i, x_i) \leq d(p, x) < \delta \leq \delta_i$. Hence, $x \in \underset{}{\mathsf{X}} \{U_i : i \in \mathbf{P}_n\}$ and we have shown that $N(p; \delta) \subset \underset{}{\mathsf{X}} \{U_i : i \in \mathbf{P}_n\}$. Thus, $\underset{}{\mathsf{X}} \{U_i : i \in \mathbf{P}_n\}$ is an open subset of (X, d).

Next suppose that U is an open subset of (X, d). Let $p \in U$. We complete the proof by showing that p is an element of a set W of the form 51.2(a) which is contained in U. Since U is open in (X, d), there is a $\delta > 0$ such that $N(p; \delta) \subset U$. Let $W = \underset{}{\mathsf{X}} \{N(p_i; n^{-\frac{1}{2}}\delta) : i \in \mathbf{P}_n\}$ and note that $p \in W$. Also, if $x \in W$, then each $d_i(x_i, p_i) < (n)^{-\frac{1}{2}}\delta$ and, hence, $d(x, p) = \left[\sum_{i=1}^{n} d_i^2(x_i, p_i) \right]^{\frac{1}{2}} < \left[\sum_{i=1}^{n} (n)^{-1}\delta^2 \right]^{\frac{1}{2}} = \delta$. Thus, $x \in W$ implies that $x \in N(p; \delta)$. We have shown therefore that $p \in W \subset N(p; \delta) \subset U$, where W is a subset of the form 51.2(a).

For the case $\mathbf{R}^2 = \mathbf{R} \times \mathbf{R}$, with the usual metric for each factor $\mathbf{R}$, the product metric gives the usual Euclidean distance formula for $\mathbf{R}^2$. Recall that for this distance formula, the ε-neighborhoods are "circular." It was pointed out in Example 48.5 that a metric g for $\mathbf{R}^2$ that is equivalent to the Euclidean metric is

$$g(x, y) = |x_1 - y_1| + |x_2 - y_2|.$$

The generalization of this metric to a finite collection of metric spaces is given in the next theorem.

51.3.   Theorem.   *Let $\{(X_i, d_i) : i \in \mathbf{P}_n\}$ be a finite collection of metric spaces. Let $X = \underset{}{\mathsf{X}} \{X_i : i \in \mathbf{P}_n\}$. For each $x = (x_1, x_2, \ldots, x_n)$ and $y = (y_1, y_2, \ldots, y_n)$ in X, define*

$$g(x, y) = \sum_{i=1}^{n} d_i(x_i, y_i).$$

Then g is a metric for X and, furthermore, g is equivalent to the product metric d as defined in 51.1.

PROOF.   It is easy to verify that $g(x, y) = 0$ if and only if $x = y$. Also it is obvious that $g(x, y) = g(y, x)$. We verify the triangle inequality as follows.

$$g(x, y) + g(y, z) = \sum_{i=1}^{n} d_i(x_i, y_i) + \sum_{i=1}^{n} d_i(y_i, z_i)$$

$$= \sum_{i=1}^{n} [d_i(x_i, y_i) + d_i(y_i, z_i)]$$

Then, since for each $i \in \mathbf{P}_n$,

$$d_i(x_i, y_i) + d_i(y_i, z_i) \geq d_i(x_i, z_i),$$

we obtain

$$g(x, y) + g(y, z) \geq \sum_{i=1}^{n} d_i(x_i, z_i)$$
$$= g(x, z).$$

We next show that the metric g is equivalent to the product metric. To do this we shall make use of Theorem 48.4. Let $\varepsilon > 0$ and $p \in X$. We shall complete the proof by finding positive numbers δ_1 and δ_2 such that

51.3(a). $N_g(p; \delta_1) \subset N_d(p; \varepsilon)$

and

51.3(b). $N_d(p; \delta_2) \subset N_g(p; \varepsilon)$

We set $\delta_1 = \varepsilon$ and $\delta_2 = n^{-1}\varepsilon$.

To verify 51.3(a), let $x \in N_g(p; \delta_1)$. Observe that

$$d(p, x) = \left[\sum_{i=1}^{n} (d_i(x_i, p_i))^2 \right]^{\frac{1}{2}}$$
$$\leq \left[\left(\sum_{i=1}^{n} d_i(x_i, p_i) \right)^2 \right]^{\frac{1}{2}}$$
$$= \sum_{i=1}^{n} d_i(x_i, p_i) = g(p, x) < \delta_1 = \varepsilon.$$

Hence, $x \in N_d(p; \varepsilon)$ and we have verified that $N_g(p; \delta_1) \subset N_d(p; \varepsilon)$. To verify 51.3(b), let $x \in N_d(p; \delta_2)$. Then $d(x, p) = \left[\sum_{i=1}^{n} (d_i(x_i, p_i))^2 \right]^{\frac{1}{2}} < \delta_2$. From this it follows that for each $i \in \mathbf{P}_n$, $d_i(x_i, p_i) < \delta_2$. But then

$$g(x, p) = \sum_{i=1}^{n} d_i(x_i, p_i) < n\delta_2 = \varepsilon.$$

Hence, we have shown that $x \in N_g(p; \varepsilon)$ and, consequently, $N_d(p; \delta_2) \subset N_g(p; \varepsilon)$. This completes the proof that d and g are equivalent metrics for X.

51.4. Definition. Projection functions. *Let X be the Cartesian product of the collection of nonempty sets $\{X_i : i \in \mathbf{P}_n\}$. For each $i \in \mathbf{P}_n$, X_i is called the ith coordinate space of X and for each $(x_1, x_2, \ldots, x_n) \in X$, x_i is called the ith coordinate. Further, the surjection $\pi_i : X \to X_i$ given by*

$$\pi_i((x_1, x_2, \ldots, x_n)) = x_i$$

is called the projection function of X onto X_i.

In terms of these projection maps, we generalize Theorem 41.2 from $\mathbf{R}^n$ to the situation with which we are presently dealing. It is instructive to observe that the proof of the following theorem makes use of Theorem 51.2, appealing to the open sets in the product and in the coordinate spaces, rather than to the product metric itself.

51.5. Theorem. *Consider the space $X = \mathsf{X}\, \{(X_i, d_i) : i \in \mathbf{P}_n)\}$. Then a sequence $(x_k)_{k=1}^{\infty}$ in X converges to a point $p \in X$ if and only if for each $j \in \mathbf{P}_n$, $(\pi_j(x_k))_{k=1}^{\infty}$ converges to $\pi_j(p)$.*

PROOF.   First assume that (x_k) is a sequence in the product space X and that $\lim (x_k) = p$. Fix j and consider the sequence $(\pi_j(x_k))_{k=1}^{\infty}$. We show next that $\lim_{k \to \infty} \pi_j(x_k) = \pi_j(p)$. Let U_j be an open set in X_j such that $\pi_j(p) \in U_j$. Note that $p \in \pi_j^{-1}[\pi_j(p)]$ and

$$\pi_j^{-1}[U_j] = X_1 \times X_2 \times \cdots \times U_j \times \cdots \times X_n$$

which by 51.2 is an open subset of X. Since $p \in \pi_j^{-1}[U_j]$ and $\lim (x_k) = p$, there is an N such that for $k \geqq N$, $x_k \in \pi_j^{-1}[U_j]$. But then for $k \geqq N$, $\pi_j(x_k) \in U_j$ and we have shown that $(\pi_j(x_k))_{k=1}^{\infty}$ converges to $\pi_j(p)$.

Next, suppose for each j, $(\pi_j(x_k))_{k=1}^{\infty}$ converges to $\pi_j(p)$. Let U be an open subset of X such that $p \in U$. By Theorem 51.2 there are open sets $U_i \subset X_i$ such that $p_i \in U_i$ and $\times \{U_i : i \in \mathbf{P}_n\} \subset U$. Since $(\pi_j(x_k))_{k=1}^{\infty}$ converges to $\pi_j(p)$, then for each j, there is an integer N_j such that for $k \geqq N_j$, $\pi_j(x_k) \in U_j$. But then for $k \geqq \max \{N_j : j \in \mathbf{P}_n\}$, $x_k \in \times \{U_j : j \in \mathbf{P}_n\} \subset U$.

EXERCISES: CARTESIAN PRODUCT OF A FINITE NUMBER OF METRIC SPACES

1.   Let $\{(X_i, d_i) : i \in \mathbf{P}_n\}$ be a collection of metric spaces. Let $X = \times \{X_i : i \in \mathbf{P}_n\}$. For each $x = (x_1, x_2, x_3, \ldots, x_n)$ and $y = (y_1, y_2, \ldots, y_n)$, define $p(x, y) = \max \{d_i(x_i, y_i) : i \in \mathbf{P}_n\}$. Show that p is equivalent to d, where d is the product metric discussed in this section.

2.   Let $\theta = (0, 0)$. For each metric ρ for $\mathbf{R}^2$, define $S(\rho) = \{x : x \in \mathbf{R}^2 \text{ and } \rho(x, \theta) = 1\}$ and $B(\rho) = \{x : x \in \mathbf{R}^2 \text{ and } \rho(x, \theta) \leqq 1\}$. The real interval $\{r : 0 \leqq r \leqq 1\}$ will be denoted by I. Also let d be the usual $\mathbf{R}^2$-metric; g and p will be the metrics for $\mathbf{R}^2$ as defined in Examples 45.3 and 45.5. Notice that $B(d) \times I$ can be perceived as a solid cylinder with radius 1 and height 1. The set $S(d) \times S(d)$ can be visualized as a torus. Describe the sets $B(d) \times S(d)$, $S(p) \times I$, $S(g) \times I$, $B(g) \times I$, and $B(g) \times S(d)$.

3.   In the proof of Theorem 51.3 it was shown that $N_d(p; n^{-1}\varepsilon) \subset N_g(p; \varepsilon)$. From the picture on page 84, we might conjecture that the larger d-neighborhood $N_d(p; n^{-\frac{1}{2}}\varepsilon)$ is contained in $N_g(p; \varepsilon)$. Prove that this conjecture is correct.

52.   CONTINUOUS MAPPINGS: INTRODUCTION

The intuitive idea behind the notion of continuity of a function is that of preserving closeness of points; that is, $f(x_0)$ can be approximated as closely as we wish by $f(x)$ provided that x is a sufficiently good approximation of x_0. Recall from the calculus that for a real-valued function defined on an open interval $(a, b) \subset \mathbf{R}$, this notion was formalized as follows:

The function f is continuous at $x_0 \in (a, b)$, provided that for each $\varepsilon > 0$, there is a $\delta > 0$ such that if $|x - x_0| < \delta$, then $|f(x) - f(x_0)| < \varepsilon$.

52.1.   Definition.   Continuous function.   *Suppose (X, d) and (Y, ρ) are metric spaces and the function $f : X \to Y$ satisfies the following condition at a point $x_0 \in X$.*

For each $\varepsilon > 0$, there is a $\delta > 0$ such that if $x \in X$ and $d(x, x_0) < \delta$, then $\rho(f(x), f(x_0)) < \varepsilon$. We then say that the function f from (X, d) to (Y, ρ) is continuous at x_0. If f is continuous at each point x in X, then f is said to be a continuous function from (X, d) into (Y, ρ).

52.2. *Example.* Let d_n and d be the Euclidean metrics for $\mathbf{R}^n$ and $\mathbf{R}$, respectively. Let a be a fixed element in $\mathbf{R}^n$ and let f be the function given by

$$f(x) = a \cdot x \quad \text{for each } x \in \mathbf{R}^n \text{ (see 34.3).}$$

Then f is a continuous function from $(\mathbf{R}^n, d_n)$ into $(\mathbf{R}, d)$. To prove this, let $x_0 \in \mathbf{R}^n$ and let $\varepsilon > 0$. If $|a| = 0$, it should be clear that f is continuous at x_0. If $|a| \neq 0$, then let $\delta = (1/|a|)\,\varepsilon$. Then, for $d_n(x, x_0) = |x - x_0| < \delta$, using the properties of the dot product (34.4 and 35.1), we obtain

$$\begin{aligned}
d(f(x), f(x_0)) &= |a \cdot x - a \cdot x_0| \\
&= |a \cdot (x - x_0)| \le |a|\,|x - x_0| \\
&\le |a|\,\delta = |a|\,\frac{1}{|a|}\,\varepsilon = \varepsilon.
\end{aligned}$$

Hence, f is continuous at x_0.

In the previous chapter when we spoke of a function f from X into Y, we used the notation $f : X \to Y$. Continuity is an example of a notion that involves the behavior of a function with respect to spaces rather than just sets. This is the reason for our using, in the definition of continuity, the expression f is a continuous function from (X, d) into (Y, ρ). We shall abbreviate such a statement by writing $f : (X, d) \to (Y, \rho)$ is continuous. However, when there is little chance for confusion, it is quite common to drop the reference to the metrics involved and refer to a "continuous mapping $f : X \to Y$" or more simply to a "continuous mapping f". Recall that we have been doing the same sort of thing when we refer to "an open set U in X" when perhaps it would be more appropriate to refer to "a d-open set U in X."

There is another remark about notation that is appropriate at this point. Sometimes in a proof one function is used in several different metric settings (see Example 52.3). In such a case it may be notationally clearer if a different name is given to the function in each case.

52.3. *Example.* Let d be the Euclidean metric for $\mathbf{R}^n$ and let m be the metric for the set $\mathbf{R}^n$ given by $m(x, y) = 0$ if $x = y$ and $m(x, y) = 1$ if $x \neq y$ (see Example 45.4). In what follows both f and g denote the identity function on the *set* $\mathbf{R}^n$. We shall show that $f : (\mathbf{R}^n, m) \to (\mathbf{R}^n, d)$ is continuous. However, the function $g : (\mathbf{R}^n, d) \to (\mathbf{R}^n, m)$ is not continuous. The point of the example is that continuity depends not only on the function but also on the metrics that are involved. We first verify that $f : (\mathbf{R}^n, m) \to (\mathbf{R}^n, d)$ is continuous. Let $a \in \mathbf{R}^n$ and $\varepsilon > 0$. We choose $\delta = 1$. Next suppose $x \in \mathbf{R}^n$ and $m(x, a) < 1$. Note that the only x that satisfies this condition is $x = a$. Hence, $d(f(x), f(a)) = d(x, a) = d(a, a) = 0 < \varepsilon$. Hence, we have shown that f is a continuous map from $(\mathbf{R}^n, m)$ into $(\mathbf{R}^n, d)$. We next consider $g : (\mathbf{R}^n, d) \to (\mathbf{R}^n, m)$, where $g(x) = x$ for each $x \in \mathbf{R}^n$. We prove that g is not continuous by contradiction. Suppose g were continuous. Let $\varepsilon = 1$. Then there is a $\delta > 0$ such that if $d(x, a) < \delta$, then $m(g(x), g(a)) < \varepsilon = 1$. Now choose an $x \in \mathbf{R}^n$ such that $d(x, a) < \delta$ and $x \neq a$ (there are such

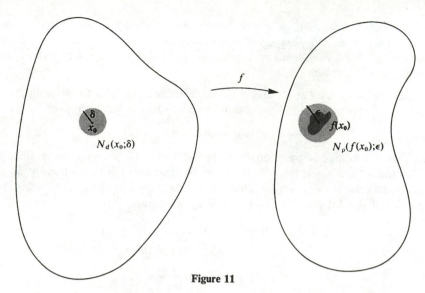

Figure 11

points). Then $m(g(x), g(a)) = m(x, a) < 1$. However, this implies that $x = a$ and we have a contradiction.

In the next theorem we give several statements, each of which is equivalent to the form of continuity given in the definition. All the forms are useful, and one of them in particular emphasizes the fact that continuity depends only on the topologies involved.

52.4. Theorem. *Each of the following properties is equivalent to continuity of $f:(X, d) \to (Y, \rho)$ at $x_0 \in X$.*

52.4(a). *For each $\varepsilon > 0$, there is a $\delta > 0$ such that*

$$f[N_d(x_0; \delta)] \subset N_\rho(f(x_0); \varepsilon).$$

52.4(b). *For each sequence (x_i) in X that converges to x_0, the sequence $(f(x_i))$ converges to $f(x_0)$.*

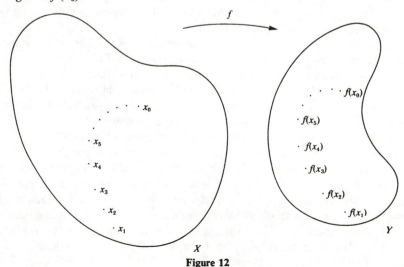

Figure 12

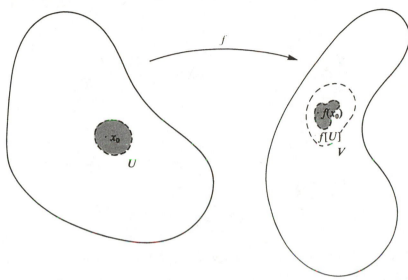

Figure 13

52.4(c). *For each open set V in Y that contains $f(x_0)$, there is an open set U in X such that $x_0 \in U$ and $f[U] \subset V$.*

PROOF. That (a) is equivalent to the continuity of f at x_0 is obvious. We assume that f is continuous at x_0 and show that (b) holds. Suppose that $\lim (x_i) = x_0$ and $\varepsilon > 0$. Then there is a $\delta > 0$ such that if $d(x, x_0) < \delta$, then $\rho(f(x), f(x_0)) < \varepsilon$. Since $\lim (x_i) = x_0$, there is a positive integer N such that for $i \geq N$, $d(x_i, x_0) < \delta$. But then for $i \geq N$, $\rho(f(x_i), f(x_0)) < \varepsilon$ and, hence, $\lim (f(x_i)) = f(x_0)$.

We next show that (b) implies (c). Assume (b) and let V be an open set in Y with $f(x_0) \in V$. Suppose that, contrary to what we wish to prove, for no open set U in X for which $x_0 \in U$ is it true that $f[U] \subset V$. Then for each positive integer n, $f[N_d(x_0; 1/n)] \cap (\sim V) \neq \varnothing$. Thus, for each positive integer n, there is an $x_n \in X$ such that $d(x_0, x_n) < \dfrac{1}{n}$ and $f(x_n) \in \sim V$. By (b), since $\lim (x_n) = x_0$, $\lim (f(x_n)) = f(x_0)$. Then for n large enough, $f(x_n) \in V$. This is a contradiction to the way the x_n's were chosen and the proof that (b) implies (c) is complete.

We next prove that (c) implies (a). Assume that (c) holds at x_0. Let $\varepsilon > 0$. The set $N_\rho(f(x_0); \varepsilon)$ is open in Y. Then by (c), there is an open set U in X such that $x_0 \in U$ and $f[U] \subset N_\rho(f(x_0); \varepsilon)$. However, there is a $\delta > 0$ such that $N_d(x_0; \delta) \subset U$ from which it follows that $f[N_d(x_0; \delta)] \subset N_\rho(f(x_0); \varepsilon)$. Hence, (c) implies (a) and the proof of the theorem has been completed.

The following theorem gives a useful characterization of continuity. The proof is left as an exercise.

52.5. Theorem. *A mapping $f:(X, d) \to (Y, \rho)$ is continuous if and only if either of the following conditions is satisfied.*

52.5(a). *For each set U that is open in Y, $f^{-1}[U]$ is open in X.*

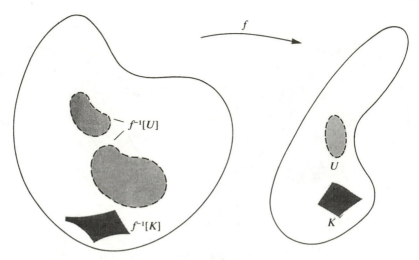

Figure 14

52.5(b). *For each set K that is closed in* Y, $f^{-1}[K]$ *is closed in* X.

The next theorem emphasizes, as does Theorem 52.5, that continuity depends only on the topologies involved and not specifically on the underlying metric. The proof follows easily from Theorem 52.5.

52.6. Theorem. *Let* d_1 *and* d_2 *be equivalent metrics for a set* X *and let* ρ_1 *and* ρ_2 *be equivalent metrics for a set* Y. *Then a mapping* $f\colon(X, d_1) \to (Y, \rho_1)$ *is continuous if and only* $f\colon(X, d_2) \to (Y, \rho_2)$ *is continuous.*

If $f\colon(X, d) \to (Y, \rho)$ is a mapping, we shall often speak of some property as being *invariant* under f or f^{-1}. For example, Theorem 52.5 tells us that if $f\colon (X, d) \to (Y, \rho)$ is continuous, then the property of being an open (closed) subset is invariant under f^{-1}; that is, if U is an open (closed) set in Y, then $f^{-1}[U]$ is an open (closed) set in X. Sometimes the expression *is preserved* is used instead of *is invariant*. For example, for one of the cases just cited we may say "the openness of sets is preserved under the inverse of a continuous function."

The following theorem is easy to verify and the proof of each part is left as an exercise.

52.7. Theorem

52.7(a). *Suppose* $f\colon(X, d) \to (Y, \rho)$ *is continuous and* $S \subset X$. *Then* $f\,|\,S\colon (S, d) \to (Y, \rho)$ *is continuous.*

52.7(b). *A function* $f\colon(X, d) \to (Y, \rho)$ *is continuous if and only if* $f\colon(X, d) \to (f[X], \rho)$ *is continuous.*

Notice that in Example 52.3 the function g is the inverse of f. Thus, the example illustrates the fact that the continuity of a bijection $f\colon(X, d) \to (Y, \rho)$ does not guarantee the continuity of $f^{-1}\colon(Y, \rho) \to (X, d)$. A continuous function that has a continuous inverse is given a special name as indicated in the next definition.

52.8. Definition. Topological mapping. *Suppose* $h:(X, d) \to (Y, \rho)$ *is a continuous bijection such that* $h^{-1}:(Y, \rho) \to (X, d)$ *is also continuous. Then* h *is called a topological mapping (from* (X, d) *onto* (Y, ρ)*). The terms homeomorphism and bicontinuous bijection are used as synonymns for topological mapping.*

It is obvious that if h is a topological mapping from (X, d) onto (Y, ρ), then h^{-1} is a topological mapping from (Y, ρ) onto (X, d). If there exists a homeomorphism h that maps (X, d) onto (Y, ρ), then (X, d) and (Y, ρ) are said to be *homeomorphic* or *topologically equivalent*.

52.9. *Example.* Let α be a positive real number. Suppose $f:(X, d) \to (Y, \rho)$ is a surjection such that $d(a, b) = \alpha\rho(f(a), f(b))$ for all x and y in X. Then f is a homeomorphism. To see this, first note that f is one-to-one. For if $f(a) = f(b)$, then $d(a, b) = \alpha \, \rho(f(a), f(b)) = 0$ and $a = b$. To see that f is continuous, let $a \in X$ and let $\varepsilon > 0$. Choose $\delta = \alpha\varepsilon$. Then if $d(a, b) < \delta$, it follows that $\rho(f(a), f(b)) = \dfrac{1}{\alpha} d(a, b) < \dfrac{1}{\alpha} \alpha\varepsilon = \varepsilon$. Hence, f is continuous. A completely similar argument that f^{-1} is continuous can be used if we make the following observation: For all c and d in Y, $\rho(c, d) = \dfrac{1}{\alpha} d(f^{-1}(c), f^{-1}(d))$.

52.10. *Example.* An example of a function that satisfies the condition imposed on the f in 52.9 is the function $f:\mathbf{R}^n \to \mathbf{R}^n$ given by $f(x) = \alpha x + b$, where $\alpha \neq 0$ and $b \in \mathbf{R}^n$.

The following theorem is often an aid in determining whether a continuous bijection is a homeomorphism. The proof is an easy consequence of Theorem 52.5 and is left as an exercise.

52.11. Theorem. *Let* $f:(X, d) \to (Y, \rho)$ *be a continuous bijection. Then* f *is a homeomorphism if and only if either of the following conditions is satisfied.*

52.11(a). $f[U]$ *is open in* Y, *if* U *is open in* X.

52.11(b). $f[K]$ *is closed in* Y, *if* K *is closed in* X.

Because of Theorems 52.5 and 52.11, we see that a bijection $f:(X, d) \to (Y, \rho)$ is a homeomorphism if and only if the openness (closedness) of sets is preserved under f and f^{-1}. Recall that in 48.3 we called a property a topological property if it could be characterized in terms of open sets. Thus, we see that a property which is possessed by a subset of a metric space is a topological property only if that property is invariant under homeomorphisms. For example, suppose X is a metric space and it has the property that every point of X is a limit point of X. The notion of limit point is a topological property and any metric space Y that is topologically equivalent to X will have the property that every point of Y is a limit point of Y. On the other hand, boundedness is not a topological property. A subset S of a metric space X can be bounded and yet under a homeomorphism $f:X \to Y$ it is possible that $f[S]$ is not bounded in Y (see Exercise 3, page 87). Topologists are often interested in determining spaces "up to a homeomorphism." To show that a space X is *not* homeomorphic to a space Y it is sufficient to find one topological property that is possessed by one of the spaces and not by the other.

The reader will recall from a study of calculus that the composition of two continuous functions is continuous. This remains true in the more general setting that we are now considering. The following lemma for composite functions will be useful in proving this fact.

52.12. Lemma. *Suppose $f: X \to Y$ and $g: Y \to Z$ are functions. Then for each $U \subset Z$, $(g \circ f)^{-1}[U] = f^{-1}[g^{-1}[U]]$.*

PROOF. Let $x \in (g \circ f)^{-1}[U]$. Then $g \circ f(x) = g(f(x)) \in U$. From the fact that $g(f(x)) \in U$, it follows that $f(x) \in g^{-1}[U]$ and, hence, $x \in f^{-1}[f(x)] \subset f^{-1}[g^{-1}[U]]$. Thus, we have shown that $(g \circ f)^{-1}[U] \subset f^{-1}[g^{-1}[U]]$. We next show that $f^{-1}[g^{-1}[U]] \subset (g \circ f)^{-1}[U]$. Let $x \in f^{-1}[g^{-1}[U]]$. Then $f(x) \in g^{-1}[U]$. But this implies that $g \circ f(x) = g(f(x)) \in g[g^{-1}[U]] \subset U$. Hence, $x \in (g \circ f)^{-1}[U]$ and the proof is complete.

52.13. Theorem. *Let $f_1 : (X_1, d_1) \to (X_2, d_2)$ and $f_2 : (X_2, d_2) \to (X_3, d_3)$ be continuous functions. Then $f_2 \circ f_1 : (X_1, d_1) \to (X_3, d_3)$ is also continuous. Further, if f_1 and f_2 are topological mappings, then so is $f_2 \circ f_1$.*

PROOF. In a proof involving composition, it is often helpful to refer to a function diagram.

$$X_1 \xrightarrow{\;\; f_2 \circ f_1 \;\;} X_3$$
$$f_1 \searrow \quad \nearrow f_2$$
$$X_2$$

Let U be open in X_3. Then since f_2 is continuous, by 52.5, $f_2^{-1}[U]$ is open in X_2. But f_1 is also continuous so that $f_1^{-1}[f_2^{-1}[U]]$ is open in X_1. Now note that by the previous lemma $(f_2 \circ f_1)^{-1}[U] = f_1^{-1}[f_2^{-1}[U]]$. Hence, since openness of sets is preserved under $(f_2 \circ f_1)^{-1}$, it follows from 52.5 that $f_2 \circ f_1$ is continuous.

The part of the proof of the theorem concerning topological mappings is left as an exercise.

Previously, we studied the notion of equivalent metrics for a set. It will follow from the next theorem that if d and ρ are equivalent metrics for a set X, then (X, d) and (X, ρ) are topologically equivalent spaces.

52.14. Theorem. *Suppose d and ρ are metrics for a set X. Then d and ρ are equivalent metrics for X if and only if the identity function on X is a topological mapping from (X, d) onto (X, ρ).*

PROOF. First we assume that the identity function $i : (X, d) \to (X, \rho)$ is a topological map. We need to show that the topologies $\mathcal{T}(d)$ and $\mathcal{T}(\rho)$ of (X, d) and (X, ρ), respectively, are the same. Let $U \in \mathcal{T}(d)$. Since i is a topological mapping, by 52.11, $i[U]$ is ρ-open in X. Hence, $U = i[U]$ is ρ-open in X and so $U \in \mathcal{T}(\rho)$. Thus, $\mathcal{T}(d) \subset \mathcal{T}(\rho)$. Next let $V \in \mathcal{T}(\rho)$. Since i is continuous, by 52.5, $i^{-1}[V]$ is d-open. But $V = i^{-1}[V]$ so that $V \in \mathcal{T}(d)$. Hence, $\mathcal{T}(\rho) \subset \mathcal{T}(d)$ and, consequently, $\mathcal{T}(d) = \mathcal{T}(\rho)$. It now remains to be proved that if d and ρ are equivalent metrics, then $i : (X, d) \to (X, \rho)$ is a topological mapping. This part also follows from 52.5 and 52.11 and the details are left as an exercise for the reader.

Suppose two metric spaces are topologically equivalent. Then the metric in one space can be used to measure distances in the other space in the following sense.

52.15. Theorem. *Suppose $f:(X, d) \to (Y, \rho)$ is a topological mapping. For each x and y in X, define $d^*(x, y) = \rho(f(x), f(y))$. Then d^* defines a metric for X that is equivalent to d.*

PROOF. That $d^*(x, y) = 0$ if and only if $x = y$ follows easily from the facts that f is one-to-one and ρ is a metric. That $d^*(x, y) = d^*(y, x)$ and the triangle inequality of d^* follow at once from the fact that ρ is a metric. We next show that d and d^* are equivalent metrics by making use of the previous theorem as follows:

$$(X, d) \xrightarrow{\quad f \quad} (Y, \rho)$$
$$g \circ f \searrow \qquad \swarrow g$$
$$(X, d^*)$$

Let $g(y) = f^{-1}(y)$ for each $y \in Y$ and consider the mapping $g:(Y, \rho) \to (X, d^*)$. Note that g is a surjection and that $\rho(c, d) = d^*(f^{-1}(c), f^{-1}(d)) = d^*(g(c), g(d))$ for all c and d in Y. Hence, it follows from Example 52.9 that g is a homeomorphism. Since f is a homeomorphism it then follows from 52.13 that $g \circ f$: $(X, d) \to (X, d^*)$ is a homeomorphism. But since $g \circ f$ is the identity function on X, we may conclude from 52.14 that d and d^* are equivalent metrics for X.

52.16. *Example.* Consider the set $\mathbf{R}_+ = \{x : x > 0\}$ and let d be the Euclidean metric. The mapping $f:(\mathbf{R}_+, d) \to (\mathbf{R}_+, d)$, given by $f(x) = 1/x$ for $x \in \mathbf{R}_+$, is a homeomorphism. For all $x > 0$ and $y > 0$, define

$$d^*(x, y) = d(f(x), f(y))$$
$$= \left| \frac{1}{x} - \frac{1}{y} \right| = \frac{1}{xy} |x - y|.$$

By 52.15, d^* is a metric for $\mathbf{R}_+$ that is equivalent to the Euclidean metric. Notice that the effect of this change of metric is to greatly expand distances between points near 0 and greatly contract distances between points far from the origin.

EXERCISES: CONTINUOUS MAPPINGS

(Recall our agreement that when we refer to $\mathbf{R}^n$ as a space, unless otherwise indicated, it is understood to be endowed with the Euclidean metric. We also make this agreement for subsets of $\mathbf{R}^n$.)

1. In each of the following examples, decide whether the mapping is continuous. If it is not, find the set of discontinuities (i.e., the set of all points in the domain at which the function is not continuous.)

(a) Let $f:\mathbf{R} \to \mathbf{R}$ be given by $f(x) = x^2$ for $x > 0$ and $f(x) = -x$ for $x \leq 0$.

(b) Let $f:\mathbf{R} \to \mathbf{R}^2$ be given by $f(x) = (x, x)$ for each $x \in \mathbf{R}$.

(c) Let $f:\mathbf{R} \to \mathbf{R}$ be given by $f(x) = 1$ for x rational and $f(x) = 2$ for x irrational.

(d) Let $f:\mathbf{R} - \{0\} \to \mathbf{R}$ be given by $f(x) = 1/x$ for $x \in \mathbf{R} - \{0\}$.

(e) Let $f:\mathbf{R} \to \mathbf{R}$ be given by $f(x) = \dfrac{1}{x}$ for $x \neq 0$ and $f(0) = 0$.

2. Suppose $f:\mathbf{R} \to \mathbf{R}$ and $g:\mathbf{R} \to \mathbf{R}$ are continuous. Let $G:\mathbf{R}^2 \to \mathbf{R}^2$ be given by $G(x, y) = (f(x), g(y))$ for each $(x, y) \in \mathbf{R}^2$. Show that $G:\mathbf{R}^2 \to \mathbf{R}^2$ is a continuous mapping.

3. Suppose $f:\mathbf{R}^2 \to \mathbf{R}$ and $g:\mathbf{R}^2 \to \mathbf{R}$ are continuous. Suppose further $G:\mathbf{R}^2 \to \mathbf{R}^2$ is given by $G(z) = (f(z), g(z))$ for each $z \in \mathbf{R}^2$. Show that $G:\mathbf{R}^2 \to \mathbf{R}^2$ is continuous.

4. Let $s:\mathbf{R}^2 \to \mathbf{R}$, $p:\mathbf{R}^2 \to \mathbf{R}$, and $q:\mathbf{R} \times (\mathbf{R} - \{0\}) \to \mathbf{R}$ be given by the rules $s(x, y) = x + y$, $p(x, y) = xy$, and $q(x, y) = x \div y$, $y \neq 0$. Prove that each of these functions is continuous. Also prove that $r:\{x:x \geq 0\} \to \mathbf{R}$, where $r(x) = \sqrt{x}$ for $x \geq 0$, is a continuous function.

5. Prove each of the following theorems: (a) 52.5; (b) 52.6; (c) 52.11; (d) the second part of 52.13; (e) the second part of 52.14.

6. Recall that a subset S in $\mathbf{R}^n$ is bounded and closed (with respect to the Euclidean metric) if and only if every sequence in S has a convergent subsequence that converges to a point in S (see Exercise 5 on page 77). Let $S \subset \mathbf{R}^n$ and suppose S is closed and bounded and $f:S \to \mathbf{R}^m$ is continuous. Prove that $f[S]$ is a closed and bounded subset of $\mathbf{R}^m$. Prove that if in addition to being continuous, f is one-to-one on the bounded and closed set S, then $f:S \to f[S]$ is a homeomorphism.

7. (Do Exercise 6 before doing this exercise.) Let S be a closed bounded nonempty set in $\mathbf{R}^n$. Suppose $f:S \to \mathbf{R}$ is continuous. Then there exists an $x_0 \in S$ and a $y_0 \in S$ such that $f(x_0) = $ g.l.b. $(f[S])$, and $f(y_0) = $ l.u.b. $(f[S])$. Thus, f attains its maximum and minimum values on S.

8. Suppose $f:\mathbf{R}^m \to \mathbf{R}^n$ satisfies the following: For each x and y in $\mathbf{R}^m$, $f(x + y) = f(x) + f(y)$. Let θ be the zero vector in $\mathbf{R}^m$. Prove that if f is continuous at θ, then f is continuous on $\mathbf{R}^m$.

9. Let $\mathcal{H}$ be a family of metric spaces. Show that "topologically equivalent to" is an equivalence relation in $\mathcal{H}$.

10. Prove that the real line $\mathbf{R}$ is homeomorphic to the open interval $(0, 1)$.

11. Let $A = \{x:x \in \mathbf{R}^n$ and $|x| < 1\}$. Is A topologically equivalent to $\mathbf{R}^n$?

12. Suppose (A_1, d_1) is topologically equivalent to (B_1, ρ_1) and (A_2, d_2) is topologically equivalent to (B_2, ρ_2). Is the product metric space $\times \{(A_i, d_i):i = 1, 2\}$ topologically equivalent to $\times \{(B_i, \rho_i):i = 1, 2\}$?

13. Consider the function $f:(X, d) \to (Y, \rho)$. Let $\mathcal{T}(d)$ and $\mathcal{T}(\rho)$ be the topologies of (X, d) and (Y, ρ), respectively. In each of the following, determine whether the statement is true.
    (a) If f is continuous, then $\mathcal{T}(d) \subset \{f^{-1}[U]: U \in \mathcal{T}(\rho)\}$.
    (b) If f is continuous, then $\{f^{-1}[U]: U \in \mathcal{T}(\rho)\} \subset \mathcal{T}(d)$.
    (c) If $\mathcal{T}(d) \subset \{f^{-1}[U]: U \in \mathcal{T}(\rho)\}$, then f is continuous.
    (d) If $\{f^{-1}[U]: U \in \mathcal{T}(\rho)\} \subset \mathcal{T}(d)$, then f is continuous.

14. Suppose that (X, d) and (Y, ρ) are metric spaces whose topologies are $\mathcal{T}(d)$ and $\mathcal{T}(\rho)$, respectively. Further assume that $f:(X, d) \to (Y, \rho)$ is a bijection for which $\mathcal{T}(\rho) = \{f[U]: U \in \mathcal{T}(d)\}$. What can you conclude about f?

53. UNIFORM CONTINUITY

Suppose $f:(X, d) \to (Y, \rho)$ is a mapping. For f to be continuous on X, we required that for each $x_0 \in X$ and $\varepsilon > 0$, there was a $\delta > 0$ (depending on x_0 and ε) such that if $x \in X$ and $d(x, x_0) < \delta$, then $\rho(f(x), f(x_0)) < \varepsilon$. The following question arises: Given an arbitrary positive ε, can a positive δ (depending only on ε) be found that satisfies the required conditions for all $x_0 \in X$. We give examples that indicate that for some functions the answer to the previous question is yes and for some functions the answer is no. The term *uniform continuity* is given to the strong type of continuity associated with the "yes" answer. As we shall see, there are certain types of metric spaces on which all continuous functions are uniformly continuous. For example, it is an important theorem in the theory of calculus that a continuous function on a closed interval $[a, b]$ is uniformly continuous. This will be a very special case of a theorem that we will prove later.

53.1. Definition. Uniform continuity. Let $f:(X, d) \to (Y, \rho)$ be a mapping. *f is said to be uniformly continuous on X provided that for each $\varepsilon > 0$, there is a $\delta > 0$ such that if $x_1 \in X$, $x_2 \in X$, and $d(x_1, x_2) < \delta$, then $\rho(f(x_1), f(x_2)) < \varepsilon$. If S is a subset of X and if $f \mid S:(S, d) \to (Y, \rho)$ is uniformly continuous, then we say that f is uniformly continuous on S.*

53.2. Example. Consider the function $f:[0, 1] \to \mathbf{R}$ given by $f(x) = x^2$ for $x \in [0, 1]$. We show that f is uniformly continuous on $[0, 1]$. Suppose x_1 and x_2 are in $[0, 1]$. Then $|f(x_1) - f(x_2)| = |x_1^2 - x_2^2| = |x_1 - x_2| \, |x_1 + x_2| \leq (|x_1| + |x_2|) \, |x_1 - x_2| \leq 2|x_1 - x_2|$. From this, we see that if $|x_1 - x_2| < \frac{\varepsilon}{2}$, then $|f(x_1) - f(x_2)| < \varepsilon$. Thus, we have found that the condition in the definition can be satisfied for $\varepsilon > 0$ by taking $\delta = \frac{1}{2}\varepsilon$.

53.3. Example. We show that the function $f:\mathbf{R} \to \mathbf{R}$ given by $f(x) = x^2$ is not uniformly continuous on $\mathbf{R}$. To do this, assume that f is uniformly continuous on $\mathbf{R}$. Then for $\varepsilon = 1$, there is a $\delta > 0$ such that if $|x_1 - x_2| < \delta$, then $|f(x_1) - f(x_2)| < 1$. Now let $x_1 > 0$ and $x_2 = x_1 + \frac{1}{2}\delta$. Then

$$|f(x_1) - f(x_2)| = |x_1 + x_2| \, |x_1 - x_2|$$
$$= |2x_1 + \tfrac{1}{2}\delta| \, |\tfrac{1}{2}\delta| = \tfrac{1}{4}(4x_1 + \delta)\delta > \tfrac{1}{4}\delta(4x_1) = x_1\delta.$$

Now note that if $x_1 = \dfrac{1}{\delta}$, then $|f(x_1) - f(x_2)| > 1$, whereas it should be that $|f(x_1) - f(x_2)| < 1$.

An interesting example of a mapping that is uniformly continuous is the metric d associated with a metric space (X, d). We prove this fact next.

53.4. Example. Let (X, d) be a metric space. Then $d : (X \times X, \rho) \to \mathbf{R}$ is uniformly continuous, where ρ is the product metric (as defined in 51.1).

PROOF. Suppose $\varepsilon > 0$. Let $\delta = \frac{1}{2}\varepsilon$. Now suppose $\rho((x_1, x_2), (y_1, y_2)) < \delta$. Then, $d(x_1, y_1) < \delta$ and $d(x_2, y_2) < \delta$. Hence, $d(x_1, x_2) \leq d(x_1, y_1) + d(y_1, y_2) + d(y_2, x_2)$ and, thus, $d(x_1, x_2) - d(y_1, y_2) \leq d(x_1, y_1) + d(x_2, y_2) < \varepsilon$. Similarly, $d(y_1, y_2) - d(x_1, x_2) < \varepsilon$ and we have that $|d(x_1, x_2) - d(y_1, y_2)| < \varepsilon$. Thus, we have shown that $d : X \times X \to \mathbf{R}$ is uniformly continuous.

In analytic geometry one derives a formula that gives the distance from a point to a line. By the distance from a given point to a line is meant the distance from the given point to the point on the line closest to it. Note that if A is an arbitrary nonempty set, say in $\mathbf{R}^2$, and p is a point in that space, then there may not be a point in A that is closest to p. However, the g.l.b. $\{d(p, x) : x \in A\}$ always exists and agrees with the distance from a point to a line if A is a line.

53.5. Example. Let $\theta = (0, 0)$. Consider the unit neighborhood $N(\theta; 1) \subset \mathbf{R}^2$ and the unit closed disc $B(\theta; 1) \subset \mathbf{R}^2$. Note that the point $(1, 0)$ is the closest point in $B(\theta; 1)$ to the point $(2, 0)$. There is no point in the open set $N(\theta; 1)$ that is closest to $(2, 0)$. However, g.l.b. $\{|p - (2, 0)| : p \in N(\theta; 1)\} = 1$. Observe also that there may not be a unique point in a set that is closest to a given point. For example, in the set $\mathbf{R}^2 - N(\theta; 1)$, every point in the circle $\{(x, y) : x^2 + y^2 = 1\}$ is a closest point to θ.

53.6. Definitions. Distance from a point to a set and distance between two sets. *Let (X, d) be a metric space and let A be a nonempty subset of X and $p \in X$. By $d(p, A)$, the distance from p to A, we shall mean the real number* g.l.b. $\{d(p, x) : x \in A\}$. *If A and B are two nonempty subsets of X, then by $d(A, B)$, the distance between A and B, we shall mean the real number* g.l.b. $\{d(a, b) : a \in A, b \in B\}$.

53.7. Example. In Example 53.5, $d(\theta, N(\theta; 1)) = 0$, $d((2, 0), N(\theta; 1)) = 1$, $d(\theta, \sim B(\theta; 1)) = 1$, and $d(N(\theta; 1), \sim B(\theta; 1)) = 0$.

53.8. Theorem. *Let (X, d) be a metric space. Let A be a nonempty subset of X. Then the mapping $d_A : X \to \mathbf{R}$ given by $d_A(x) = d(x, A)$ is a uniformly continuous real-valued function.*

PROOF. Let $\varepsilon > 0$. We shall show that if $d(x_1, x_2) < \varepsilon$, then $|d_A(x_1) - d_A(x_2)| < \varepsilon$. Let $a \in A$. Then $d(x_1, a) \leq d(x_2, a) + d(x_1, x_2)$. Hence, g.l.b. $\{d(x_1, a) : a \in A\} < $ g.l.b. $\{d(x_2; a) : a \in A\} + \varepsilon$ and $d(x_1, A) < d(x_2, A) + \varepsilon$. By a symmetric argument we can also show that $d(x_2, A) < d(x_1, A) + \varepsilon$. Hence, it follows that $|d(x_1, A) - d(x_2, A)| < \varepsilon$ and the proof is complete.

53.9. Definition. Isometry. *If $f : (X, d) \to (Y, \rho)$ maps X onto Y in such a way that distances are preserved (i.e., $d(x, y) = \rho(f(x), f(y)))$, then f is called an isometry.*

It is to be noted that if f is an isometry, then it is both uniformly continuous and a homeomorphism.

53.10. Examples. Each of the following are isometries on $\mathbf{R}^2$.

53.10(a). The reflection in $\mathbf{R}^2$ through a line L. For example, let $r : \mathbf{R}^2 \to \mathbf{R}^2$

be the reflection through the "x-axis" given by $r(x, y) = (x, -y)$ for each $(x, y) \in \mathbf{R}^2$.

53.10(b). The translation $T:\mathbf{R}^2 \to \mathbf{R}^2$ given for a fixed b by $T(x) = x + b$, for each $x \in \mathbf{R}^2$.

53.10(c). Any rotation in $\mathbf{R}^2$.

Let f be a function from X into Y. Suppose d and d^* are equivalent metrics for X, ρ and ρ^* are equivalent metrics for Y, and $f:(X, d) \to (Y, \rho)$ is uniformly continuous. This is not enough to assure us that $f:(X, d^*) \to (Y, \rho^*)$ is uniformly continuous. This fact is illustrated in the next example.

53.11. *Example.* Let d be the Euclidean metric for $\mathbf{R}_+$ and let d^* be the metric for $\mathbf{R}_+$ given by $d^*(x, y) = \left| \dfrac{1}{x} - \dfrac{1}{y} \right| = \dfrac{1}{xy} |x - y|$. Recall from Example 52.16 that d and d^* are equivalent metrics. Consider the functions $f:(\mathbf{R}_+, d) \to (\mathbf{R}_+, d)$ and $f^*:(\mathbf{R}_+, d) \to (\mathbf{R}_+, d^*)$ where f and f^* both denote the identity function on $\mathbf{R}_+$. Since $d(f(x), f(y)) = d(x, y)$ for each $(x, y) \in \mathbf{R}_+ \times \mathbf{R}_+$, it is obvious that $f:(\mathbf{R}_+, d) \to (\mathbf{R}_+, d)$ is uniformly continuous. However, the fact that $f^*:(\mathbf{R}_+, d) \to (\mathbf{R}_+, d^*)$ is not uniformly continuous is seen by the following argument: Suppose $f^*:(\mathbf{R}_+, d) \to (\mathbf{R}_+, d^*)$ were uniformly continuous. Then there is a $\delta > 0$ such that for all $x > 0$ and $y > 0$,

$$|x - y| < \delta \text{ implies } d^*(f^*(x), f^*(y)) = \frac{1}{xy} |x - y| < 1.$$

However, this leads to a contradiction. For let $\beta = \min \{\frac{1}{2}, \frac{1}{2}\delta\}$. Then $d(2\beta, \beta) = \beta < \delta$. However, $d^*(f^*(2\beta), f^*(\beta)) = \dfrac{|2\beta - \beta|}{2\beta^2} = \dfrac{1}{2\beta} \geq 1$.

EXERCISES: UNIFORM CONTINUITY AND MISCELLANEOUS EXERCISES

1. Let $f:\mathbf{R} \to \mathbf{R}$ be given by $f(x) = 3x$. Is f uniformly continuous? (Recall our convention is that, unless otherwise specified, the metric for $\mathbf{R}^n$ is assumed to be the Euclidean metric.)

2. Let $X = \{x:0 < x\}$ and suppose $f(x) = 1/x$ for $x \in X$. Is f uniformly continuous on X?

3. Suppose $f:(X, d) \to (Y, \rho)$ is uniformly continuous and $Z \subset X$. Is f uniformly continuous on Z?

4. Let $f:\mathbf{R} \to \mathbf{R}$ be given by $f(x) = x^3$. Is f uniformly continuous on $\mathbf{R}$?

5. Prove the following. Suppose (X, d) is a metric space and $S \subset X$. Then $d(x, S) = 0$ if and only if $x \in \text{cl}(S)$. Hence, if $x \in X - S$ and S is closed, $d(x, S) > 0$.

6. Use 53.8 and Exercise 5 to prove the following: Suppose S is a closed set in (X, d) and $x \in X - S$. Then there exists a pair of disjoint open sets U and V such that $x \in U$ and $S \subset V$.

7. Let A be a nonempty closed subset of a metric space (X, d). Let $d_A : X \to \mathbf{R}$ be as in Theorem 53.8. Show that $A = \bigcap \left\{ d_A^{-1} \left[\left\{ y : y < \frac{1}{n} \right\} \right] : n \in \mathbf{P} \right\}$. Give a counterexample to show that the conclusion need not hold if A is not closed.

8. Let A be a closed subset of a metric space. Show that A is the intersection of a countable collection of open subsets. Show that if U is an open subset of X, then U is the union of a countable collection of closed subsets.

9. Suppose A is a nonempty closed subset of $\mathbf{R}^n$. Show that if $x \notin A$, then there is a point $a \in A$ such that $d(x, a) = d(x, A)$.

10. Give an example of a pair of disjoint closed subsets A and B such that $d(A, B) = 0$.

11. Suppose that $f : (X, d) \to (Y, \rho)$ and $g : (Y, \rho) \to (Z, d^*)$ are uniformly continuous mappings. Is $g \circ f : (X, d) \to (Z, d^*)$ necessarily uniformly continuous?

12. In each of the following, determine if the function is uniformly continuous.
    (a) The function f in Example 52.2.
    (b) The function f in Example 52.3.
    (c) The function f in Example 52.9.
    (d) The function f in Example 52.10.

13. Let $\mathbf{Z}$ be the set of all integers and let $g : \mathbf{Z} \times \mathbf{Z} \to \mathbf{R}$ be given by the following:

$$g(i, j) = \left| \frac{1}{i} - \frac{1}{j} \right| \quad \text{if} \quad i \neq 0 \quad \text{and} \quad j \neq 0$$

$$g(0, 0) = 0$$

$$g(0, i) = g(i, 0) = \left| \frac{1}{i} \right| \quad \text{if} \quad i \neq 0$$

Is g a metric for $\mathbf{Z}$? If the answer is yes, does the sequence (a_i) converge, where $a_i = i$ for $i \in \mathbf{P}$?

14. Suppose that $f : (X, d) \to (Y, \rho)$ is a function for which the following is given: If (x_i) converges in X, then $(f(x_i))$ converges in Y. (Note that we are not assuming that $\lim (x_i) = x$ implies that $\lim (f(x_i)) = f(x)$.) Is f necessarily continuous?

15. Suppose that (X, d) is a metric space. Let $d^* : X \times X \to \mathbf{R}$ be given by:
$$d^*(a, b) = d(a, b)(1 + d(a, b))^{-1}$$

Is d^* a metric for X? If the answer is yes, determine whether d^* is equivalent to d.

4

Metric Spaces: Special Properties and Mappings on Metric Spaces

There are a number of useful properties that are possessed by all metric spaces. We shall show, for example, that metric spaces are sufficiently rich in open sets so that each metric space has the following topological property known as *normality*. If A and B are disjoint closed sets, then there exist disjoint open sets U_A and U_B that contain A and B, respectively. Some of the other properties that we shall study in this chapter, although not possessed by all metric spaces, are possessed by important special classes. Our previous study of $\mathbf{R}^n$ and certain types of subsets of $\mathbf{R}^n$ will serve as a foundation for our study of various properties. For example, we proved in 42.1 that a sequence in $\mathbf{R}^n$ converges if and only if it satisfies the *Cauchy criterion*. Metric spaces for which convergence is characterized by the Cauchy criterion are called *complete* metric spaces and are very important for many applications. Such spaces enjoy some useful properties that metric spaces, in general, do not have. For example, an important theorem that can be regarded as a generalization of the nested interval theorem can be proved for complete metric spaces.

In this chapter we shall also define the concept of *connectedness* and prove that the real line $\mathbf{R}$ is connected. By making use of this fact, we will then be able to prove that $\mathbf{R}^n$ is connected. Connectedness is an extremely important concept with useful implications in the application of topology to analysis. For example, we shall see that the intermediate value theorem in calculus depends upon the fact that connectedness is preserved under continuous mappings. In Chapter 2 we saw that closed and bounded subsets of $\mathbf{R}^n$ possessed some very strong properties. For example, if S is a closed and bounded subset of $\mathbf{R}^n$, then every sequence in S has a convergent subsequence that converges to a point in S (see 41.4). This property is called *sequential compactness*. We will prove that it is equivalent to the *finite subcovering property* of the classic Heine-Borel theorem (43.4). Recall from calculus that if a real-valued function is continuous on a closed interval then it is uniformly continuous on that interval. We give a generalization of this theorem for metric spaces provided the domain is compact. Recall also the Lindelöf theorem for $\mathbf{R}^n$ (43.3). Not all metric spaces satisfy the countable subcovering property of that theorem. However, we shall show that the Lindelöf theorem can be extended to a class of metric spaces known as *separable metric spaces*.

As mentioned in the last chapter, to show that two spaces are not homeomorphic, it is necessary only to show that one of the spaces possesses a topological property that the other space does not have. For this as well as other reasons, as we study various special properties, it will be of interest to note which are topological properties. For those properties that are topological, it is useful to determine which are invariant under continuous mappings. Since homeomorphisms are continuous, nontopological properties cannot be invariant under all continuous mappings. However, for nontopological properties it is useful to know if the property is invariant under uniformly continuous mappings.

The reader is probably familiar with the notion of uniformly convergent sequences of real-valued functions. In this chapter this notion will be extended to sequences of metric-valued functions. For example, it will be proved that the limit function of a uniformly convergent sequence of continuous functions $(f_i:(X, d) \to (Y, \rho))$ is itself continuous.

In the material that follows, properties will be stated for metric spaces. If we then say that a subset S of a space (X, d) has that property, this will be taken to mean that the subspace (S, d) has the property.

54. SEPARATION PROPERTIES

Suppose x and y are distinct points in a metric space (X, d). Then x can be "separated" from y by an open set U such that $x \in U$ and $y \notin U$. Notice that the effect of this property is to make the complement of $\{y\}$ open. Thus, a set consisting of a single point is a closed set. This is a rather weak separation property. A little stronger separation property is that for every two points x and y there exist disjoint open sets U and V such that $x \in U$ and $y \in V$. This follows from the fact that if $x \neq y$ and $d = \frac{1}{2}d(x, y)$, then $N(x; \frac{1}{2}d) \cap N(y; \frac{1}{2}d) = \varnothing$. We shall show in Theorem 54.1 that metric spaces have a rather strong separation property called *normality*, i.e., for every pair of disjoint closed sets A and B, there exists a pair of disjoint open sets U and V such that $A \subset U$ and $B \subset V$. Since sets consisting of a single point are closed, the property stated in Exercise 6, page 105, is a special case of normality for metric spaces.

54.1. Theorem.   *Let (X, d) be a metric space. Let A and B be disjoint closed subsets of X. Then there exist disjoint open sets U_A and U_B such that $A \subset U_A$ and $B \subset U_B$.*

PROOF.   The conclusion obviously holds if at least one of the sets is empty. We shall assume that $A \neq \varnothing$ and $B \neq \varnothing$. For each $a \in A$, a is not a limit point of B. Hence, we may choose an $\varepsilon(a) > 0$ such that $N(a; \varepsilon(a)) \cap B = \varnothing$. Similarly, for each $b \in B$ we may choose an $\varepsilon(b) > 0$ such that $N(b; \varepsilon(b)) \cap A = \varnothing$. Note that for each $a \in A$ and $b \in B$, $N(a; \frac{1}{2}\varepsilon(a)) \cap N(b; \frac{1}{2}\varepsilon(b)) = \varnothing$. For suppose

$z \in N(a; \frac{1}{2}\varepsilon(a)) \cap N(b; \frac{1}{2}\varepsilon(b))$. Then

$$d(a, b) \leq d(a, z) + d(z, b) < \tfrac{1}{2}\varepsilon(a) + \tfrac{1}{2}\varepsilon(b).$$

Suppose $\varepsilon(b) \leq \varepsilon(a)$. Then $d(a, b) < \varepsilon(a)$. This gives a contradiction because $N(a; \varepsilon(a)) \cap B = \varnothing$. We get a similar contradiction if $\varepsilon(a) < \varepsilon(b)$. Next let

$$U_A = \bigcup \{N(a; \tfrac{1}{2}\varepsilon(a)) : a \in A\}$$

and

$$U_B = \bigcup \{N(b; \tfrac{1}{2}\varepsilon(b)) : b \in B\}$$

and notice that these two sets are open. Obviously, $A \subset U_A$ and $B \subset U_B$. Furthermore, it should be clear that $U_A \cap U_B = \varnothing$ since for $a \in A$ and $b \in B$,

$$N(a; \tfrac{1}{2}\varepsilon(a)) \cap N(b; \tfrac{1}{2}\varepsilon(b)) = \varnothing.$$

Note that in the previous proof for $a \in A$ to have a neighborhood $N(a; \varepsilon(a))$ that is disjoint from B, it is merely required that the point a not be a point of B or a limit point of B. Thus, if $A \cap \mathrm{cl}\,(B) = \varnothing$, then there exists for each $a \in A$, a neighborhood $N(a; \varepsilon(a))$ such that $N(a; \varepsilon(a)) \cap B = \varnothing$. Similarly if $B \cap \mathrm{cl}\,(A) = \varnothing$, there exists for each $b \in B$, a neighborhood $N(b; \varepsilon(b))$ such that $N(b; \varepsilon(b)) \cap A = \varnothing$. Hence, the proof of 54.1 yields the following result.

54.2. Theorem. *Suppose A and B are subsets of X such that $\mathrm{cl}\,(A) \cap B = A \cap \mathrm{cl}\,(B) = \varnothing$. Then there exist disjoint open subsets U_A and U_B such that $A \subset U_A$ and $B \subset U_B$.*

The property for metric spaces given in the previous theorem is referred to as *complete normality*. A pair of sets A and B that satisfies the condition stated in the hypothesis of the theorem is given a special name as indicated next.

54.3. Definitions. Mutually separated sets and separation of a set. *Suppose X is a metric space. Two subsets A and B of X are said to be mutually separated in X provided that*

$$\mathrm{cl}\,(A) \cap B = A \cap \mathrm{cl}\,(B) = \varnothing.$$

If $S \subset X$ is the union of a pair of nonempty mutually separated sets A and B, then $\{A, B\}$ is said to be a separation of S (in X).

It should be noted that any pair of sets A and B such that $\mathrm{cl}\,(A) \cap \mathrm{cl}\,(B) = \varnothing$ are mutually separated. However, this condition is not necessary for A and B to be mutually separated as the following example indicates.

54.4. Example. Using the notation $B(p; r)$ for a closed disc in $\mathbf{R}^2$ with center at p and radius r, consider the following two subsets of $\mathbf{R}^2$:

$$C = B((0, 0); 1) - \{(1, 0)\}, \; D = B((2, 0); 1) - \{(1, 0)\}.$$

The sets C and D are mutually separated sets.

In the definition of mutually separated sets the condition, $\mathrm{cl}\,(A) \cap B = A \cap \mathrm{cl}\,(B) = \varnothing$, is given with reference to the closure operation for the containing space X. However, whether A and B are actually mutually separated depends only on the topology of the subspace $A \cup B$. To see this it is necessary only to observe the following fact.

54.5. *Remark.* Suppose X is a metric space. Then subsets A and B are mutually separated subsets of X if and only if A and B are both closed (or equivalently both open) in $A \cup B$ and are disjoint.

There is another useful separation property possessed by metric spaces. Nonempty disjoint closed sets can be "separated" by a continuous function in the sense of the following theorem.

54.6. **Theorem.** *Suppose A and B are closed nonempty disjoint subsets of a metric space (X, d). Then there exists a continuous function f from X into the real, closed interval $[0, 1]$ such that $f[A] = \{0\}$ and $f[B] = \{1\}$.*

Proof. Let $f: X \to \mathbf{R}$ given by

$$f(x) = d(x, A)[d(x, A) + d(x, B)]^{-1}.$$

The proof that this function has the required properties is left as an exercise.

54.7. *Remark.* It is instructive to note that the function given by the previous theorem suggests an alternate proof that a metric space possesses the normality property. Suppose that A and B are nonempty disjoint closed subsets of a metric space X. By 54.6, there exists a continuous function $f: X \to \mathbf{R}$ such that $f[A] = \{0\}$ and $f[B] = \{1\}$. Let $U_A = f^{-1}[y: y \in \mathbf{R}$ and $y < \frac{1}{2}]$ and let $U_B = f^{-1}[y: y \in \mathbf{R}$ and $y > \frac{1}{2}]$. Recall that, by Theorem 52.5, openness is preserved under the inverse of a continuous mapping. Finally, observe that $A \subset U_A$, $B \subset U_B$, and $U_A \cap U_B = \varnothing$.

EXERCISES: SEPARATION PROPERTIES

1. Let $\{A_i : i \in \mathbf{P}_n\}$ be a collection of n pairwise disjoint closed subsets of a metric space. Show that there exists a collection $\{U_i : i \in \mathbf{P}_n\}$ of open subsets such that $\{\text{cl } (U_i) : i \in \mathbf{P}_n\}$ is pairwise disjoint and for each $i \in \mathbf{P}_n$, $A_i \subset U_i$.

2. Suppose $\{A, B\}$ is a separation of an open set U in a space X. Prove that A and B are both open in X.

3. Suppose $\{A, B\}$ is a separation of a closed set F in a space X. Prove that A and B are both closed in X.

4. Prove 54.5.

5. Complete the proof of 54.6.

6. Suppose A and B are two nonempty closed disjoint subsets of a metric space X. Let a and b be real numbers with $a < b$. Prove that there exists a continuous real-valued function $f: X \to [a, b]$ such that $f(x) = a$ if and only if $x \in A$ and $f(x) = b$ if and only if $x \in B$.

55. CONNECTEDNESS IN METRIC SPACES

An extremely important topological property is that of connectedness. As we shall see, it is essentially this property that makes the intermediate value theorem hold for continuous real-valued functions defined on intervals. Intuitively we would like to say that a set is connected if it "hangs together" in one piece. Whatever definition we choose for *connected space*, we would certainly wish, for example, the real line **R** to have that property. Notice that there are many ways in which the real line can be decomposed into a pair of nonempty disjoint subsets. However, it can be shown that **R** is *not* the union of two nonempty disjoint *open* sets. This is one of the facts to be verified in this section.

55.1. Definition. Connected space. *Let (X, d) be a metric space. Then (X, d) is connected provided X is not the union of two nonempty disjoint open sets. A subset S of X is said to be connected provided (S, d) is a connected space.*

If a set is not connected we shall say that it is *disconnected*.

The following theorem gives some properties that are equivalent to connectedness. The proof is left as an exercise for the reader.

55.2. Theorem. *Let X be a metric space. Then the following statements are equivalent.*

55.2(a). *X is connected.*

55.2(b). *X is not the union of two nonempty disjoint closed sets.*

55.2(c). *There exists no separation of X.*

55.2(d). *A subset S of X is both open and closed if and only if $S = \varnothing$ or $S = X$.*

Note that connectedness is a topological property. Also the connectedness of a subset S of a space depends only on the space S and not on the containing space. Nevertheless, in investigating the possible connectedness of a subset S of a metric space X, often we will find it convenient to deal directly with the topology of the containing space X rather than explicitly with the open subsets of the subspace S.

We accomplish this by making use of the notion of *separation* of a set defined in 54.3 as follows.

55.3. Theorem. *Suppose X is a metric space and S is a subset of X. Then S is connected if and only if S is not the union of sets A and B such that*

$$A \neq \varnothing, \quad B \neq \varnothing, \quad A \cap \operatorname{cl}(B) = B \cap \operatorname{cl}(A) = \varnothing$$

(i.e., there exists no separation of S in X).

PROOF. The condition that $A \neq \varnothing$, $B \neq \varnothing$, and $A \cap \operatorname{cl}(B) = B \cap \operatorname{cl}(A) = \varnothing$ is equivalent to the condition that A and B are disjoint nonempty open subsets of $S = A \cup B$ (see 54.5). The theorem now follows from the definition of connected set.

If what we have given is a reasonable definition of connectedness, we should be able to prove, for example, that an interval in **R** is connected, and this we shall do in this section. Subsequently the reader will be given exercises in which he will be asked to prove that certain other sets that we would want to call connected are indeed connected.

55.4. Theorem. *Let $[a, b]$ be a closed interval in **R**. Then $[a, b]$ is connected.*

PROOF. Suppose $[a, b]$ is not connected. Then there is a separation $\{A, B\}$ of $[a, b]$. Note that $[a, b] = A \cup B$ and $[a, b]$ is closed. Then since B has no limit points of A, A must be closed. Similarly B is closed. We may assume that $a \in A$. Let $\alpha = $ l.u.b. (A) and since A is closed $\alpha \in A$. Suppose $\alpha < b$. Then α would be a limit point of B and since B is closed we have a contradiction. So we have that $\alpha = b$. Then $B \subset [a, b] - \{a, b\}$. Now let $\beta = $ g.l.b. (B). Then since B is closed $\beta \in B$ and $a < \beta < b$. However, since $a < \beta$, β is a limit point of A and we have arrived at a contradiction. Summarizing, assuming that $[a, b]$ had a separation $\{A, B\}$ with $a \in A$, we found that each of the statements l.u.b. $(A) < b$ and l.u.b. $(A) = b$ led to a contradiction. We therefore are forced to conclude that $[a, b]$ has no separation and consequently $[a, b]$ is connected.

The following two propositions will prove to be quite helpful in our dealing with the concept of connectedness.

55.5. Theorem. *Suppose $\{A, B\}$ is a separation of a set S (note that S is not connected). Let C be a connected subset of S. Then $C \subset A$ or $C \subset B$.*

PROOF. Suppose that the conclusion is false. Then $C \cap A \neq \varnothing$ and $C \cap B \neq \varnothing$. Further, $\{C \cap A, C \cap B\}$ is a separation of C. To see this, note that $C = (C \cap A) \cup (C \cap B)$. Moreover, if $p \in (C \cap A)$ and p is a limit point of $C \cap B$, then it is also a limit point of B. But this would give a contradiction since we are assuming that $\{A, B\}$ is a separation of S. Similarly we would get a contradiction if a point of $C \cap B$ were a limit point of $C \cap A$. Thus, we have found a separation of C and, since C is connected, we have arrived at a contradiction.

55.6. Theorem. *Suppose $\mathscr{H}$ is a collection of connected subsets of a space such that if $C_1 \in \mathscr{H}$ and $C_2 \in \mathscr{H}$, then $C_1 \cap C_2 \neq \varnothing$. Then $\bigcup \mathscr{H}$ is connected.*

PROOF. Suppose $\bigcup \mathscr{H}$ is not connected. Let $\{A, B\}$ be a separation for $\bigcup \mathscr{H}$. Let $a \in A$ and $b \in B$. There is a $C_a \in \mathscr{H}$ and $C_b \in \mathscr{H}$ such that $a \in C_a$ and $b \in C_b$. But from the hypothesis, $C_a \cap C_b \neq \varnothing$. Let $z \in C_a \cap C_b$. By Theorem 55.5, $C_a \subset A$ since a point of C_a is in A. Similarly $C_b \subset B$. However, since $z \in C_a \cap C_b$, $z \in A \cap B$. But this is a contradiction. Hence, $\bigcup \mathscr{H}$ is connected.

We next make use of the previous theorem in proving that the real line is connected.

55.7. Theorem. *The real line is connected.*

PROOF. For each $n \in \mathbf{P}$, by 55.4, the closed interval $[-n, n]$ is connected. Observe that $\mathbf{R} = \bigcup \{[-n, n] : n \in \mathbf{P}\}$ and that $0 \in [-n, n]$ for each $n \in \mathbf{P}$. Hence, by 55.6, $\mathbf{R}$ is connected.

EXERCISES: CONNECTEDNESS IN METRIC SPACES

1. Prove 55.2.

2. Suppose (X, d) is a metric space.
    (a) Is $\varnothing$ a connected subset of (X, d)?
    (b) Let $x \in X$. Is $\{x\}$ connected?
    (c) Suppose $\{x_1, x_2, \ldots, x_n\}$ is a finite subset of X, $n > 1$. Is this set disconnected?
    (d) Suppose S is a countably infinite subset of X. Is S necessarily disconnected?

3. Recall the definition of an interval in **R**. Prove that each interval (not necessarily closed) in **R** is connected. Prove that if S is a connected subset of **R** then S is an interval (possibly empty).

4. Prove the following important proposition: If S is a connected subset of a metric space and C is a subset of the space such that $S \subset C \subset \operatorname{cl}(S)$, then C is connected.

5. Suppose m is the metric for **R** given by $m(x, y) = 0$ if $x = y$ and $m(x, y) = 1$, otherwise. Is the space $(\mathbf{R}, m)$ a connected space?

6. Does there exist in $\mathbf{R}^2$ a decreasing sequence of connected sets (C_i) such that $\bigcap \{C_i : i \in P\}$ is not connected? If the answer is yes, give an example. If the answer is no, justify with a proof as to why no such example exists.

7. If your answer to the previous question was yes, add to the requirement on the collection $\{C_i : i \in P\}$ that each C_i is a closed and bounded subset of $\mathbf{R}^n$ and repeat the question.

56. THE INVARIANCE OF CONNECTEDNESS UNDER CONTINUOUS MAPPINGS

The fact that certain properties are invariant under various kinds of mappings is of fundamental interest in topology and the application of topology to various branches of analysis. In the case of invariance of connectedness that we consider here, classic forms of the intermediate value theorem will be recognized as immediate corollaries.

56.1. Theorem. *Suppose that (X, d) is a connected metric space and $f: (X, d) \to (Y, \rho)$ is a continuous surjection. Then (Y, ρ) is connected.*

PROOF. Suppose that Y is not connected. Let $\{A, B\}$ be a separation of Y. Then A and B are nonempty open subsets of Y with an empty intersection. We shall show that $\{f^{-1}[A], f^{-1}[B]\}$ is a separation of X, thus arriving at a contradiction. Note first that $X = f^{-1}[A] \cup f^{-1}[B]$. Since A and B are disjoint, $f^{-1}[A]$ and $f^{-1}[B]$ are also disjoint. Because A and B are open and f is continuous, it follows that $f^{-1}[A]$ and $f^{-1}[B]$ are open in X. Since f is a surjection, $f^{-1}[A]$ and $f^{-1}[B]$ are nonempty. Hence, $\{f^{-1}[A], f^{-1}[B]\}$ is a separation of X and the proof is complete.

56.2. Corollary. Intermediate value theorem. *Let $f: X \to \mathbf{R}$ be a real-valued continuous function defined on a metric space X. Suppose S is a connected subset of X. Then if a and b are elements of S and $f(a) < c < f(b)$, there is an $x_0 \in S$ such that $f(x_0) = c$.*

56.3. Classic intermediate value theorem. *Suppose $f: [a, b] \to \mathbf{R}$ is a continuous real-valued mapping from a closed real interval $[a, b]$. If $f(a) \neq f(b)$ and c is a number between $f(a)$ and $f(b)$, then there is a number $x_0 \in (a, b)$ such that $f(x_0) = c$.*

EXERCISES: THE INVARIANCE OF CONNECTEDNESS UNDER
           CONTINUOUS MAPPINGS

1. Review the definition of line and line segment in $\mathbf{R}^n$. Prove that lines and line segments are connected subsets of $\mathbf{R}^n$.

2.   (a) Show that $\mathbf{R}^n$ is connected.
     (b) Show that if $n > 1$ and S is a countable subset of $\mathbf{R}^n$, then $\mathbf{R}^n - S$ is connected.
     (c) Show that in $\mathbf{R}^n$ open spheres $N(p; \varepsilon)$ and closed balls $B(p; \varepsilon)$ are connected subsets of $\mathbf{R}^n$.
     (d) Suppose S is an open subset of $\mathbf{R}^n$ and C is a connected subset of S which has the property that no other connected subset of S contains it (hence, it is a maximal connected subset of S). Prove that C is an open subset of $\mathbf{R}^n$.

3. Let $f: \mathbf{R}^2 \to \mathbf{R}$ be continuous. Suppose for $a \in \mathbf{R}^2$ and $b \in \mathbf{R}^2$, $f(a) < 0$ and $f(b) > 0$. Show that there is an uncountable collection of points $z \in \mathbf{R}^2$ such that $f(z) = 0$.

4. Let (X, d) be a space. Suppose it is known that there exists no continuous mapping $f:(X, d) \to \mathbf{R}$ such that $f[X] = \{0, 1\}$. Does this condition imply that X is connected?

5. Let $f:[a, b] \to \mathbf{R}$ be a real-valued continuous nonconstant mapping defined on a closed real interval $[a, b]$. The range is a closed interval $[\alpha, \beta]$. (Why?) Suppose $c \in (\alpha, \beta)$. Show that there exists a $\delta > 0$ such that if $g:[a, b] \to \mathbf{R}$ is continuous and

$$|g(x) - f(x)| < \delta$$

on $[a, b]$, then the equation $g(x) = c$ has at least one solution in $[a, b]$. Also show by an example that the conclusion is not true for $c = \alpha$ or $c = \beta$.

6. Let $f:[a, b] \to \mathbf{R}$ be a continuous real-valued mapping defined on a real closed interval $[a, b]$. Prove that the graph

$$\{(x, f(x)): x \in [a, b]\}$$

is a bounded closed and connected subset of $\mathbf{R}^2$ that is topologically equivalent to $[a, b]$.

7. Let A and B be the following subsets of $\mathbf{R}^2$:
$$A = \{(x, y): x = 0, \quad 0 \le y \le 1\},$$
$$B = \{(x, y): y = \sin \frac{1}{x} \quad \text{for } x > 0\}.$$
Sketch $A \cup B$. Is $A \cup B$ connected?

57. POLYGONAL CONNECTEDNESS

A subset S of $\mathbf{R}^n$ is said to be a polygon if there exist points $x_0, x_1, x_2, \ldots, x_k$, not necessarily distinct, such that $S = \bigcup \{L(x_{i-1}, x_i): i \in P_k\}$, where $L(x_{i-1}, x_i)$ is the closed line segment joining x_{i-1} to x_i.

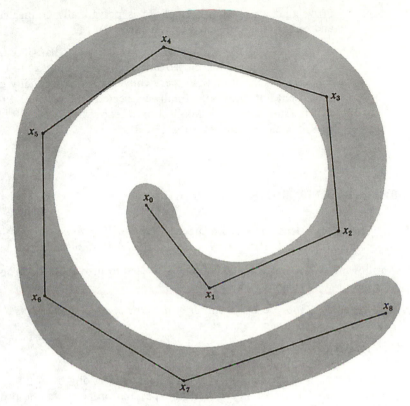

Figure 15

57.1. Definition. Polygonal connectedness. *A subset S of* $\mathbf{R}^n$ *is said to be polygonally connected provided that each two points in the set can be joined by a polygon that is contained in the set.*

As we shall see, this type of connectedness is stronger than the concept of connectedness that was introduced earlier. In an exercise the reader will be asked to show that polygonal connectedness implies connectedness. However, note that the graph of the sine function is connected but not polygonally connected. It is natural to ask under what conditions these concepts are the same. Open subsets of $\mathbf{R}^n$ are connected if and only if they are polygonally connected, a fact that the reader will be asked to verify in the next set of exercises.

EXERCISES: POLYGONAL CONNECTEDNESS

1. Prove that a polygon in $\mathbf{R}^n$ is connected.

2. Suppose S is a subset of $\mathbf{R}^n$ such that S is polygonally connected. Prove that S is connected.

3. Let A be an open subset of $\mathbf{R}^n$. Let $\alpha \in A$ and let $U = \{x : x \in A$ and x can be connected to α by a polygon contained in $A\}$. Show that U is an open subset of $\mathbf{R}^n$. Show that $A - U$ is an open subset of $\mathbf{R}^n$.

4. Prove that an open subset of $\mathbf{R}^n$ is polygonally connected if and only if it is connected (see Exercise 3).

5. A subset S of $\mathbf{R}^n$ is said to be convex provided that, if x_0 and x_1 are points in S, the closed line segment joining x_0 to x_1 is contained in S. Note that every convex set is polygonally connected but not every polygonally connected set is convex. Prove that the intersection of a collection of convex subsets of $\mathbf{R}^n$ is a (possibly empty) convex subset of $\mathbf{R}^n$.

58. SEPARABLE METRIC SPACES

58.1.   Definition. Dense subset of a space. *Suppose X is a metric space and D is a subset of X such that* cl $(D) = X$. *Then D is called a dense subset of X.*

The following characterization of denseness is easy to prove and its proof is left as an exercise.

58.2.   Theorem. *Suppose D is a subset of a metric space X. Then D is dense in X if and only if every nonempty open subset of X intersects D.*

58.3.   *Examples*
(a) The set $\mathbf{Q}$ of rational numbers is a dense subset of $\mathbf{R}$.
(b) The set $\{(x, y): x \neq 0, y \neq 0\} \subset \mathbf{R}^2$ is a dense subset of $\mathbf{R}^2$.
(c) The space (X, m) in Example 45.4 has the peculiar property that no proper subset of X is a dense subset of X, since each subset of X is closed.

Recall from Theorem 43.1 that there is a countable set $D \subset \mathbf{R}^n$ that is dense in $\mathbf{R}^n$. Not all spaces have this property. However, spaces with this property occur with sufficient frequency to warrant their study.

58.4.   Definition. Separable metric space. *Let X be a metric space. If there is a countable set in X that is dense in X, then X is said to be separable.*

It turns out that separability is invariant under continuous mappings. We may prove this by first proving the following stronger result.

58.5.   Theorem. *Let X and Y be metric spaces and suppose $f: X \to Y$ is a continuous surjection. If D is dense in X, then $f[D]$ is dense in Y.*

58.6.   Theorem. *Suppose $f: X \to Y$ is a continuous surjection. Then if X is separable, so is Y.*

The reader will be asked to prove the previous two theorems in the next set of exercises.

The next theorem gives an important property that is equivalent to separability for metric spaces.

58.7.   Theorem. *Suppose X is a metric space. Then the following property is equivalent to separability.*

58.7(a).   *There exists a countable collection $\mathscr{K}$ of open subsets of X such that*

each open subset of X can be expressed as the union of a subcollection of $\mathscr{K}$ (see
43.2). (Such a collection is called a countable base for the topology of (X, d).)

PROOF. First suppose that X is separable. Then there is a countable dense
subset $\{c_i : i \in \mathbf{P}\}$ in X. Let $\{r_i : i \in \mathbf{P}\}$ be the collection of positive rational numbers,
also indexed by the set $\mathbf{P}$. Let $\mathscr{K} = \{N(c_i ; r_j) : (i, j) \in \mathbf{P} \times \mathbf{P}\}$. We claim that $\mathscr{K}$
satisfies the conditions imposed on $\mathscr{K}$ in 58.7(a). To show this it will suffice to
show that if W is open and $x \in W$, then there is a $U \in \mathscr{K}$ such that $x \in U \subset W$.
We show this next.

Suppose W is open and $x \in W$. Then there is an $\varepsilon > 0$ such that $N(x; \varepsilon) \subset W$.
Since $\{c_i : i \in \mathbf{P}\}$ is dense in X, there is a c_i such that $d(c_i, x) < \frac{1}{3}\varepsilon$. Next, note that
$N(c_i ; \frac{2}{3}\varepsilon) \subset W$. Now, let r_j be a positive rational such that $\frac{1}{3}\varepsilon < r_j < \frac{2}{3}\varepsilon$. We
complete the proof of this part by noting that $x \in N(c_i ; \frac{1}{3}\varepsilon) \subset N(c_i ; r_j) \subset$
$N(c_i ; \frac{2}{3}\varepsilon) \subset W$. Thus, separability implies 58.7(a).

To show that 58.7(a) implies separability, let $\{U_i : i \in \mathbf{P}\}$ be a countable collection of open subsets with the property stated in 58.7(a). We may assume that each
of the U_i is nonempty. For each $i \in \mathbf{P}$, select $c_i \in U_i$. To see that $D = \{c_i : i \in \mathbf{P}\}$
is dense, let W be a nonempty open subset of X. Let $x \in W$. There is a U_i such that
$x \in U_i \subset W$. But then $c_i \in W$ so $W \cap D \neq \varnothing$. Hence, by 58.2, D is dense in X.

58.8. Definition. Covering. *Suppose S is a subset of X. If $\mathscr{K}$ is a collection of*
subsets of X such that $S \subset \bigcup \mathscr{K}$, then $\mathscr{K}$ is said to be a covering of S. ($\mathscr{K}$ is also
said to cover S.) If each element of $\mathscr{K}$ is an open set in X, then $\mathscr{K}$ is called an
open covering of S.

If $\mathscr{K}$ is a covering of S and $\mathscr{H}$ is a subcollection of $\mathscr{K}$ that covers S, then $\mathscr{H}$
is sometimes referred to as a *subcovering*.

We next use the previous theorem to generalize the classic Lindelöf theorem
for $\mathbf{R}^n$ to the setting of a separable metric space. The reader who proved 43.3
should have a proof that essentially carries over to this case. We include a proof
at this point.

58.9. Theorem (Lindelöf). *Let X be a separable metric space. Suppose $\mathscr{K}$*
is an open covering of a subset S of X. Then there exists a countable subcollection
of $\mathscr{K}$ that also covers S.

PROOF. Suppose $\mathscr{K}$ is an open covering of S. By 58.7(a), there exists a countable collection $\mathscr{U} = \{U_i : i \in \mathbf{P}\}$ of open subsets of X such that each open subset of
X is the union of a subcollection of $\mathscr{U}$. Let $J = \{i : i \in \mathbf{P}$ and $U_i \subset W$ for at least
one $W \in \mathscr{K}\}$. First we shall show that $S \subset \bigcup \{U_i : i \in J\}$. Suppose $x \in S$; then
there is a $W \in \mathscr{K}$ such that $x \in W$. Since W is the union of a subcollection of $\mathscr{U}$,
there is a $k \in \mathbf{P}$ such that $x \in U_k \subset W$. Hence, $k \in J$ and $x \in \bigcup \{U_i : i \in J\}$. Next,
for each $j \in J$, choose one $W \in \mathscr{K}$ such that $U_j \subset W$ and call it W_j. It should now
be clear that $S \subset \bigcup \{U_i : i \in J\} \subset \bigcup \{W_i : i \in J\}$. Observe that $\{W_i : i \in J\}$ is a
countable subcollection of $\mathscr{K}$.

The next theorem is easy to prove by making use of Theorem 58.7. The proof
is left as an exercise.

58.10. Theorem. *Let (X, d) be a separable metric space. Suppose (S, d)*
is a subspace of (X, d). Then (S, d) is separable also.

EXERCISES: SEPARABLE METRIC SPACES

1. Prove 58.2.

2. Let (X, d) be a metric space. Then D is dense in X if and only if the following condition holds: If $x \in X$ and $\varepsilon > 0$, there is a $z \in D$ such that $d(z, x) < \varepsilon$.

3. Prove Theorems 58.5 and 58.6.

4. Consider the metric space (X, m) of Example 45.4, with X taken to be an uncountable set. Show that this space is not separable. Also show that this space does not satisfy the conclusion of the Lindelöf theorem.

5. Prove Theorem 58.10.

6. Prove that if (X, d) is a metric space, then separability is equivalent to the following property:
    For each $\varepsilon > 0$ there is a countable subset $C \subset X$ such that $X = \bigcup\{ N(c; \varepsilon) : c \in C\}$ (or equivalently for each $\varepsilon > 0$, each point of X is within ε distance of some point in C).

7. Let (X, d) be a metric space. Suppose X satisfies the conclusion of the Lindelöf theorem 58.9. Prove that (X, d) is separable.

8. A point p is a condensation point of a set S in a metric space if each open subset containing p contains an uncountable subset of S. Prove that if (X, d) is a separable metric space and M is the set of condensation points of X, then $X - M$ is countable. (Notice that there exist countably infinite subsets of $\mathbf{R}^n$ that have no limit points. But, in view of the above proposition, if S is an uncountable subset of $\mathbf{R}^n$, then in most places the points must be so "crowded" that there is an uncountable collection of condensation points of S in S.

9. Suppose (X, d) is a metric space and (Y, d) is a subspace of (X, d). Suppose that (Y, d) is separable. Is $(\mathrm{cl}_X(Y), d)$ also separable?

10. Prove the following proposition: Suppose that $f:(X, d) \to (Y, \rho)$ and $g:(X, d) \to (Y, \rho)$ are continuous and there exists a dense subset $D \subset X$ such that the restrictions $f \mid D$ and $g \mid D$ are equal. Then $f = g$.

11. In the proof of Theorem 58.9, is $\bigcup \{W_i : i \in J\} = \bigcup \mathcal{K}$?

59. TOTALLY BOUNDED METRIC SPACES

In Exercise 6, this page, we saw that a metric space S is separable if for each $\varepsilon > 0$, there is a countable set D_ε such that $S = \bigcup \{N(z; \varepsilon) : z \in D_\varepsilon\}$. If we alter this property so that the words "countable set D_ε" are replaced by "finite set D_ε," we obtain a stronger property. This stronger property is of considerable importance in the study of certain metric spaces and applications.

In a sense, it is precisely because of this property that bounded subsets of $\mathbf{R}^n$ enjoy the nice properties that they do.

59.1. Definition. Totally bounded spaces. *A metric space (X, d) is said to be totally bounded provided the following condition is satisfied. For each $\varepsilon > 0$, there is a finite set F_ε (called an ε-net) contained in X such that*

$$X = \bigcup \{N(z; \varepsilon): z \in F_\varepsilon\}$$

(or equivalently each $x \in X$ is within ε distance from at least one point in F_ε).

Based on Exercise 6, page 118, if a space is totally bounded it is separable. However, we include a proof of this fact.

59.2. Theorem. *If (X, d) is totally bounded, then it is separable.*

PROOF. Suppose X is totally bounded. For each positive integer i, let D_i be a $\frac{1}{i}$-net for X. Then $D = \bigcup \{D_i : i \in \mathbf{P}\}$ is a countable subset of X. (Note that in proving the statement in Exercise 6, page 118, we could take D_i as countable but not necessarily finite.) We complete the proof by showing that D is dense in X. To see this, let $x \in X$ and $\varepsilon > 0$ (see Exercise 2, page 118). Let i be an integer such that $0 < \frac{1}{i} < \varepsilon$. Then there is a $z \in D_i$ such that $d(z, x) < \frac{1}{i} < \varepsilon$ and the proof that D is dense in X is complete.

In Theorem 58.10 it is stated that a subspace of a separable metric space is separable. Because of the similarity of the concepts of separability and total boundedness, one might guess that a subset of a totally bounded space is totally bounded. We shall prove that such is the case.

59.3. Theorem. *Suppose that S is a subset of a totally bounded metric space. Then S is totally bounded.*

PROOF. Let $\varepsilon > 0$. We wish to find a finite set $H \subset S$ that is an ε-net for S. Since X is totally bounded, there is a finite subset F of X that is an $\frac{\varepsilon}{2}$-net for X. Let $F^* = \left\{ x : x \in F \text{ and } d(x, S) < \frac{\varepsilon}{2} \right\}$. For each $z \in F^*$, choose a point $a_z \in S$ such that $d(z, a_z) < \frac{\varepsilon}{2}$. Let H be the finite set of all a_z chosen as in the previous sentence. We claim that H is an ε-net for S. To see this, let $y \in S$. Then there is an $x \in F$ such that $d(x, y) < \frac{\varepsilon}{2}$. But then $x \in F^*$. Now the point a_x in H, chosen to correspond to x, satisfies $d(a_x, x) < \frac{\varepsilon}{2}$. Thus, $d(a_x, y) < \varepsilon$ and we have shown that H is an ε-net for S.

59.4. Theorem. *Let (X, d) be a metric space. Suppose S is a subset of X. Then S is totally bounded if and only if cl (S) is totally bounded.*

PROOF. Suppose first that S is totally bounded. Let $\varepsilon > 0$. We find an ε-net for cl (S) as follows: Let F be an $\frac{\varepsilon}{2}$-net for S. Then F is an ε-net for cl (S).

To verify this, we show that if $z \in$ cl (S), then there is a $p \in F$ such that $d(p, z) < \varepsilon$. Proceeding with the verification, let $z \in$ cl (S). Then there exists a $q \in S$ such that $d(q, z) < \dfrac{\varepsilon}{2}$. But since F is an $\dfrac{\varepsilon}{2}$-net for S and $q \in S$, there is a $p \in F$ such that $d(q, p) < \dfrac{\varepsilon}{2}$. Thus, $d(z, p) < \varepsilon$ and we have shown that cl (S) is totally bounded.

Next, suppose that cl (S) is totally bounded. Then the fact that S is totally bounded follows from Theorem 59.3.

59.5. Definitions. Diameter of a set, bounded subsets of a metric space, and bounded functions. *Suppose that S is a subset of a metric space (X, d). Then S is said to be a bounded set provided that there exists a positive number M such that $d(x, y) \leq M$ for all x and y in S. The diameter of S $(diam(S))$ is defined as follows:*

> *$diam (\varnothing) = 0$;*
> *$diam (S) =$ l.u.b. $\{d(x, y) : x \in S, y \in S\}$, if S is a nonempty bounded set;*
> *$diam (S) = \infty$, if S is an unbounded set.*

If $f : (X, d) \to (Y, \rho)$ is such that $f[X]$ is a bounded set, then we say that f is a bounded function.

We shall show in the next theorem that every totally bounded subset of a metric space is bounded. On the other hand, not all bounded subsets of a metric space are totally bounded. For example, consider the space $(\mathbf{R}, m)$ of Exercise 5, page 113. $\mathbf{R}$ is a bounded set in $(\mathbf{R}, m)$ but it is easy to see that there is no $\frac{1}{2}$-net for $\mathbf{R}$ in $(\mathbf{R}, m)$. We shall show, however, that every bounded subset of $\mathbf{R}^n$ is totally bounded with respect to the Euclidean metric.

59.6. Theorem. *If S is a totally bounded subset of a metric space (X, d) then S is bounded. Moreover, if S is a subset of $\mathbf{R}^n$, then S is bounded if and only if it is totally bounded.*

PROOF.   First suppose S is a totally bounded subset of a metric space (X, d). Let $F = \{x_1, x_2, \dots, x_n\}$ be a 1-net for S. Let $M =$ max. $\{d(x_i, x_j) : x_i$ and x_j in $F\} + 2$. We show that diam $(S) \leq M$. To see this, let x and z be points in S. Then there is an $x_i \in F$ and an $x_j \in F$ such that $d(x, x_i) < 1$ and $d(z, x_j) < 1$. But then $d(x, z) \leq d(x, x_i) + d(x_i, x_j) + d(x_j, z) < 2 + d(x_i, x_j) \leq M$.

Next, let S be a bounded subset of $\mathbf{R}^n$. Recall that we proved in 41.4 that if (a_i) is a bounded sequence in $\mathbf{R}^n$, then (a_i) has a convergent subsequence. We next assume that S is not totally bounded and show that we can find a sequence in S which has no convergent subsequence. This will be a contradiction. If S is not totally bounded, then for some $\varepsilon > 0$ there is no ε-net for S. Let $x_1 \in S$. Since there is no ε-net there must be an $x_2 \in S$ such that $d(x_1, x_2) \geq \varepsilon$. Assume now $x_1, x_2, \dots, x_n$ have been chosen so that if $i \neq j$, $d(x_i, x_j) \geq \varepsilon$. Since $\{x_1, x_2, \dots, x_n\}$ is not an ε-net for S there is an $x_{n+1} \in S$ such that $d(x_{n+1}, x_i) \geq \varepsilon$ for $i \in \mathbf{P}_n$. We may thus inductively define a sequence (x_i) such that $d(x_i, x_j) \geq \varepsilon$ for $i \neq j$. Obviously no subsequence of (x_i) can converge so we have a contradiction and S is totally bounded.

That total boundedness is not a topological property is shown in the example that follows.

59.7. *Example.* Let $X = \{x : 0 < x < 1\}$ and $Y = \{y : 1 < y\}$, both endowed with the Euclidean metric. The bijection $f : X \to Y$ given by $f(x) = \dfrac{1}{x}$ for all $x \in X$ is a homeomorphism. Now it follows from 59.6 that X is totally bounded and Y is not.

In the next section we shall discuss a concept, sequential compactness, which, as we shall see, is somewhat stronger than total boundedness. In this connection it is of interest to note that total boundedness is sometimes called precompactness.

EXERCISES: TOTALLY BOUNDED METRIC SPACES

1. Let $f : (X, d) \to (Y, \rho)$ be a uniformly continuous surjection. Prove that if (X, d) is totally bounded, so is (Y, ρ).

2. Give an example of a metric $d*$ for $\mathbf{R}^2$ that is equivalent to the Euclidean metric d and which is such that $(\mathbf{R}^2, d*)$ is totally bounded.

3. Prove that if (X, d) is not separable, then for no metric $d*$ equivalent to d is it true that $(X, d*)$ is totally bounded.

60. SEQUENTIAL COMPACTNESS FOR METRIC SPACES

Sets in $\mathbf{R}^n$ that are both bounded and closed have some very strong and useful properties. However, the property of being both bounded and closed is not a topological property. To see this, let $d*$ be a metric for $\mathbf{R}^n$ that is bounded and equivalent to the Euclidean metric d (see Exercise 3, page 87). Then $\mathbf{R}^n$ is a closed and bounded set in $(\mathbf{R}^n, d*)$ but in $(\mathbf{R}^n, d)$, although it is closed, it is not bounded.

Recall that we characterized the property of a set being a closed and bounded subset of $\mathbf{R}^n$ (with respect to the Euclidean metric!) by the property that every sequence in the set has a convergent subsequence that converges to a point in the set (see Exercise 5, page 77). As we shall show, this latter property, which is called *sequential compactness*, is a topological property. In this and in the next several sections we shall study this concept in the setting of metric spaces. In the course of our study we shall show that, for metric spaces, sequential compactness is equivalent to several other useful properties. When we study more general spaces, we shall see that some of the properties which are equivalent to sequential compactness in metric spaces are not equivalent in general and each will merit study on its own.

60.1. Definition. Sequentially compact space. *If X is a metric space, then X is said to be sequentially compact provided every sequence in X has a subsequence that converges in X.*

It is useful to know what kind of subsets of a sequentially compact space are sequentially compact. The reader should think about this before reading the answer in the next theorem, the proof of which is left as an exercise.

60.2. Theorem. *Suppose X is a sequentially compact metric space and S is a subset of X. Then S is sequentially compact if and only if S is a closed subset of X.*

We next give an example and a theorem which point out one of the ways in which sequential compactness affects the action of a continuous mapping.

60.3. *Example.* Consider the mapping f defined on the nonsequentially compact set of real numbers $S = \{x : 0 \leq x < 1\}$ and given by $f(x) = (\cos 2\pi x, \sin 2\pi x)$. f takes the half open interval S onto a unit circle and is one-to-one. However, f^{-1} is not continuous. Notice that if we set $a_n = \left(\cos 2\pi\left(1 - \frac{1}{n}\right), \sin 2\pi\left(1 - \frac{1}{n}\right)\right)$, then $\lim (a_n) = (1, 0)$. If f^{-1} were continuous, $\lim (f^{-1}(a_n)) = f^{-1}(1, 0)$. However, $(f^{-1}(a_n)) = \left(1 - \frac{1}{n}\right)$ and so $\lim (f^{-1}(a_n)) = 1$, whereas actually $f^{-1}(1, 0) = 0$.

This previous example points out the importance of the sequential compactness of the domain in the following theorem.

60.4. **Theorem.** *Suppose X and Y are metric spaces, X is sequentially compact, and $f : X \to Y$ is a continuous bijection. Then $f^{-1} : Y \to X$ is continuous.*

Proof. Suppose (y_i) is a sequence in Y and (y_i) converges to y. Consider the sequence (x_i), $x_i = f^{-1}(y_i)$. We need to show that $\lim (x_i) = f^{-1}(y)$. If the opposite is true, then there is an $\varepsilon > 0$ and a subsequence (x_{n_i}) such that $x_{n_i} \in X - N(f^{-1}(y); \varepsilon)$. However, $X - N(f^{-1}(y); \varepsilon)$ is a closed subset of a sequentially compact space and is therefore sequentially compact. Hence, a subsequence (z_i) of (x_{n_i}) converges to a point $z \in X - N(f^{-1}(y); \varepsilon)$. Since f is continuous $(f(z_i))$ converges to $f(z)$. But $(f(z_i))$ is a subsequence of $(f(x_{n_i}))$ and, hence, of (y_i). So $(f(z_i))$ converges to y. So $f(z) = y$. Note that $z \neq f^{-1}(y)$ and $f(f^{-1}(y)) = y$. Thus, two different points map onto y and we have a contradiction to the one-to-oneness of f.

The next proposition is easy to prove using Theorem 51.5 and the reader will be asked to prove it in the next set of exercises. It is a special case of a more general theorem which we shall prove later.

60.5. **Theorem.** *Let $\{(X_i, d_i) : i \in \mathbf{P}_n\}$ be a collection of metric spaces and let (X, d) be the product space for that collection. Then (X, d) is sequentially compact if and only if each (X_i, d_i) is sequentially compact.*

EXERCISES: SEQUENTIAL COMPACTNESS FOR METRIC SPACES

1.   (a) Let (x_i) be a convergent sequence in a metric space. Let x be its limit and let R be the range of (x_i). Is $R \cup \{x\}$ sequentially compact?

    (b) Show that every finite subset of a metric space is sequentially compact.

2.   Prove Theorem 60.2.

3.   Let (x_i) be a sequence in a sequentially compact metric space (X, d). Suppose the sequence satisfies the following condition called the Cauchy condition. For each $\varepsilon > 0$, there is an integer M such that for $m \geq M$ and $n \geq M$, $d(x_m, x_n) < \varepsilon$. Show that (x_i) converges. Show by an example that this

condition is not sufficient for convergence in all metric spaces but recall that it is sufficient in $\mathbf{R}^n$ (see 42.1).

4. Prove that if (X, d) is sequentially compact, then it is totally bounded and, hence, separable.

5. Suppose that (S_i) is a sequence of nonempty sequentially compact subsets of a metric space and $S_{i+1} \subset S_i$ for $i \in \mathbf{P}$. Prove that $\bigcap \{S_i : i \in \mathbf{P}\}$ is a nonempty subset of the space. Prove also that if $\lim (\text{diam} (S_i)) = 0$, then the intersection has exactly one point.

6. In Exercise 5 add to the hypothesis that each S_i is connected and show that the intersection is connected. Show by an example in $\mathbf{R}^2$ that if the sequential compactness hypothesis is removed then the intersection need not be connected.

7. Suppose f is a continuous mapping from a sequentially compact metric space X onto a metric space Y. Prove that Y is sequentially compact.

8. Let $f:(X, d) \to (Y, \rho)$ be a continuous bijection. Suppose f has the property that for each sequentially compact subset, $Z \subset Y$, $f^{-1}[Z]$ is sequentially compact. Show that f is a homeomorphism.

9. Let (X, d) be a metric space. Prove each of the following:
   (a) If H and K are nonempty sequentially compact subsets of X, then there are points $x \in H$ and $y \in K$ such that $d(x, y) = d(H, K)$.
   (b) If H is a nonempty sequentially compact subset of X and $x \in X$, then there is at least one point $z \in H$ which is closest to x.
   (c) If H is a nonempty sequentially compact set, then there exist points x and y in H such that $d(x, y) = \text{diam} (H)$.

10. Prove Theorem 60.5.

61. THE BOLZANO-WEIERSTRASS PROPERTY

The Bolzano-Weierstrass theorem, 40.4, states that every bounded infinite subset of $\mathbf{R}^n$ has at least one limit point. Thus, a closed bounded infinite subset S in $\mathbf{R}^n$ has at least one limit point in S. A sequentially compact set in a metric space has this attribute, which we shall refer to as the *Bolzano-Weierstrass property*.

61.1. Theorem. *Suppose S is sequentially compact. Then every infinite subset of S has at least one limit point in S.*

PROOF. Suppose S is sequentially compact and Q is an infinite subset of S. Then there exists an infinite sequence (a_i) in Q such that $a_i \neq a_j$ if $i \neq j$. Since S

is sequentially compact, there is a subsequence (a_{n_i}) of (a_i) that converges to a point $a_0 \in S$. It is easy to see that a_0 is a limit point of the range of (a_i) and, hence, is a limit point of Q. Thus, S has the Bolzano-Weierstrass property.

Now if the reader will refer back to 41.4, he will recall that if (a_i) is a bounded sequence in $\mathbf{R}^n$, then there is a subsequence of (a_i) that converges. A review of 41.4 shows that the proof depended upon the fact that each bounded infinite subset of $\mathbf{R}^n$ has at least one limit point. This observation suggests the theorem that follows.

61.2. Theorem. *Suppose S has the Bolzano-Weierstrass property (i.e., every infinite subset of S has a limit point in S). Then S is sequentially compact.*
The proof is left to the next set of exercises.

Note that the equivalence of sequential compactness and the Bolzano-Weierstrass properties for metric spaces follows from the previous two theorems.

EXERCISES: THE BOLZANO-WEIERSTRASS PROPERTY

1. Prove that if a subset of a metric space has the Bolzano-Weier-strass property, then it is a closed subset of the space. Do this without using the equivalence of the property to sequential compactness.
2. Prove Theorem 61.2.
3. Point out why it is that sequentially compact metric spaces satisfy the conclusion of the Lindelöf theorem.

62. COMPACTNESS OR FINITE SUBCOVERING PROPERTY

Recall that one version of the Lindelöf theorem asserts that for a separable metric space X, if $\mathcal{K}$ is an open covering of X, then some countable subcollection of $\mathcal{K}$ also covers X. A much stronger property, known as *compactness*, requires that if $\mathcal{K}$ is an open covering of X, then some finite subcollection of $\mathcal{K}$ also covers X. The Heine-Borel theorem in classic analysis asserts that closed bounded subsets of real numbers have this property (see Theorem 43.4). As the reader who proved Theorem 43.4 probably discovered, the proof of the Heine-Borel theorem can be made to depend on the Lindelöf theorem for $\mathbf{R}^n$ and on the fact that closed and bounded subsets of $\mathbf{R}^n$ are sequentially compact (or perhaps the reader used the Bolzano-Weierstrass property). Recall that since a sequentially compact metric space is separable it possesses the Lindelöf property. Thus, we should be able to extend the Heine-Borel theorem to sequentially compact metric spaces. We shall give a proof of this extension.

62.1. Definition. Compact space. *A metric space X is said to be compact provided it has the following property: If $\mathcal{K}$ is an open covering of X, then $\mathcal{K}$ contains a finite subcollection that also covers X.*

62.2. Theorem. *If X is a sequentially compact metric space, then every open covering of X contains a finite subcovering (i.e., X is compact).*

PROOF. Suppose X is sequentially compact. Let $\mathcal{K}$ be an open covering of X. By Exercise 4, page 123, X is separable and, hence, by the Lindelöf theorem for separable metric spaces (58.9), some countable subcollection $\{U_1, U_2, U_3, \ldots\}$ of $\mathcal{K}$ also covers X. The claim is made that for some n, $\{U_1, U_2, \ldots, U_n\}$ also covers X. Otherwise, for each positive integer n, we may choose an

$$x_n \in X - \bigcup \{U_i : i \in \mathbf{P}_n\}.$$

Since X is sequentially compact, the sequence (x_n) has a convergent subsequence (x_{n_i}) that converges to a point $x_0 \in X$. But $\{U_i : i \in \mathbf{P}\}$ is a covering of X so for some j_0, $x_0 \in U_{j_0}$. From the way in which the x_i's were chosen, there are only a finite number of integers n_i for which $x_{n_i} \in U_{j_0}$. This is a contradiction.

62.3. Theorem. *If X is compact, then it has the Bolzano-Weierstrass property.*

PROOF. Suppose X is compact and K is a subset of X that does not have a limit point in X. We shall complete the proof by showing that K is finite. This will imply that every infinite subset does have a limit point. If K has no limit point in X, then for each $x \in X$, x is in an open set U_x such that $(U_x \cap K) - \{x\} = \varnothing$. Then since $\{U_x : x \in X\}$ is an open covering of X, some subcollection $\{U_{x_i} : i \in \mathbf{P}_n\}$ also covers X. But each U_{x_i} has in it at most one point of K. Hence, K is a finite set.

Reviewing the last two sections (61 and 62), we see that we have shown that for metric spaces *sequential compactness*, *compactness*, and the *Bolzano-Weierstrass* properties are equivalent. In proving propositions about sequentially compact metric spaces it may be that one of the other properties is easier to work with than sequential compactness. When we study spaces more general than metric spaces, we shall see that these properties are not equivalent in every setting.

In Exercise 4, page 123, the reader was asked to prove that sequentially compact metric spaces are totally bounded. Although this exercise was instructive, the total boundedness follows at once from the finite covering property of compactness as is shown in the next proof.

62.4. Theorem. *If (X, d) is compact, then it is totally bounded.*

PROOF. Let $\varepsilon > 0$. The collection $\mathcal{U} = \{N(x; \varepsilon) : x \in X\}$ is an open covering for X. Since X is compact, there is a finite subcollection $\{N(x_i; \varepsilon) : i \in \mathbf{P}_n\}$ that covers X. Then $F_\varepsilon = \{x_i : i \in \mathbf{P}_n\}$ is obviously a finite ε-net for X.

The reader may recall from calculus that if a real-valued function is continuous on a closed interval, then it is uniformly continuous. This fact follows from the following more general theorem.

62.5. Theorem. *Suppose $f : (X, d) \to (Y, \rho)$ is continuous. If (X, d) is compact, then f is uniformly continuous.*

PROOF. Let $\varepsilon > 0$. For each $x \in X$, choose a $\delta(x) > 0$ such that if $d(y, x) < \delta(x)$, then $\rho(f(y), f(x)) < \dfrac{\varepsilon}{2}$. $\{N(x; \frac{1}{2}\delta(x)) : x \in X\}$ is an open covering of X. Hence, there is a finite subcollection $\{N(x_i; \frac{1}{2}\delta(x_i)) : i \in \mathbf{P}_n\}$ that also covers X. Let $\delta = $ min. $\{\frac{1}{2}\delta(x_i) : i \in \mathbf{P}_n\}$. We complete the proof by showing that if a and b are elements of X such that $d(a, b) < \delta$, then $d(f(a), f(b)) < \varepsilon$. For any two such points a

and b, $a \in N(x_j; \frac{1}{2}\delta(x_j))$ for some positive integer j. Then $d(b, x_j) \leq d(a, x_j) - d(a, b) < \frac{1}{2}\delta(x_j) + \delta \leq \delta(x_j)$. From the definition of $\delta(x_j)$ it then follows that $\rho(f(a), f(x_j)) < \frac{\varepsilon}{2}$ and $\rho(f(b), f(x_j)) < \frac{\varepsilon}{2}$; so $\rho(f(a), f(b)) < \varepsilon$.

In Exercise 7, page 123, the reader was asked to prove that sequential compactness is invariant under continuous mappings. Since compactness and sequential compactness are equivalent for metric spaces, it follows that compactness is invariant under continuous mappings. However, it is instructive to include a proof of the invariance of compactness that is independent of the corresponding result for sequential compactness.

62.6. Theorem. *Suppose X and Y are metric spaces, X is compact, and $f: X \to Y$ is a continuous surjection. Then Y is compact.*

PROOF. Let $\mathcal{U} = \{U_\lambda : \lambda \in \Lambda\}$ be an open covering of Y. Then since f is continuous, $\mathcal{K} = \{f^{-1}[U_\lambda] : \lambda \in \Lambda\}$ is an open covering of X. Since X is compact, $\mathcal{K}$ contains a finite subcollection $\{f^{-1}[U_{\lambda(i)}] : i \in \mathbf{P}_n\}$ that covers X. Since f is a surjection, for each $\lambda(i)$, $f[f^{-1}[U_{\lambda(i)}]] = U_{\lambda(i)}$. Hence,

$$f[\bigcup \{f^{-1}[U_{\lambda(i)}] : i \in \mathbf{P}_n\}]$$
$$= \bigcup \{f[f^{-1}[U_{\lambda(i)}]] : i \in \mathbf{P}_n\}$$
$$= \bigcup \{U_{\lambda(i)} : i \in \mathbf{P}_n\} = Y.$$

Thus, $\mathcal{U}$ contains a finite subcollection $\{U_{\lambda(i)} : i \in \mathbf{P}_n\}$ that covers Y. The space Y is therefore compact.

EXERCISES: COMPACTNESS OR FINITE SUBCOVERING PROPERTY

1. Suppose (X, d) is a metric space and S is a subset of X. Prove that S is compact if and only if every covering $\mathcal{K}$ of S (i.e., $S \subset \bigcup \mathcal{K}$) consisting of open subsets of X contains a finite subcovering of S.

2. Prove that $\mathbf{R}^n$ can be expressed as the union of a countable collection $\{K_i : i \in \mathbf{P}\}$ of compact sets such that $K_i \subset K_{i+1}$ for $i \in \mathbf{P}$.

3. Let $\mathcal{K}$ be an open covering for a compact metric space (X, d). Prove that there is an $\varepsilon > 0$ such that if $d(x, y) < \varepsilon$, then there is a $U \in \mathcal{K}$ such that x and y both belong to U. (Such an ε is called a Lebesgue number for the covering $\mathcal{K}$.) Use this result to give an alternate proof of Theorem 62.5.

4. Suppose X is a subset of $\mathbf{R}^n$ that has the following property: There is an $\varepsilon > 0$ such that if x and y are two distinct points in X, then $|x - y| \geq \varepsilon$. Prove that X is finite or unbounded.

5. Suppose $f: X \to X$ is a continuous mapping from a compact metric space into itself. Let $X_1 = f[X]$ and for each i, $X_{i+1} = f[X_i]$. Define A to be $\bigcap \{X_i : i \in \mathbf{P}\}$. Prove that $f[A] = A$.

6. Suppose f is a continuous mapping from a compact metric space X into a metric space Y. Prove that if F is a closed subset of X, then $f[F]$ is a closed subset of Y.

7. Suppose f is a one-to-one continuous mapping from the compact metric space X onto the metric space Y. Prove that if U is an open subset of X, then $f[U]$ is an open subset of Y. Explain why this shows that f^{-1} is continuous.

8. Let f be a continuous mapping from a compact metric space onto a metric space Y. Suppose it is known that for each $y \in Y, f^{-1}[y]$ is connected. Let K be a connected subset of Y. Show that $f^{-1}[K]$ is connected. (See [33], (2.2) page 138.)

9. Re-prove Theorem 62.5 by using the sequential compactness of X rather than the equivalent compactness property.

10. Suppose $f: \mathbf{R}^m \to \mathbf{R}^n$ is continuous. Is f necessarily uniformly continuous on every bounded subset of $\mathbf{R}^m$?

63. COMPLETE METRIC SPACES

63.1. Definition. Cauchy sequences. *Let (X, d) be a metric space. If (a_n) is a sequence in X that satisfies the following condition, then (a_n) is said to be a Cauchy sequence (C.S.) in (X, d).*

For each $\varepsilon > 0$, there is a positive integer N such that if $m \geq N$ and $n \geq N$, then $d(a_m, a_n) < \varepsilon$.

Recall that in 42 we proved that a sequence in $\mathbf{R}^n$ converges if and only if it is a Cauchy sequence. So, in $\mathbf{R}^n$, the class of Cauchy sequences and the class of convergent sequences are the same. For metric spaces, every convergent sequence is a Cauchy sequence. (The first part of the proof of 42.1(a) is a direct verification of this fact.) However, it is not true for every metric space that Cauchy sequences necessarily converge. To verify this fact, refer to Exercise 3, page 122. Those metric spaces for which every Cauchy sequence converges have many other useful properties because of this important property.

63.2. Definition. Complete metric space. *A metric space (X, d) is complete provided every Cauchy sequence in (X, d) converges.*

It should be clear to the reader that a closed subspace of a complete metric space is complete. Also suppose that (X, d) is a metric space and Y is a subset of X that is not closed. Then there is a sequence (y_i) in Y that converges to a point $x_0 \in X - Y$. Thus, (y_i) is a Cauchy sequence in Y that does not converge in (Y, d). Hence, (Y, d) is not complete. These remarks are summarized in the following statement.

63.3. Theorem. *Suppose that X is a metric space and $S \subset X$. If S is a complete subset of X, then S is closed. Further, if X is complete and S is closed, then S is complete.*

It is important to notice that *completeness is not a topological property*. Recall that the real line **R** is homeomorphic to the open interval $(0, 1) \subset$ **R** (Exercise 10, page 102). Although **R** is complete, it follows from Theorem 63.3 that the real open interval $(0, 1)$ is not complete.

The results of Exercise 3, page 122, in terms of completeness, together with Theorem 62.4, give us the following.

63.4.   Theorem.   *If (X, d) is a compact metric space, then it is complete and totally bounded.*

Note that **R**^{n} is complete but not compact. Thus, for metric spaces the property of completeness is weaker than the property of compactness. Nevertheless, one might suspect that some of the theorems for compact metric spaces would suggest possible results for complete metric spaces. Also, because of 63.4, one might wonder if completeness taken together with total boundedness implies compactness. If it does, it is very interesting since taken as separate properties neither total boundedness nor completeness is a topological property, whereas compactness is a topological property. We shall pursue this matter further in the exercises and in a subsequent section.

The proofs of the following two theorems involving uniform continuity are left to the reader in the next set of exercises.

63.5.   Theorem.   *If $f:(X, d) \to (Y, \rho)$ is uniformly continuous and (x_i) is a Cauchy sequence in (X, d), then $(f(x_i))$ is a Cauchy sequence in (Y, ρ).*

63.6.   Theorem.   *Suppose (X, d) and (Y, ρ) are metric spaces and (Y, ρ) is complete. Let D be a dense subset of X. If $f:(D, d) \to (Y, \rho)$ is uniformly continuous, then there is a unique continuous extension $f^*:(X, d) \to (Y, \rho)$ of f (i.e., $f^* \mid D = f$). Furthermore, f^* is uniformly continuous on X. Moreover, if f is an isometry, so is f^*.*

EXERCISES: COMPLETE METRIC SPACES

1.   Let X be a metric space. Suppose (x_i) is a Cauchy sequence in X.

   (a) Prove that if the range of (x_i) is not an infinite set, then for some N, $x_N = x_{N+h}$ for each $h \in$ **P**.

   (b) Prove that the range of (x_i) is a bounded set.

   (c) Prove that if a subsequence of (x_i) converges, then (x_i) also converges.

2.   Suppose (X, d) is a complete and totally bounded metric space. Is (X, d) necessarily compact?

3.   Prove Theorem 63.5.

4.   Prove Theorem 63.6.

64.   NESTED SEQUENCES OF SETS FOR COMPLETE SPACES

In view of the fact that sequentially compact spaces are complete, the following is a generalization of the statement in Exercise 5, page 123.

64.1. **Theorem.** *Let (X, d) be a complete metric space. Suppose (A_i) is a decreasing sequence of nonempty closed subsets of X such that $\lim (\text{diam } (A_i)) = 0$. Then there exists one and only one point x in $\bigcap \{A_i\}$.*

PROOF. Let $x_i \in A_i$. We shall show that (x_i) is a Cauchy sequence and that its limit is the point in $\bigcap \{A_i\}$.

Let $\varepsilon > 0$. There is a positive integer N such that $\text{diam } (A_N) < \varepsilon$. Since the sequence (A_i) is a decreasing sequence of sets, for $m \geq N$ and $n \geq N$, x_m and x_n are elements of A_N and we have $d(x_m, x_n) < \varepsilon$. Thus, (x_i) is a Cauchy sequence and has a limit x_0. Now for each $j \in \mathbf{P}$, the sequence $(x_{j+i})_{i=1}^{\infty}$ is a sequence in A_j such that $\lim_{i \to \infty} x_{j+i} = x_0$. Since each A_j is closed, $x_0 \in A_j$ and, thus, $x_0 \in \bigcap \{A_j : j \in \mathbf{P}\}$. It is clear that if there were another point y in the intersection, then $\text{diam } (A_j) \geq d(x_0, y) > 0$ and we would have a contradiction to the fact that $\lim (\text{diam } (A_j)) = 0$.

As an application of the previous theorem, we prove next a theorem that guarantees that, under certain conditions, the intersection of a collection of sets is not only nonempty but dense.

64.2. **Theorem (Baire).** *Let (X, d) be a complete metric space. Let $\{D_i : i \in \mathbf{P}\}$ be a countable collection of open subsets of X, each dense in X. Then $\bigcap \{D_i : i \in \mathbf{P}\}$ is dense in X.*

PROOF. Let $p \in X$ and $r > 0$. To show that $\bigcap \{D_i : i \in \mathbf{P}\}$ is dense in X, it will be sufficient to show that $N(p; r) \cap [\bigcap \{D_i : i \in \mathbf{P}\}] \neq \varnothing$. We will show that we can find a nested sequence of balls $(B(p_i; r_i))$ such that

64.2(a). $\qquad\qquad B(p_i; r_i) \subset N(p; r)$

64.2(b). $\qquad\qquad r_i < 1/i$

64.2(c). $\qquad\qquad B(p_i; r_i) \subset D_i.$

The fact that the sequence of closed sets $(B(p_i; r_i))$ is a decreasing sequence and $r_i < \dfrac{1}{i}$ will allow us to use the previous theorem to conclude that there is a $z \in \bigcap \{B(p_i; r_i)\}$. Then 64.2(a) and 64.2(c) will imply that $z \in [\bigcap \{D_i\}] \cap N(p; r)$. We complete the proof by showing by induction that such a sequence of balls exists.

Since D_1 is dense and open, $D_1 \cap N(p; r)$ is a nonempty open subset. Hence, there is a p_1 and $r_1 < 1$ such that

$$B(p_1; r_1) \subset N(p; r) \cap D_1$$

so that 64.2(a) and 64.2(c) are satisfied by this ball.

Next assume that for $i \in \mathbf{P}_h$, $B(p_i; r_i)$ has been chosen to satisfy 64.2(b) and 64.2(c) and so that $B(p_1; r_1) \supset B(p_2; r_2) \supset \cdots \supset B(p_h; r_h)$. Then, since D_{h+1} is open and dense, $D_{h+1} \cap N(p_h; r_h)$ is a nonempty open subset. Thus, there is a p_{h+1} and an $r_{h+1} < \dfrac{1}{h+1}$ such that the closed ball $B(p_{h+1}; r_{h+1}) \subset D_{h+1} \cap N(p_h; r_h)$. Hence, 64.2(b) and 64.2(c) are satisfied for $i = h + 1$ and, further, $B(p_1; r_1) \supset B(p_2; r_2) \supset \cdots \supset B(p_{h+1}; r_{h+1})$. By induction there is a nested sequence $(B(p_i; r_i))$

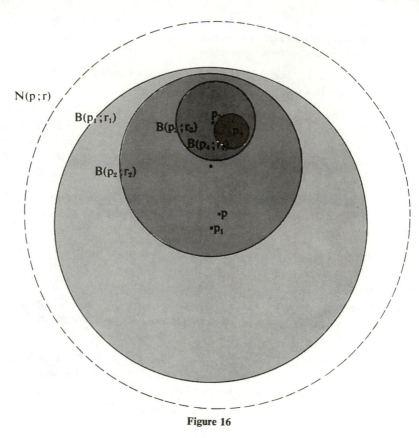

Figure 16

satisfying 64.2(b) and 64.2(c); recall that 64.2(a) is also satisfied. The proof has thus been completed.

EXERCISES: NESTED SEQUENCES OF SETS FOR COMPLETE SPACES

1. Apply the Baire theorem to prove the following: Suppose X is a nonempty complete metric space and $X = \bigcup \{F_i : i \in \mathbf{P}\}$, where each F_i is closed. Then at least one of the F_i's contains a nonempty open subset of X.

2. Suppose X is a complete metric space and W is a nonempty open subset of X. If $\{F_i : i \in \mathbf{P}\}$ is a countable collection of closed subsets of X such that $W = \bigcup \{F_i : i \in \mathbf{P}\}$, then at least one of the F_i's contains a nonempty open subset of X. (The proposition in this exercise is one form of a theorem known as the Baire Category theorem.)

3. In 64.1 it was proved that a complete metric space had the following property: If (A_i) is a decreasing sequence of nonempty closed subsets of the space such that $\lim (\operatorname{diam} (A_i)) = 0$, then $\bigcap \{A_i : i \in \mathbf{P}\} \neq \varnothing$. Prove that if a metric space has this property, then the space is complete.

65. ANOTHER CHARACTERIZATION OF COMPACT METRIC SPACES

We have already seen that a compact metric space is complete and totally bounded (see 63.4). Thus, a necessary condition for compactness of a metric space is that it be both totally bounded and complete. We show next that this is also a sufficient condition. This gives an affirmative answer to the question raised in Exercise 2, page 128 .

65.1. Theorem. *A necessary and sufficient condition that a metric space be compact is that it be complete and totally bounded.*

PROOF. The necessity is given by Theorem 63.4. We shall prove the sufficiency by assuming that X is a totally bounded and complete metric space and by showing that it has the Bolzano-Weierstrass property. (Our proof will be based on the proof of the Bolzano-Weierstrass theorem for $\mathbf{R}^n$ (40.3). Instead of using the nested interval theorem as in the proof of 40.3, we shall use the nested subset theorem (64.1) for complete metric spaces.)

Let Z be an infinite subset of X. Since X is totally bounded, X is the union of a finite collection of (closed) balls each of whose radius is less than $\frac{1}{2}$. For one of these balls, B_1, it must be true that $B_1 \cap Z$ is an infinite set. Assume that closed sets $B_1, B_2, \ldots, B_h$ have been chosen so that $B_i \supset B_{i+1}$, diam $(B_i) < \frac{1}{i}$ and $B_i \cap Z$ is an infinite set. Now a finite collection of balls each of whose diameter is less than $\dfrac{1}{h+1}$ covers X. Since $B_h \cap Z$ is infinite, one of these balls B^*_{h+1} must be such that $B^*_{h+1} \cap B_h \cap Z$ is infinite. Let $B_{h+1} = B^*_{h+1} \cap B_h$. Note that B_{h+1} is closed, diam $(B_{h+1}) < \dfrac{1}{h+1}$, $B_{h+1} \cap Z$ is infinite, and $B_h \supset B_{h+1}$. Thus, by induction, we find an infinite sequence of closed sets (B_i) such that $B_i \supset B_{i+1}$, diam $(B_i) <$ $1/i$ and $B_i \cap Z$ is infinite. By Theorem 64.1, $\bigcap \{B_i : i \in \mathbf{P}\}$ has in it a single element z. As in the proof of the Bolzano-Weierstrass theorem for $\mathbf{R}^n$, it is now easy to show that z is a limit point of Z. We leave this detail to the reader.

EXERCISE: ANOTHER CHARACTERIZATION OF COMPACT METRIC SPACES

1. Recall that subsets of $\mathbf{R}^n$ that are closed and bounded are compact. The property of closedness together with boundedness does not characterize compactness in general (see the introductory paragraph in 60).
    (a) Give an example of a complete metric space for which there exists a closed and bounded subset that is not compact.
    (b) Prove that in a complete metric space, a set is compact if and only if it is closed and totally bounded.

66. COMPLETION OF A METRIC SPACE

In 53.9 we defined an isometry to be a surjection $f : (X, d) \rightarrow (Y, \rho)$ such that the distance between points is preserved under f, i.e., $\rho(f(x), f(y)) = d(x, y)$

for all x and y in X. It should be clear that if f is an isometry from (X, d) onto (Y, ρ), then f^{-1} is an isometry from (Y, ρ) onto (X, d). If (X, d) and (Y, ρ) are metric spaces for which there exists an isometry that maps (X, d) onto (Y, ρ), then we shall speak of (X, d) and (Y, ρ) as being *isometric*. In such a case, we shall sometimes refer to one of the spaces as being an *isometric copy* of the other. If Ψ maps (X_1, d_1) isometrically onto (X_2, d_2), then we may regard (X_1, d_1) as essentially the same space as (X_2, d_2) with the name of a point x in X_1 being changed to $\Psi(x)$. For example, the real line $\mathbf{R}$ can be mapped isometrically onto the x-axis of the plane with the mapping $\Psi : \mathbf{R} \to \mathbf{R} \times \mathbf{R}$ given by $\Psi(x) = (x, 0)$. We can then think of the x-axis as the real line by identifying each point $(x, 0)$ in the plane with the corresponding real number x.

66.1. Definition. Isometric embedding. *A mapping* $f:(X, d) \to (Y, \rho)$ *is said to be an isometric embedding of* (X, d) *in* (Y, ρ) *provided* $f:(X, d) \to (f[X], \rho)$ *is an isometry.*

We see from this definition that if $f:(X, d) \to (Y, \rho)$ is an isometric embedding, then Y contains an isometric copy of X, namely, $f[X]$. Isometric embeddings will be of special interest to us in the discussion of the *completion process*.

We have seen examples of spaces that are not complete. For example, a sub-space of a metric space cannot be complete if it is not closed in the space that contains it. We shall show that although a metric space (X, d) may not be complete, there exists a complete metric space (X^*, d^*) which contains a dense isometric copy of (X, d). We shall do this by first defining an equivalence relation E for the d-Cauchy sequences of X. A suitable metric d^* will then be defined for the collection X^* of all E-equivalence classes. Finally, we shall define an isometric embedding map $\Psi:(X, d) \to (X^*, d^*)$ such that $\Psi[X]$ is dense in X^*. Thus, we shall be able to think of a metric space (X, d) as being a dense subset of some complete metric space (X^*, d^*).

66.2. Definition. Completion of a metric space. *Suppose that* (Y, ρ) *is a complete metric space and* $\Psi : (X, d) \to (Y, \rho)$ *is an isometric embedding of* (X, d) *in* (Y, ρ) *such that* $\Psi[X]$ *is a dense subset of* Y. *Then the pair* $(\Psi, (Y, \rho))$ *is called a completion of* (X, d).

66.3. Theorem. *Let* (X, d) *be a metric space. Then there exists a completion* $(\Psi, (X^*, d^*))$ *of* (X, d).

The proof of Theorem 66.3 will be broken up into a number of parts. The verification of each part for which a proof is not given should be considered as an exercise to be done before going on to the next part. Some of the steps are designed more to help the reader to get a feel for the procedure than to provide a necessary part of the development.

In the statements that follow (X, d) is a metric space and C.S. is an abbreviation for Cauchy sequence. The symbol $\mathscr{S}$ will be used to denote the collection of all d-Cauchy sequences in X. The reader should notice the dependence of the completion of (X, d) on the metric d.

66.4. If (α_i) and (β_i) are convergent sequences in X, then $\lim (\alpha_i) = \lim (\beta_i)$ if and only if $\lim (d(\alpha_i, \beta_i)) = 0$.

66.5. If (α_i) is a C.S. (not necessarily convergent) and (β_i) is a sequence such that $\lim (d(\alpha_i, \beta_i)) = 0$, then (β_i) is also a C.S.

66.6. If (α_i) is a C.S. in X and $(\alpha_{n(i)})$ is a subsequence of (α_i), then $(\alpha_{n(i)})$ is also a C.S. and $\lim (d(\alpha_i, \alpha_{n(i)})) = 0$.

66.7. If (α_i) and (β_i) are Cauchy sequences in X, then $(d(\alpha_i, \beta_i))$ is a C.S. of nonnegative real numbers and, hence, has a limit, and the limit is nonnegative.

66.8. Define the relation E in the set $\mathscr{S}$ of all Cauchy sequences in X as follows: For $(\alpha_i) \in \mathscr{S}$ and $(\beta_i) \in \mathscr{S}$, $(\alpha_i) E (\beta_i)$ if and only if $\lim (d(\alpha_i, \beta_i)) = 0$. The relation E is an equivalence relation in $\mathscr{S}$.

66.9. Let (α_i) be a convergent sequence in X with the limit x. Let (β_i) be the sequence each of whose terms is x. Then $(\alpha_i) E (\beta_i)$.

66.10. Suppose that (α_i), $(\tilde{\alpha}_i)$, (β_i), and $(\tilde{\beta}_i)$ are in $\mathscr{S}$. If $(\alpha_i) E (\tilde{\alpha}_i)$ and $(\beta_i) E (\tilde{\beta}_i)$, then $\lim (d(\alpha_i, \beta_i)) = \lim (d(\tilde{\alpha}_i, \tilde{\beta}_i))$.

66.11. Let X^* be the collection of all E-equivalence classes. For each $\alpha = (\alpha_i) \in \mathscr{S}$, $E[\alpha]$ denotes the E-equivalence class that has α as an element. On the basis of 66.10 and some of the previous remarks, we can now make the following definition:

For $E[\alpha]$ and $E[\beta]$ in X^*, let

66.11(a). $$d^*(E[\alpha], E[\beta]) = \lim (d(\alpha_i, \beta_i)).$$

Then d^* as defined in 66.11(a) is a metric for X^*.

66.12. For each $x \in X$, define $\gamma(x) \in \mathscr{S}$ as the C.S. in X given by $\gamma(x)_i = x$ for each $i \in \mathbf{P}$, i.e., $\gamma(x)$ is the sequence each of whose terms is x. Define $\Psi : (X, d) \to (X^*, d^*)$ by

$$\Psi(x) = E[\gamma(x)].$$

The mapping Ψ is an isometry; that is, $d(x, y) = d^*(E[\gamma(x)], E[\gamma(y)])$ for x and y in X.

66.13. With $\Psi : (X, d) \to (X^*, d^*)$ as in 66.12, $\Psi[X]$ is dense in X^*.

PROOF. Let $E[\beta] \in X^*$. We shall prove the proposition by showing that there is a sequence in $\Psi[X]$ that converges to $E[\beta]$. This will show that $E[\beta] \in \mathrm{cl}\,(\Psi[X])$.

For each term β_i of β, let $\gamma(\beta_i)$ be the element in $\mathscr{S}$ given, as in 66.12, by $\gamma(\beta_i)_j = \beta_i$ for each $j \in \mathbf{P}$. We complete the proof by showing that

$$\lim (d^*(\Psi(\beta_i), E[\beta])) = 0.$$

To see this, first note that $d^*(\Psi(\beta_i), E[\beta]) = d^*(E[\gamma(\beta_i)], E[\beta])$.

Let $\varepsilon > 0$. Since (β_i) is a C.S. in X, there is a positive integer I such that for $i \geq I$, and each $h \in \mathbf{P}$,

$$d(\beta_i, \beta_{i+h}) < \frac{\varepsilon}{2}.$$

Let $i \geq I$. Now i is fixed. From the definition of d^*, there is an integer J such that for $j \geq J$

$$d^*(E[\gamma(\beta_i)], E[\beta]) < d(\gamma(\beta_i)_j, \beta_j) + \frac{\varepsilon}{2}.$$

Thus, if we choose $j \geq \max \{I, J\}$,

$$d^*(E[\gamma(\beta_i)], E[\beta]) < d(\gamma(\beta_i)_j, \beta_j) + \frac{\varepsilon}{2}$$

$$= d(\beta_i, \beta_j) + \frac{\varepsilon}{2}$$

$$< \frac{\varepsilon}{2} + \frac{\varepsilon}{2} = \varepsilon.$$

Thus, we have shown that for $i \geq I$, $d^*(E[\gamma(\beta_i)], E[\beta]) < \varepsilon$ and, consequently, we have shown that $\lim (d^*(E[\gamma(\beta_i)], E[\beta])) = 0$. Hence, $\lim (d^*(\Psi(\beta_i), E[\beta])) = 0$.

66.14. The space (X^*, d^*) is complete.

PROOF. Let (Λ_m) be a C.S. in X^*. For each $m \in \mathbf{P}$ we choose a representative $x_m \in \Lambda_m$ so that we may write $\Lambda_m = E[x_m]$. Notice that for each positive integer m, x_m is a C.S. in X so that we may write $x_m = (x_{m,i})_{i=1}^\infty$. We shall prove that (X^*, d^*) is complete by exhibiting a $z \in \mathscr{S}$ such that $\lim (E[x_m]) = E[z]$. We shall choose z by a modified diagonal process as follows:

For each m, $(x_{m,j})_{j=1}^\infty$ is a C.S. in X. Then there is an integer $N(m)$ for each $m \in \mathbf{P}$ such that for $j \geq N(m)$,

66.14(a).

$$d(x_{m,j}, x_{m,N(m)}) < \frac{1}{m} .$$

Define $z_m = x_{m,N(m)}$ and set $z = (z_m)_{m=1}^\infty$. We shall show next that z is a C.S. in X and finally complete the proof by proving that $\lim (E[x_m]) = E[z]$.

To show that $(z_m) \in \mathscr{S}$, let $\varepsilon > 0$. We may choose a positive integer N^* such that

66.14(b).

$$\frac{1}{N^*} < \frac{\varepsilon}{4}$$

and such that for $m \geq N^*$ and $n \geq N^*$

66.14(c).

$$d^*(E[x_m], E[x_n]) < \frac{\varepsilon}{4} .$$

Next let $m \geq N^*$ and $n \geq N^*$. We show that $d(z_m, z_n) < \varepsilon$ so that (z_m) is a C.S. in X. With m and n now fixed, because of 66.14(c) and the definition of d^* we may choose an i so large that

$$d(x_{m,i}, x_{n,i}) < \frac{\varepsilon}{2}$$

and we also choose the i to satisfy

$$i \geq N(m) \quad \text{and} \quad i \geq N(n).$$

Then, from 66.14(a),

$$d(z_m, z_n) = d(x_{m,N(m)}, x_{n,N(n)})$$

$$\leq d(x_{m,N(m)}, x_{m,i}) + d(x_{m,i}, x_{n,i}) + d(x_{n,i}, x_{n,N(n)})$$

$$< \frac{1}{m} + \frac{\varepsilon}{2} + \frac{1}{n}$$

$$< \frac{\varepsilon}{4} + \frac{\varepsilon}{2} + \frac{\varepsilon}{4} = \varepsilon.$$

Now that we have shown that (z_m) is a C.S. in X, it follows that $E[z] \in X^*$ and we show next that $\lim (E[x_m]) = E[z]$. To do this, let $\varepsilon > 0$. Since z is a C.S. in X, there is a positive integer M such that

66.14(d).

$$d(z_m, z_n) < \frac{\varepsilon}{2} \quad \text{for} \quad m \geq M, n \geq M.$$

Next, let N be an integer such that

66.14(e).

$$N > \max \left\{M, \frac{4}{\varepsilon}\right\}.$$

We complete the proof by showing that for $m \geq N$, $d^*(E[x_m], E[z]) < \varepsilon$.

Let $m \geq N$. From the definition of d^*, there is an integer I such that for $i \geq I$,

66.14(f).

$$d^*(E[x_m], E[z]) < d(x_{m,i}, z_i) + \frac{\varepsilon}{4}.$$

Choose a positive integer i such that

66.14(g). $i \geq \max \{I, N(m), M\}.$

Since $i \geq I$, from 66.14-(f),

$$d^*(E[x_m], E[z]) < d(x_{m,i}, z_i) + \frac{\varepsilon}{4}$$

$$= d(x_{m,i}, x_{i,N(i)}) + \frac{\varepsilon}{4}$$

$$\leq d(x_{m,i}, x_{m,N(m)}) + d(x_{m,N(m)}, x_{i,N(i)}) + \frac{\varepsilon}{4}$$

$$= d(x_{m,i}, x_{m,N(m)}) + d(z_m, z_i) + \frac{\varepsilon}{4}$$

Then since $i \geq N(m)$, $m \geq M$, and $i \geq M$ from 66.14(a) and 66.14(d), it follows that we may write

$$d^*(E[x_m], E[z]) < \frac{1}{m} + \frac{\varepsilon}{2} + \frac{\varepsilon}{4}.$$

Finally, since $m \geq N$, from 66.14(e) we get

$$d^*(E[x_m], E[z]) < \frac{\varepsilon}{4} + \frac{\varepsilon}{2} + \frac{\varepsilon}{4} = \varepsilon.$$

Hence, (X^*, d^*) is complete.

Reviewing what we have done so far, we see that by 66.12 and 66.14, the mapping Ψ is an isometric embedding of (X, d) in the complete space (X^*, d^*). Furthermore, by 66.13, $\Psi[X]$ is dense in X^*. Hence, $(\Psi, (X^*, d^*))$ is a completion of (X, d) and the proof of Theorem 66.3 is complete.

One might wonder if there are other completions of (X, d). In a certain sense, any other completion is essentially the same. If $(\Psi_1, (X_1, d_1))$ and $(\Psi_2, (X_2, d_2))$ are two completions of a space (X, d), it can be shown that there exists an isometry h from (X_1, d_1) onto (X_2, d_2) that relates Ψ_1 to Ψ_2 as indicated by the following commutative diagram:

These remarks are summarized in the next statement, the proof of which is left as an exercise.

66.15. Theorem. *Suppose that $(\Psi_1, (X_1, d_1))$ and $(\Psi_2, (X_2, d_2))$ are completions of (X, d). Then there exists an isometry h from (X_1, d_1) onto (X_2, d_2) such that $h \circ \Psi_1 = \Psi_2$.*

Hint for the proof of 66.15: Before trying to prove the theorem, review the statement of Theorem 63.6. Define h first on $\Psi_1[X]$ in a way that will force $h \circ \Psi_1$ to equal Ψ_2. Then extend h by making use of 63.6.

It is interesting to note that if (X, d) is itself complete, then $(i, (X, d))$ is a completion of (X, d), where $i: (X, d) \to (X, d)$ is the identity mapping. Hence, by 66.15, the space (X^*, d^*), constructed in the proof of Theorem 66.3, is isometric to (X, d).

EXERCISES: COMPLETION OF A METRIC SPACE

1. Give proofs for items 66.4 through 66.12.

2. Prove Theorem 66.15.

3. Suppose that (X, d) is a metric space. Prove that (X, d) is complete if and only if (X, d) is a closed subset of every metric space in which it can be isometrically embedded (i.e., for every isometric embedding $\Psi: (X, d) \to (Y, \rho)$, $\Psi[X]$ is a closed subset of (Y, ρ)).

67. SEQUENCES OF MAPPINGS INTO A METRIC SPACE

Recall from the study of calculus that for each $x \in \mathbf{R}$, $\sin x = \sum\limits_{n=0}^{\infty} \dfrac{(-1)^n x^{2n+1}}{(2n+1)!}$.

What this means is that the sine function is the limit of a sequence of polynomials

$(S_n)_{n=0}^{\infty}$, where $S_n(x) = \sum_{j=0}^{n} \dfrac{(-1)^j x^{2j+1}}{(2j+1)!}$. The reader may recall also that the sequence (S_n) converges to the sine function uniformly on every closed and bounded interval. In this section we will deal with the notions of convergent and of uniformly convergent sequences of functions with values in a metric space. In doing this we will be able to generalize many familiar phenomena where the functions have their ranges in **R**.

67.1. **Definition.** *Let $(f_n:X \to (Y, d))$ be a sequence of maps defined on a set X, with functional values in a metric space (Y, d). Suppose for some $x \in X$, $\lim (f_n(x))$ exists. We then say that (f_n) converges at x. If the sequence of maps (f_n) converges for each $x \in X$, and we let $f(x) = \lim (f_i(x))$ for each $x \in X$, then $f(x)$ defines a map $f:X \to (Y, d)$. In that case we say that $(f_n:X \to (Y, d))$ converges pointwise (or converges) to the map $f:X \to (Y, d)$.*

The reader should note that the ranges of the mappings f_i and f are contained in a common metric space. However, the common domain X need not be a metric space but only a set.

Next suppose $(f_n:(X, d) \to (Y, \rho))$ is a sequence of continuous maps that converges pointwise to a map $f:(X, d) \to (Y, \rho)$. A natural question is whether f inherits the continuity from the f_n's. The following example answers the question in the negative.

67.2. *Example.* For each $i \in \mathbf{P}$, let $f_i:[0, 1] \to \mathbf{R}$ be given by $f_i(x) = x^i$. Note that the limit exists and is given by $f:[0, 1] \to \mathbf{R}$, where $f(x) = 0$ if $x \in [0, 1)$ and $f(1) = 1$. We see that f is not continuous.

There is a stronger kind of convergence, uniform convergence, that is strong enough to guarantee that the continuity of f is inherited from that of the f_i's. We define that concept next. (See Figure 17.)

67.3. **Definition. Uniform convergence.** *Let $(f_n:X \to (Y, \rho))$ be a sequence of maps. Suppose (f_n) converges pointwise to f and for each $\varepsilon > 0$, there is an integer N such that for $n \geq N$, $\rho(f_n(x), f(x)) < \varepsilon$ for all $x \in X$. Then (f_n) is said to converge uniformly to f on X.*

We see that pointwise convergence requires only that for each x and $\varepsilon > 0$ a suitable N be determined such that for $n \geq N$, $\rho(f_n(x), f(x)) < \varepsilon$, whereas for uniform convergence the N can be chosen independent of the $x \in X$. We should notice also that if $(f_n:X \to (Y, \rho))$ converges uniformly to $f:X \to (Y, \rho)$, then for each subset X^* of X it is true that $(f_n \,|\, X^*:X^* \to (Y, \rho))$ converges uniformly to $f \,|\, X^*:X^* \to (Y, \rho)$.

In the next theorem convergence of a sequence of continuous functions is considered. Hence, the theorem is stated in a setting in which both the domain and range are metric spaces.

67.4. **Theorem.** *Let $(f_n:(X, d) \to (Y, \rho))$ be a sequence of continuous mappings that converges uniformly to $f:(X, d) \to (Y, \rho)$. Then f is continuous.*

Proof. Let $x_0 \in X$ and let $\varepsilon > 0$. By uniform convergence, there is an integer n such that $\rho(f(x), f_n(x)) < \dfrac{\varepsilon}{3}$ for all $x \in X$. Then, by the continuity of f_n there is an open set U containing x_0 such that $\rho(f_n(x), f_n(x_0)) < \dfrac{\varepsilon}{3}$ for each $x \in U$.

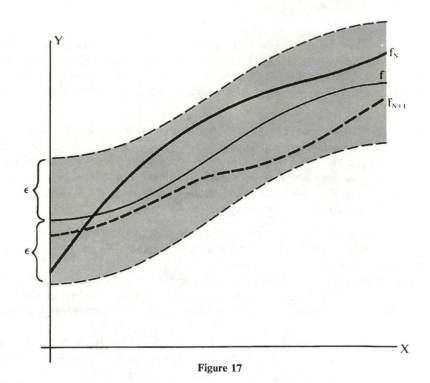

Figure 17

We complete the proof by showing that if $x \in U$, then $\rho(f(x), f(x_0)) < \varepsilon$. To see this, notice that if $x \in U$, then

$$\rho(f(x), f(x_0)) \leq \rho(f(x), f_n(x)) + \rho(f_n(x), f_n(x_0)) + \rho(f_n(x_0), f(x_0))$$

$$< \frac{\varepsilon}{3} + \frac{\varepsilon}{3} + \frac{\varepsilon}{3} = \varepsilon.$$

The following theorems, whose proofs are left to the reader in the next set of exercises, are quite useful in analysis. They furnish sufficient conditions that imply uniform convergence; the first also gives a necessary condition. Note that in the first theorem the domain need not be a space.

67.5. Theorem. Cauchy criterion for uniform convergence. *Suppose (Y, ρ) is a complete metric space and X is a set. Then the sequence $(f_i : X \to (Y, \rho))$ converges uniformly to some function if and only if the following condition is satisfied:*
For each $\varepsilon > 0$, there is a positive integer M such that for $m \geq M$ and $n \geq M$, $\rho(f_m(x), f_n(x)) < \varepsilon$ for all $x \in X$.

67.6. Theorem. *Suppose $(f_i : (X, d) \to (Y, \rho))$ is a sequence of continuous mappings that converges pointwise to a continuous mapping f. Suppose (X, d) is compact and the convergence is monotone in the following sense: For all $x \in X$ and $i \in \mathbf{P}$, $\rho(f_{i+1}(x), f(x)) \leq \rho(f_i(x), f(x))$. Then the convergence is uniform.*

The following important theorem is an easy corollary to the previous theorem.

Figure 18

67.7. **Theorem.** **(Dini).** *Suppose $(f_i:(X, d) \to \mathbf{R})$ is a sequence of continuous real-valued functions defined on a compact space (X, d). Suppose the sequence is monotonic increasing; i.e., $f_i(x) \leqq f_{i+1}(x)$ for each $i \in \mathbf{P}$. If (f_i) converges pointwise to a continuous function f, then the convergence is uniform.*

EXERCISES: SEQUENCES OF MAPPINGS INTO A METRIC SPACE

1. Suppose that $(f_i:(X, d) \to (Y, \rho))$ converges uniformly to a function $f:(X, d) \to (Y, \rho)$ and for sufficiently large i, f_i is uniformly continuous. Prove that f is uniformly continuous.

2. Suppose that the sequence $(f_i:(X, d) \to (Y, \rho))$ converges pointwise to a map $f:(X, d) \to (Y, \rho)$. Let $x_0 \in X$. Suppose that there exists an open subset $U \subset X$ that contains x_0 such that for sufficiently large i, the restriction $f_i \mid U$ is continuous. Prove that if $(f_i \mid U)$ converges uniformly to $f \mid U$, then the function f is continuous at x_0.

3. For each positive integer n, let f_n be given by $f_n(x) = \dfrac{1}{n} x^2$ for $x \in \mathbf{R}$. Show that $\lim (f_n)$ is continuous. Is the convergence uniform on $\mathbf{R}$? Does the situation fall under that covered in the previous exercise?

4. Suppose $(f_i:(X, d) \to (Y, \rho))$ converges uniformly to a continuous mapping $f:(X, d) \to (Y, \rho)$. Prove that for each sequence (x_i) in X and $x \in X$, $\lim (x_i) = x$, implies $\lim (f_i(x_i)) = f(x)$.

5. Prove Theorem 67.5 and Theorem 67.6. A hint for the proof of 67.6 is given here.

    Let $\varepsilon > 0$. For each $x \in X$, there exists a positive integer $N(x)$ such that $\rho(f_{N(x)}(x), f(x)) < \dfrac{\varepsilon}{3}$. Choose an open set $U(x)$ that contains x and has the property that for $z \in U(x)$, $\rho(f_{N(x)}(z), f_{N(x)}(x)) < \dfrac{\varepsilon}{3}$ and $\rho(f(z), f(x)) < \dfrac{\varepsilon}{3}$. The collection of open sets $\{U(x) : x \in X\}$ is an open covering of X and, hence, has a finite subcovering. Use the finite subcovering to find an appropriate positive integer N needed to prove uniform convergence. The monotone convergence property will play a crucial role in showing that the appropriate N will work. (If one chooses to prove the theorem by an indirect proof, the sequential compactness property can be used to good advantage.)

6. Show that Theorem 67.7 follows from 67.6. Show how 67.6 can be obtained from 67.7 by considering the real-valued functions g_i given by $g_i(x) = \rho(f_i(x), f(x))$.

68. REVIEW EXERCISES

The set of exercises below provides a review of the last two chapters. Some of the exercises will serve as motivation for several of the concepts to be taken up later.

1. Define the term separability for metric spaces. Give two other properties that are equivalent to separability for metric spaces.

2. Is total boundedness invariant under a continuous mapping? If the answer is no, is total boundedness invariant under uniformly continuous mappings?

3. Define the concepts of sequential compactness, compactness, and that of the Bolzano-Weierstrass property for metric spaces. Prove that they are equivalent for metric spaces.

4. Suppose $f: \mathbf{R}^n \to \mathbf{R}^m$ is continuous and $S \subset \mathbf{R}^n$ is bounded. Is $f[S]$ necessarily bounded?

5. Suppose $f: S \to \mathbf{R}^m$ is continuous where S is a bounded subset of $\mathbf{R}^n$. Is $f[S]$ necessarily bounded?

6. (a) Let S be an open subset of $\mathbf{R}^n$. Let $\{U_\alpha : \alpha \in A\}$ be a collection of nonempty pairwise disjoint open subsets such that $S = \bigcup \{U_\alpha : \alpha \in A\}$. Show that the collection $\{U_\alpha : \alpha \in A\}$ is a countable collection.
    (b) Show by an example that the statement proved in part (a) is not true for metric spaces in general.
    (c) Is the result of part (a) true for separable metric spaces?

7. Suppose S is an open subset of $\mathbf{R}^n$. Let $x \in S$ and $C_x = \bigcup \{Q : Q$ is connected and $x \in Q \subset S\}$. Show that C_x is an open connected subset of $\mathbf{R}^n$.

8. Let (X, d) be the subspace of $\mathbf{R}^2$, where

$$X = \{(0, y): -1 \leq y \leq 1\} \cup \left\{\left(x, \sin \frac{1}{x}\right): 0 < x \leq 1\right\}.$$

   (a) Is (X, d) connected?
   (b) Show that (X, d) does not have the property possessed by $\mathbf{R}^n$ described in the previous exercise; i.e., show that there is a point $x \in X$ and an open subset S of X such that $x \in S$ and for which $C_x = \bigcup \{Q : Q$ is connected and $x \in Q \subset S\}$ is not open in X.

9. Let (X, d) be the product metric space formed from the collection $\{(X_i, d_i) : i \in \mathbf{P}_n\}$. Let $\pi_i : (X, d) \to (X_i, d_i)$ be the projection onto (X_i, d_i) for each $i \in \mathbf{P}_n$. As usual, for $x \in X$, use x_i to denote $\pi_i(x)$.
   (a) Is π_i uniformly continuous?
   (b) For each $i \in \mathbf{P}_n$ and $p \in X$, let $X_i^*(p) = \{x : x \in X$ and $x_j = p_j$ for $j \neq i\}$. Show that $X_i^*(p)$ is a closed subset of X containing p and that $\pi_i \mid X_i^*(p) : X_i^*(p) \to X_i$ is an isometry so that $X_i^*(p)$ is an "isometric copy" of X_i.
   (c) For each of the following properties, show that (X, d) has the property if and only if each (X_i, d_i) has the property: (i) separable; (ii) compact; (iii) totally bounded; (iv) connected; (v) complete.

10. Prove that in $\mathbf{R}^n$ every open set is the union of a countable collection of pairwise disjoint open connected sets. Use an example to show that this property does not hold for metric spaces in general.

11. Suppose $(f_n : (X, d) \to (Y, \rho))$ is a sequence of bounded mappings that converges uniformly to a mapping $f : (X, d) \to (Y, \rho)$.
   (a) Is f necessarily a bounded mapping?
   (b) Show that there is a uniform bound for the sequence (i.e., show that there is a number M such that diam $(f_n[X]) \leq M$ for each $n \in \mathbf{P}$).

12. Suppose (X, d) is a metric space. A nonempty connected subset S of X is said to be a *component* provided that it is not a proper subset of any other connected subset of X.
   (a) Show that in Exercise 7 the set C_x is a component of the subspace S.
   (b) Give an example of a space in which the components are not all open.
   (c) Are components of a space necessarily closed?
   (d) Suppose $f : (X, d) \to (Y, \rho)$ is a continuous surjection and S is a component of Y. Show that $f^{-1}[S]$ is the union of a collection of components of X.

13. Suppose (X, d) is a metric space and Y is a set that has the same cardinality as does X. Show that there is a metric ρ for Y and a mapping $h:(X, d) \rightarrow (Y, \rho)$ that is an isometry of (X, d) onto (Y, ρ).

14. Let $B(0; 1)$ be the unit ball in $\mathbf{R}^n$ with center at the origin. Define $f:\mathbf{R}^n \rightarrow B(0; 1)$ as follows: For each $x \in \mathbf{R}^n$, let
$$f(x) = x \text{ if } |x| \leq 1 \text{ and } f(x) = \frac{x}{|x|} \text{ for } |x| > 1.$$
    (a) Show that f is continuous.
    (b) Is f uniformly continuous on $\mathbf{R}^n$?

15. (This exercise is intended for the reader who has some familiarity with the theory of functions of a complex variable. In particular, recall that every complex polynomial is continuous. Also recall that from the fundamental theorem of algebra, it follows that if f is a nonconstant complex polynomial and Z is the complex plane, then $f[Z] = Z$.) Let $f:Z \rightarrow Z$ be a nonconstant complex polynomial. Think of the complex plane Z as $\mathbf{R}^2$.
    (a) Show that f maps bounded subsets of Z onto bounded subsets of Z.
    (b) By making use of the algebraic form of complex polynomials and your knowledge of the complex number system show that f maps unbounded subsets of Z onto unbounded subsets of Z.
    (c) Show that for each compact subset K of Z, $f^{-1}[K]$ is compact.
    (d) Show that if F is a closed subset of Z, then $f[F]$ is closed.

5

Metric Spaces: Some Examples and Applications

In Chapter 2 we studied $\mathbf{R}^n$ in some detail. We learned that $\mathbf{R}^n$ had both an algebraic and a metric structure. Recall that an *inner product* (34.3) was defined for $\mathbf{R}^n$, and the magnitude function was defined in terms of the inner product. The Euclidean distance between two points x and y was then defined as the magnitude of their vector difference. We proved that $\mathbf{R}^n$, endowed with the Euclidean metric, is a complete separable metric space. The space $\mathbf{R}^n$ is a special case of a larger class of vector spaces. Generally a nonempty set on which there is defined a *vector addition* and a *scalar multiplication* that satisfy the properties listed in 34.2 is called a *real linear space*. If a *norm* or *magnitude* is defined on a real linear space, then it is called a *real normed linear space* and a metric can be given for the space in terms of its norm. If the normed linear space is complete, it is called a *Banach* space. If the norm of a Banach space is induced by an inner product, the space is called a *Hilbert* space. Thus, we see that $\mathbf{R}^n$, together with the appropriate vector addition, scalar multiplication and inner product, is a real Hilbert space.

In this chapter we shall study two important examples of Banach spaces, one of which we shall prove is a Hilbert space. In one example, each point in the space is a real-valued sequence. In the other example, each point is a real-valued continuous mapping defined on a closed interval. Like $\mathbf{R}^n$, both spaces turn out to be separable and connected. In these two examples, however, closed and bounded subsets are not necessarily compact.

In the last part of this chapter we shall consider *contraction mappings* on complete metric spaces and present a proof of the very important Banach fixed point theorem. This theorem will then be used to prove an existence theorem for a class of first order differential equations.

69. LINEAR OR VECTOR SPACES

Although we shall primarily be interested in the metric and topological properties of the examples in this chapter, it is useful to consider first the algebraic structure of the spaces involved.

69.1. Definition. Real linear or vector space. *A real linear or vector space is a nonempty set of objects V on which there is defined a binary operation* $+$ *called vector addition and a scalar multiplication (multiplication of the element by a real number) that satisfies the following properties.*

69.1(a). Properties of vector addition. *To each x and y in V, there corresponds a unique element $x + y$ in V called the vector sum of x and y. Furthermore this operation* $+$ *satisfies the following properties*:

(i) $x + y = y + x$ *for all x and y in V.*

(ii) $x + (y + z) = (x + y) + z$ *for all x, y, and z in V.*

(iii)   *There exists a unique element $\theta \in V$ (called the zero element) such that $x + \theta = x$ for all $x \in V$.*

(iv)   *To each $x \in V$ there corresponds a unique element $-x \in V$ (called the negative of x) such that $x + (-x) = \theta$.*

69.1(b). Properties of scalar multiplication. *To each x in V and real number α, there corresponds a unique αx in V. Furthermore, for all real numbers α and β and all x and y in V:*

(i) $\alpha(\beta x) = (\alpha\beta)x$

(ii) $(\alpha + \beta)x = \alpha x + \beta x$

(iii) $\alpha(x + y) = \alpha x + \alpha y$

(iv) $1x = x$

Suppose V is a real linear space and S is a nonempty subset such that S is closed under vector addition and scalar multiplication. That is,

$$\text{if} \quad x \in S \quad \text{and} \quad y \in S, \quad \text{then} \quad x + y \in S$$

and

$$\text{if} \quad x \in S \quad \text{and} \quad \alpha \in \mathbf{R}, \quad \text{then} \quad \alpha x \in S.$$

We then say that S is a *linear subspace* of V. This name is appropriate since it is easy to see that these conditions make S a vector space with respect to the vector addition and scalar multiplication for V restricted to S. For example, the set $\{(x, 0) : x \in \mathbf{R}\}$ is a subspace of $\mathbf{R}^2$.

Suppose V_1 and V_2 are real linear spaces and $f : V_1 \to V_2$ is a mapping such that $f(\alpha x + \beta y) = \alpha f(x) + \beta f(y)$ for all x and y in V_1 and all α and β in $\mathbf{R}$. Then f is called a *linear mapping*. If, in addition, f is a bijection, f is said to be a *vector isomorphism* from V_1 onto V_2. If there exists a vector isomorphism that carries one space onto another, the spaces are said to be *isomorphic*. For example, the vector subspace $S = \{(x, 0) : x \in \mathbf{R}\}$ of $\mathbf{R}^2$ is isomorphic to $\mathbf{R}$. To see this, simply examine the mapping $h : \mathbf{R}^2 \to \mathbf{R}$ given by $h((x, 0)) = x$ for each $(x, 0) \in S$.

69.2. Definition. Norm. *Let V be a real linear space. A real-valued function defined on V is called a norm for the space V provided it satisfies the following properties ($\|x\|$ will be the symbol for norm of x):*

For all x and y in V and real α,
(a) $\|x\| > 0$ *if* $x \neq \theta$ *and* $\|\theta\| = 0$
(b) $\|\alpha x\| = |\alpha| \, \|x\|$
(c) $\|x + y\| \leq \|x\| + \|y\|$ *(triangle inequality)*

A real linear space V for which a norm is defined is said to be a *real normed linear space*.

Just as in $\mathbf{R}^n$, it follows from the triangle inequality that for all x, y, and z in V, $\|x - y\| + \|y - z\| \geq \|x - z\|$ and $\|x - y\| = \|y - x\|$. Hence, using this together with 69.2(a), we see that $d: V \times V \to \mathbf{R}$ given by

$$d(x, y) = \|x - y\|$$

defines a metric for V. When we speak of a normed linear space as a metric space it will always be with reference to this metric.

69.3. Definition. Inner product. *Suppose V is a real linear space. A real-valued function defined on $V \times V$ that satisfies the following properties is called an inner product for V. We shall use the notation $\langle x, y \rangle$ for the inner product of x and y.*
For all x, y, and z in V and real numbers α and β,
(a) $\langle x, y \rangle = \langle y, x \rangle$
(b) $\langle x, \alpha y + \beta z \rangle = \alpha \langle x, y \rangle + \beta \langle x, z \rangle$
(c) $\langle x, x \rangle > 0$ *if* $x \neq \theta$ *and* $\langle \theta, \theta \rangle = 0$

If V is an inner product space and if for each $x \in V$ we let

$$\|x\| = [\langle x, x \rangle]^{\frac{1}{2}},$$

then this formula defines a norm for V. Properties (a) and (b) of 69.2 are easy to check by making use of the properties of the inner product. The details are left as an exercise. To prove the triangle inequality, recall that the proof of this inequality for $\mathbf{R}^n$ made use of the Cauchy-Schwarz inequality (35.1). It is left as an exercise to check that the proof of the Cauchy-Schwarz inequality given in 35.1 (with appropriate change in notation) holds for a real inner product space. Thus, we have the following.

69.4. Theorem. Cauchy-Schwarz inequality for a real inner product space. *Let V be a real inner product space. Then for each x and y in V, the following holds.*

$$|\langle x, y \rangle| \leq \|x\| \, \|y\|.$$

Because of the Cauchy-Schwarz inequality, the proof of the triangle inequality as presented in 35.2 carries over to the present more general setting. The reader should check to see that this is so.

Whenever we speak of the norm for an inner product space, it will be understood that reference is made to the norm just discussed.

69.5. Definitions. Banach and Hilbert spaces. *If a real normed linear space is complete, it is called a real Banach space. If a real inner product space is complete it is called a real Hilbert space.*

Recall that in $\mathbf{R}^n$ a line segment with endpoints a and b is defined as the set $\{x : x = (1 - t)a + tb, 0 \leq t \leq 1\}$. Also, if a and b are distinct points in $\mathbf{R}^n$, then the set $\{x : x = (1 - t)a + tb, -\infty < t < \infty\}$ is a line in $\mathbf{R}^n$. By making use of the fact that lines in $\mathbf{R}^n$ are connected, we were able to show that $\mathbf{R}^n$ is connected. Notice that these definitions are meaningful in a real linear space. It will be left as an exercise to show that line segments and lines in a normed linear space are connected and to use this information to prove that normed linear spaces are connected.

In this section we have been dealing with linear spaces for which the scalars are allowed to range over the real field $\mathbf{R}$. For this reason the spaces considered are called *real linear spaces* or *linear spaces over the real field*. If the scalars are allowed to range over the complex field, the properties of the inner product have to be modified to give a satisfactory theory (See for example, [5], [21], or [30].). We will deal only with real linear spaces.

EXERCISES: LINEAR OR VECTOR SPACES

1. Give the details of the proof of Theorem 69.4.

2. Prove that line segments and lines in a real normed linear space are connected.

3. Let V be a real normed linear space. A subset $S \subset V$ is said to be convex if for every pair of points a and b in S the line segment with a and b as endpoints is contained in S. Prove that if S is a convex subset of a real normed linear space, then S is connected. Prove that all real normed linear spaces are connected. Prove that all neighborhoods $N(p; \varepsilon)$ and balls $B(p; \varepsilon)$ in a real normed linear space are connected.

4. Prove that if V_1 and V_2 are real linear spaces and h is a vector isomorphism from V_1 onto V_2, then h^{-1} is a vector isomorphism from V_2 onto V_1.

5. Show that for each $j \in \mathbf{P}_n$, $\mathbf{R}^n$ contains a vector subspace that is isomorphic to $\mathbf{R}^j$. Thus, for example, $\mathbf{R}^3$ contains isomorphic copies of $\mathbf{R}^1$, $\mathbf{R}^2$, and $\mathbf{R}^3$.

70. THE HILBERT SPACE ℓ^2

Let S be the set of all real-valued sequences. For all $(x_i) \in S$, $(y_i) \in S$, and $\alpha \in \mathbf{R}$, define

$$(x_i) + (y_i) = (x_i + y_i) \quad \text{and} \quad \alpha(x_i) = (\alpha x_i).$$

Furthermore, let Θ be the element of S, each of whose elements is 0. Also define $-(x_i) = (-x_i)$ for all $(x_i) \in S$. Taking these definitions as vector addition and scalar multiplication, together with the 0 element and the negative of an element as just defined, we can easily verify that S is a real linear space. This linear space S contains isomorphic copies of $\mathbf{R}^n$ for each positive integer n. To see this, for each positive integer n, let

$$\tilde{R}^n = \{(x_i) : (x_i) \in S \quad \text{and} \quad x_i = 0 \quad \text{for} \quad i > n\}.$$

The mapping that assigns $(x_1, x_2, x_3, \ldots, x_n) \in \mathbf{R}^n$ to the point $(x_1, x_2, x_3, \ldots, x_n, 0, 0, 0, \ldots)$ is clearly a vector isomorphism. The inner product that we have been using for $\mathbf{R}^n$ suggests that an inner product for $\tilde{R}^n$ is given by

$$\langle x, y \rangle = \sum_{i=1}^{\infty} x_i y_i = \sum_{i=1}^{n} x_i y_i.$$

It is easy to see that this formula does indeed give an inner product for $\tilde{R}^n$. If we try to extend this definition to all of S, we have an obvious convergence problem since $\sum_{i=1}^{\infty} x_i y_i$ does not necessarily converge for all x and y in S. However, we can find a subspace of S that is small enough so that $\sum_{i=1}^{\infty} x_i y_i$ converges for all (x_i) and (y_i) in the space and yet large enough to contain an isomorphic copy of each $\mathbf{R}^n$. We define such a subspace next.

We define ℓ^2 as that subset of S, consisting of all $(x_i) \in S$ for which $\sum_{i=1}^{\infty} x_i^2$ converges.

At this point a few preliminary observations concerning the set ℓ^2 will be helpful. Suppose $(x_i) \in \ell^2$ and $(y_i) \in \ell^2$. Then for each n, $(|x_1|, |x_2|, \ldots, |x_n|)$ and $(|y_1|, |y_2|, \ldots, |y_n|)$ are elements of $\mathbf{R}^n$. Hence, from the Cauchy-Schwarz inequality for $\mathbf{R}^n$ (35.1), we obtain

$$\sum_{i=1}^{n} |x_i|\,|y_i| \leq \left[\sum_{i=1}^{n} x_i^2\right]^{\frac{1}{2}} \left[\sum_{i=1}^{n} y_i^2\right]^{\frac{1}{2}}.$$

Then, since $(x_i) \in \ell^2$ and $(y_i) \in \ell^2$, we have

$$\sum_{i=1}^{n} |x_i|\,|y_i| \leq \left[\sum_{i=1}^{\infty} x_i^2\right]^{\frac{1}{2}} \left[\sum_{i=1}^{\infty} y_i^2\right]^{\frac{1}{2}}.$$

Thus, the finite sum on the left (which is increasing with n) is bounded above so that we have

$$\sum_{i=1}^{\infty} |x_i|\,|y_i| \leq \left[\sum_{i=1}^{\infty} x_i^2\right]^{\frac{1}{2}} \left[\sum_{i=1}^{\infty} y_i^2\right]^{\frac{1}{2}}.$$

Thus, $\sum_{i=1}^{\infty} x_i y_i$ converges absolutely and, hence, converges. Furthermore, observe that $\left|\sum_{i=1}^{\infty} x_i y_i\right| \leq \sum_{i=1}^{\infty} |x_i|\,|y_i|$. We summarize our results in the following statement.

70.1. Theorem. *If $(x_i) \in \ell^2$ and $(y_i) \in \ell^2$, then $\sum_{i=1}^{\infty} x_i y_i$ converges and*

$$\left|\sum_{i=1}^{\infty} x_i y_i\right| \leq \left[\sum_{i=1}^{\infty} x_i^2\right]^{\frac{1}{2}} \left[\sum_{i=1}^{\infty} y_i^2\right]^{\frac{1}{2}}.$$

By making use of the first part of the previous theorem, we can now prove the following important fact.

70.2. Theorem. *ℓ^2 is a subspace of the real linear space S of all real-valued sequences.*

PROOF. It is sufficient to show that ℓ^2 is closed under vector addition and scalar multiplication. It is clear that if $\sum_{i=1}^{\infty} x_i^2$ converges, then so does $\sum_{i=1}^{\infty} \alpha^2 x_i^2$.

Consequently, $x \in \ell^2$ implies that $\alpha x \in \ell^2$. Next assume that $x \in \ell^2$ and $y \in \ell^2$. Since $\sum_{j=1}^{\infty} x_i^2$ converges and $\sum_{i=1}^{\infty} y_i^2$ converges, it follows from 70.1 that $\sum_{i=1}^{\infty} x_i y_i$ converges. Next, notice that

$$\sum_{i=1}^{n} (x_i + y_i)^2 = \sum_{i=1}^{n} x_i^2 + 2 \sum_{i=1}^{n} x_i y_i + \sum_{i=1}^{n} y_i^2.$$

Since the three sequences of partial sums on the right converge, so does the one on the left. Hence, $\sum_{i=1}^{\infty} (x_i + y_i)^2$ converges and, consequently, $x + y \in \ell^2$.

70.3. Theorem. *The real-valued function defined on $\ell^2 \times \ell^2$ by the following formula is an inner product for ℓ^2. For each $x = (x_i) \in \ell^2$ and $y = (y_i) \in \ell^2$, let*

$$\langle x, y \rangle = \sum_{i=1}^{\infty} x_i y_i.$$

It follows from 70.1 that $\sum_{i=1}^{\infty} x_i y_i$ is defined for all x and y in ℓ^2. The proof that 69.3(a), (b), and (c) are satisfied is left as an easy exercise for the reader.

We now take as our norm for ℓ^2,

$$\|x\| = [\langle x, x \rangle]^{\frac{1}{2}} = \left[\sum_{i=1}^{\infty} x_i^2 \right]^{\frac{1}{2}}.$$

It is interesting to note that if we write out the Cauchy-Schwarz inequality for ℓ^2 in series form, we get the inequality in 70.1. To see this, observe that

$$|\langle x, y \rangle| \leq \|x\| \, \|y\|$$

is equivalent for ℓ^2 to

$$\left| \sum_{i=1}^{\infty} x_i y_i \right| \leq \left[\sum_{i=1}^{\infty} x_i^2 \right]^{\frac{1}{2}} \left[\sum_{i=1}^{\infty} y_i^2 \right]^{\frac{1}{2}}.$$

Notice next that the metric given by this norm takes the form

$$d(x, y) = \left[\sum_{i=1}^{\infty} (x_i - y_i)^2 \right]^{\frac{1}{2}}.$$

We should notice further that ℓ^2 contains each $\tilde{R}^n$. Furthermore, the metric d restricted to $\tilde{R}^n \times \tilde{R}^n$ becomes $d(x, y) = \left[\sum_{i=1}^{n} (x_i - y_i)^2 \right]^{\frac{1}{2}}$. Thus, we may look at the metric given here for ℓ^2 as a generalization of the Euclidean metric.

70.4. Theorem. *ℓ^2 is complete.*

PROOF. Let (q_m) be a Cauchy sequence in ℓ^2, where for each m,

$$q_m = (q_{m,1}, q_{m,2}, \ldots, q_{m,n}, \ldots).$$

We prove the proposition by showing: (a) For each j, $(q_{m,j})_{m=1}^{\infty}$ is a Cauchy sequence in **R**. Call its limit z_j. (b) $z = (z_1, z_2, \ldots) \in \ell^2$. (c) $\lim (q_m) = z$.

Proof of (a). Fix j and let $\varepsilon > 0$. There is a positive integer N such that for $m \geq N$ and $n \geq N$, $d(q_m, q_n) < \varepsilon$. But then for $m \geq N$ and $n \geq N$,

$$|q_{m,j} - q_{n,j}| \leq \left[\sum_{i=1}^{\infty} (q_{m,i} - q_{n,i})^2 \right]^{\frac{1}{2}} = d(q_m, q_n) < \varepsilon.$$

Hence, for each $j \in \mathbf{P}$, $(q_{m,j})_{m=1}^{\infty}$ is a C.S. of real numbers and there is a $z_j = \lim_{m \to \infty} (q_{m,j})$. We now set $z = (z_j)_{j=1}^{\infty}$.

Proof of (b). We wish to show that $z \in \ell^2$ where z is defined as in part (a). We shall do this in the following way: Using the fact that (q_m) is a C.S. in ℓ^2, we will exhibit a positive integer N such that $(q_{N,i} - z_i)_{i=1}^{\infty} \in \ell^2$. But this will be sufficient, for $(z_i) = (q_{N,i}) - (q_{N,i} - z_i)$ and the difference of two elements of ℓ^2 is an element of ℓ^2.

Let $h > 0$. Since (q_m) is a C.S., there is a positive integer N such that for each $p \in \mathbf{P}$, $d(q_N, q_{N+p}) < h$. Then $\sum_{i=1}^{\infty} (q_{N,i} - q_{N+p,i})^2 < h^2$ and, hence, for each positive integer n we have $\sum_{i=1}^{n} (q_{N,i} - q_{N+p,i})^2 < h^2$. Then from the fact that $\lim_{p \to \infty} q_{N+p,i} = z_i$, we obtain

$$\lim_{p \to \infty} \sum_{i=1}^{n} (q_{N,i} - q_{N+p,i})^2 = \sum_{i=1}^{n} (q_{N,i} - z_i)^2 \leqq h^2.$$

Then since $\left(\sum_{i=1}^{n} (q_{N,i} - z_i)^2 \right)_{n=1}^{\infty}$ is a monotonic increasing sequence of real numbers bounded above, $\sum_{i=1}^{\infty} (q_{N,i} - z_i)^2$ converges and $(q_{N,i} - z_i)_{i=1}^{\infty} \in \ell^2$. The result now follows from the preliminary remarks at the beginning of the proof.

Proof of (c). We prove that $\lim (q_n) = z$. (The proof is very similar to that for (b).)

Let $\varepsilon > 0$. Then there is a positive integer N such that for $n \geqq N$, $d(q_n, q_{n+p}) < \varepsilon$ for all $p \in \mathbf{P}$. Let $n \geqq N$. Then

$$\sum_{i=1}^{\infty} (q_{n,i} - q_{n+p,i})^2 < \varepsilon^2$$

and, hence for each $k \in \mathbf{P}$,

$$\sum_{i=1}^{k} (q_{n,i} - q_{n+p,i})^2 < \varepsilon^2.$$

Therefore, since $\lim_{p \to \infty} q_{n+p,i} = z_i$, we get

$$\sum_{i=1}^{k} (q_{n,i} - z_i)^2 \leqq \varepsilon^2$$

and, thus,

$$\sum_{i=1}^{\infty} (q_{n,i} - z_i)^2 \leqq \varepsilon^2.$$

From this we see that $d(q_n, z) \leqq \varepsilon$ for $n \geqq N$. This completes the proof.

Now that we have shown that ℓ^2 is a complete inner product space, we have verified that it is a Hilbert space. In the next theorem we shall show that ℓ^2 is also separable. The idea behind the proof is that for each $x_0 \in \ell^2$ and $\varepsilon > 0$, there is a positive integer n such that the subset $D_n = \{x : x_i$ is rational for $i \in \mathbf{P}_n$ and $x_i = 0$ for $i \geqq n + 1\}$ is within ε distance from x_0.

70.5. Theorem. ℓ^2 *is separable.*

PROOF. We construct a countable dense subset for ℓ^2 as follows: For each $n \in \mathbf{P}$, let $D_n = \{(x_i) : (x_i) \in \ell^2, x_i$ is rational for $i \in \mathbf{P}_n$ and $x_i = 0$ for $i \geqq n + 1\}$.

Note that each $D_n \subset \ell^2$ and is countable. Let $D = \bigcup \{D_n : n \in \mathbf{P}\}$ and note that D is countable. We next show that D is dense in ℓ^2. Let $p = (p_i) \in \ell^2$ and let $\varepsilon > 0$. We shall complete the proof by showing that there is a $z \in D$ such that $d(z, p) < \varepsilon$. Since $p \in \ell^2$, $\sum\limits_{i=1}^{\infty} p_i^2$ converges. Choose N such that

$$\sum_{i=N+1}^{\infty} p_i^2 < \frac{\varepsilon^2}{2}$$

Next for $i = 1, 2, \ldots, N$ choose a rational number r_i such that

$$|r_i - p_i| < ((2N)^{-\frac{1}{2}})\varepsilon.$$

Let $z = (r_1, r_2, \ldots, r_N, 0, 0, 0, \ldots)$. Note that $z \in D$ and

$$d(z, p) = \left[\sum_{i=1}^{N} (r_i - p_i)^2 + \sum_{i=N+1}^{\infty} p_i^2 \right]^{\frac{1}{2}}$$

$$< \left[N(2N)^{-1}\varepsilon^2 + \frac{\varepsilon^2}{2} \right]^{\frac{1}{2}} = \varepsilon.$$

Recall that the closed balls in $\mathbf{R}^n$ are compact since closed and bounded subsets of $\mathbf{R}^n$ are compact. We shall show that this is not the case in ℓ^2. We shall prove next that the unit ball $B(\Theta; 1)$ in ℓ^2 is not compact. It then follows easily that no closed ball is compact in ℓ^2. To see this consider the ball $B(p; r)$. The mapping $f : \ell^2 \to \ell^2$ given by $f(x) = rx + p$ is a homeomorphism; furthermore, it is easy to verify that $f[B(\Theta; 1)] = B(p; r)$. Hence, if $B(\Theta; 1)$ is not compact, neither is $B(p; r)$.

70.6. Theorem.   *For each $r > 0$ and $p \in \ell^2$, the ball $B(p; r)$ is not compact.*

PROOF.   From the previous remarks it is sufficient to prove the theorem for the unit ball. We shall prove that $B(\Theta; 1)$ is not compact by exhibiting a sequence in ℓ^2 with no convergent subsequence. For each $i \in \mathbf{P}$, let q_i be the real-valued sequence whose ith term is 1 and all the other terms are 0. Thus, $q_1 = (1, 0, 0, 0, \ldots)$, $q_2 = (0, 1, 0, 0, 0, \ldots)$, etc. It is clear that $q_i \in \ell^2$ for each $i \in \mathbf{P}$. Now (q_i) is a sequence in ℓ^2 such that for each $i \neq j$, $d(q_i, q_j) = 2^{\frac{1}{2}}$. Obviously no subsequence of (q_i) could converge. Hence, $B(\Theta; 1)$ is not compact.

Some spaces possess the important property that for each point x in the space there exists an open set U_x and a compact set K_x such that $x \in U_x \subset K_x$. (Such a set K_x is called a compact neighborhood of x.) A space with this property is called a *locally compact* space. Note that $\mathbf{R}^n$ is locally compact. However, the fact that closed balls in ℓ^2 are not compact implies that ℓ^2 is not locally compact.

EXERCISES: THE HILBERT SPACE ℓ^2

1.   Prove Theorem 70.3.

2.   In the discussion in this section it was remarked that the function $f : \ell^2 \to \ell^2$ given by $f(x) = rx + p$, where $r > 0$, is a homeomorphism. Prove this. Also verify that for each $p \in \ell^2$ and $r > 0$, $f[B(\Theta; 1)] = B(p; r)$.

3. For each $n \in P$, let $\tilde{R}^n$ be as defined in this section. Let $A = \bigcup \{\tilde{R}^n : n \in P\}$. Is A a connected subset of ℓ^2? Does A contain any nonempty open subsets of ℓ^2?

4. Verify the following statement which was made in this section: The fact that closed balls in ℓ^2 are not compact implies that ℓ^2 is not locally compact.

71. THE HILBERT CUBE

In the last section we showed that the unit ball $\{x : \|x\| \leq 1\}$ in ℓ^2 is not compact. Because of this, it is an interesting fact that the following subset of ℓ^2, called the *Hilbert cube* and denoted by I^∞, is a compact subset of ℓ^2:

$$I^\infty = \left\{ x : |x_i| \leq \frac{1}{i} \right\}.$$

Not only shall we prove that the Hilbert cube is compact but, in a later chapter, this fact will be put to important use.

For each $j \in P$, let us define $\pi_j : I^\infty \to \left[-\frac{1}{j}, \frac{1}{j} \right]$ as follows: For every $x = (x_1, x_2, x_3, \ldots, x_i, \ldots)$, let $\pi_j(x) = x_j$. Hence, π_j selects the jth coordinate of x. (This is an extension of what we have already done for finite-sequences in 51.4.) We can characterize convergence of sequences (x_i) in I^∞ in terms of the sequences $(\pi_j(x_i))_{i=1}^\infty$ in $\left[-\frac{1}{j}, \frac{1}{j} \right]$, $j \in P$.

(See 41.2 and 51.5 for motivation for the characterization of convergence given next.) In reading the proof of the next theorem, it should be kept in mind that if (x_i) is a sequence in I^∞, then each term x_i is a sequence of real numbers such that the jth coordinate, $\pi_j(x_i) \in \left[-\frac{1}{j}, \frac{1}{j} \right]$.

71.1. Theorem. *Let (x_i) be a sequence in I^∞ and let $x_0 \in I^\infty$. Then $\lim_{i \to \infty} x_i = x_0$ if and only if for each $j \in P$, $\lim_{i \to \infty} \pi_j(x_i) = \pi_j(x_0)$.*

PROOF. First assume that for each $j \in P$, $\lim \pi_j(x_i) = \pi_j(x_0)$. Let $\varepsilon > 0$. Choose a positive integer N such that

$$\sum_{j=N+1}^\infty \frac{1}{j^2} < \frac{\varepsilon^2}{8}.$$

For each $j \in P_N$ let N_j be chosen so that for $i \geq N_j$, $|\pi_j(x_i) - \pi_j(x_0)| < (2N)^{-\frac{1}{2}}\varepsilon$. Next let $M = \max. \{N_j : j \in P_N\}$. Then for $i \geq M$,

$$d(x_i, x_0) = \left[\sum_{j=1}^N (\pi_j(x_i) - \pi_j(x_0))^2 + \sum_{j=N+1}^\infty (\pi_j(x_i) - \pi_j(x_0))^2 \right]^{\frac{1}{2}}$$

$$< \left[N(2N)^{-1}\varepsilon^2 + \sum_{j=N+1}^\infty \frac{4}{j^2} \right]^{\frac{1}{2}} < \left[\frac{\varepsilon^2}{2} + \frac{\varepsilon^2}{2} \right]^{\frac{1}{2}} = \varepsilon.$$

Hence, (x_i) converges to x_0.

Suppose next that (x_i) is a sequence in I^∞ and $\lim (x_i) = x_0 \in I^\infty$. Then for each $j \in \mathbf{P}$ and $i \in \mathbf{P}$,

$$|\pi_j(x_i) - \pi_j(x_0)| \leq d(x_i, x_0).$$

It is now clear that since $\lim (d(x_i, x_0)) = 0$,

$$\lim_{i \to \infty} |\pi_j(x_i) - \pi_j(x_0)| = 0.$$

Hence,

$$\lim_{i \to \infty} (\pi_j(x_i)) = \pi_j(x_0).$$

Notice that the last paragraph of the previous proof holds also for ℓ^2. However, a sequence (x_i) in ℓ^2 does not necessarily converge just because $(\pi_j(x_i))_{i=1}^\infty$ converges for each $j \in \mathbf{P}$. To see this, refer back to the sequence (q_i) in 70.6. It was pointed out in that proof that (q_i) could not converge. However, notice that for each $j \in \mathbf{P}$, $\lim_{i \to \infty} (\pi_j(x_i)) = 0$.

71.2.   Remark. I^∞ is a closed subset of ℓ^2 and, hence, is complete.

71.3.   Theorem.   *The Hilbert cube I^∞ is a compact subset of ℓ^2.*

PROOF.   Since I^∞ is complete, we will prove the theorem by showing that I^∞ is totally bounded. To do this we let $\varepsilon > 0$ and find an ε-net for I^∞ as follows: Choose an N such that $\sum_{i=N+1}^\infty \frac{1}{i^2} < \frac{\varepsilon^2}{2}$. Observe that for each positive integer i, the closed interval $\left[-\frac{1}{i}, \frac{1}{i}\right]$ is totally bounded. For each $i \in \mathbf{P}_N$, let F_i be a $(2N)^{-\frac{1}{2}}\varepsilon$-net for $\left[-\frac{1}{i}, \frac{1}{i}\right]$.   Next, define

$$F = \{x : x_i \in F_i \quad \text{for} \quad i \in \mathbf{P}_N \quad \text{and} \quad x_i = 0 \quad \text{for} \quad i \geq N+1\}.$$

We complete the proof by showing that F is an ε-net for I^∞. To verify this, let $y \in I^\infty$. For each $i \in \mathbf{P}_N$, choose an $x_i \in F_i$ such that $|x_i - y_i| < (2N)^{-\frac{1}{2}}\varepsilon$. Let $x_i = 0$ for $i \geq N+1$. Then $x = (x_i) \in F$. Next we compute $d(x, y)$.

$$\begin{aligned}
d(x, y) &= \left[\sum_{i=1}^\infty |x_i - y_i|^2\right]^{\frac{1}{2}} \\
&\leq \left[\sum_{i=1}^N |x_i - y_i|^2 + \sum_{i=N+1}^\infty \frac{1}{i^2}\right]^{\frac{1}{2}} \\
&< \left[N\frac{\varepsilon^2}{2N} + \sum_{i=N+1}^\infty \frac{1}{i^2}\right]^{\frac{1}{2}} \\
&< \left[\frac{\varepsilon^2}{2} + \frac{\varepsilon^2}{2}\right]^{\frac{1}{2}} = \varepsilon.
\end{aligned}$$

Hence, we have verified that F is an ε-net for I^∞.

The notion of the Cartesian product of a finite number of metric spaces was defined in 51. In the last chapter of the text we shall extend the notion to include the Cartesian product of an infinite collection of spaces. The space I^∞ will turn out to be the product space of an infinite collection of closed intervals. Hence, the symbol I^∞ is appropriate.

EXERCISES: THE HILBERT CUBE

1. Carry out the details of the proof of the statement in 71.2 that I^∞ is a closed subset of ℓ^2.

2. One might be tempted to guess that the set

$$\left\{ x : x \in \ell^2 \text{ and } |x_i| < \frac{1}{i} \right\}$$

is an open subset of ℓ^2. Show that this is false. Does I^∞ contain any nonempty open subset of ℓ^2?

72. THE SPACE $\mathscr{C}([a, b])$ OF CONTINUOUS REAL-VALUED MAPPINGS ON A CLOSED INTERVAL $[a, b]$

An interesting and important question arises in connection with convergent sequences of mappings. Suppose $\mathscr{F}$ is a collection of mappings from one space into another. Suppose for sequences (f_i) in $\mathscr{F}$ we consider some mode $\mathscr{M}$ of convergence (e.g., pointwise, uniform). Can a metric ρ be introduced on $\mathscr{F}$ so that a sequence of mappings (f_i) converges in mode $\mathscr{M}$ to a mapping f if and only if (f_i) as a sequence of points in $(\mathscr{F}, \rho)$ converges to f? Notice that in the previous question we are looking at (f_i) from two points of view. On one hand, we are looking at (f_i) as a sequence of mappings as we did in 67; on the other hand, we are looking at (f_i) as a sequence of points in a metric space $(\mathscr{F}, \rho)$. In this section we shall give an example of the introduction of a metric that makes the two types of convergence equivalent.

In the remainder of this section we shall be concerned with the collection $\mathscr{C}([a, b])$ of all continuous functions defined on a fixed closed interval $[a, b]$. We introduce a vector addition and real scalar multiplication for $\mathscr{C}([a, b])$ as follows: For each f and g in $\mathscr{C}([a, b])$ and $\alpha \in \mathbf{R}$, let $f + g$ and αf be the elements of $\mathscr{C}([a, b])$ given by

$$(f + g)(x) = f(x) + g(x) \quad \text{for all} \quad x \in [a, b]$$
$$(\alpha f)(x) = \alpha(f(x)) \quad \text{for all} \quad x \text{ in } [a, b].$$

(Recall this was precisely what was meant by $f + g$ and αf in calculus.) Let Θ be the identically 0 function on $[a, b]$.

The reader should verify that $\mathscr{C}([a, b])$ together with addition and scalar multiplication as defined here is a real linear space. We next introduce a norm for $\mathscr{C}([a, b])$.

72.1. Definition. Uniform norm for $\mathscr{C}([a, b])$. *For each $f \in \mathscr{C}([a, b])$, define*

$$\|f\| = \text{l.u.b. } \{|f(x)| : x \in [a, b]\}.$$

(Notice that in the situation that we are considering each $f \in \mathscr{C}([a, b])$ is continuous on $[a, b]$ so that l.u.b. $\{|f(x)| : x \in [a, b]\} = \text{max. } \{|f(x)| : x \in [a, b]\}$.)

It is left as an exercise to show that this is indeed a norm for $\mathscr{C}([a, b])$. We take note that the metric ρ for $\mathscr{C}([a, b])$ induced by this norm is given by

$$\rho(f, g) = \text{l.u.b. } \{|f(x) - g(x)| : x \in [a, b]\}.$$

It can be shown that $(\mathscr{C}([a, b]), \rho)$ is a complete metric space and, hence, that $\mathscr{C}([a, b])$ is a Banach space. Like all real normed linear spaces, $\mathscr{C}([a, b])$ is connected and furthermore ε-neighborhoods and balls are connected. Also like ℓ^2, balls in $\mathscr{C}([a, b])$ are not compact. The next set of exercises will provide the reader with an opportunity to verify these facts. Also he will be asked to show that for $f_i \in \mathscr{C}([a, b])$ and $f \in \mathscr{C}([a, b])$, (f_i) converges uniformly to f on $[a, b]$ if and only if (f_i) converges to f in the metric space $(\mathscr{C}([a, b]), \rho)$.

Mathematicians have long been interested in problems related to convergence of functions. The fact that metric spaces can include such examples as the one discussed in this section served as a motivation for the study of metric spaces. Not all convergence questions of interest to mathematicians can be placed in the setting of metric spaces. However, the study of spaces more general than metric spaces has also proved to be useful in connection with convergence problems.

EXERCISES: THE SPACE $\mathscr{C}([a, b])$ OF CONTINUOUS REAL-VALUED MAPPINGS ON A CLOSED INTERVAL $[a, b]$

1. Verify that with the vector addition and scalar multiplication as given in this section, $\mathscr{C}([a, b])$ is a real normed linear space.

2. Verify that $\mathscr{C}([a, b])$ is complete and, hence, that it is a Banach space.

3. Prove that a sequence (f_i) in $\mathscr{C}([a, b])$ converges uniformly on $[a, b]$ to a mapping $f \in \mathscr{C}([a, b])$ if and only if $\lim (f_i) = f$ in the space $(\mathscr{C}([a, b]), \rho)$.

4. Let $\mathscr{B}$ be the collection of all finite-sequences of *rational* numbers. For each $\beta \in \mathscr{B}$ we construct a continuous function whose graph is the union of a finite number of line segments as follows: Let n be the number of terms in β so that we may write $\beta = (\beta_1, \beta_2, \ldots, \beta_n)$. Choose n numbers $\alpha_1, \alpha_2, \ldots, \alpha_n$ in $[a, b]$ such that $a = \alpha_1 < \alpha_2 < \alpha_3 < \cdots < \alpha_n = b$ and
$$\alpha_{i+1} - \alpha_i = \frac{1}{n-1}(b-a) \text{ for } i \in \mathbf{P}_{n-1}. \text{ For each } i \in \mathbf{P}_{n-1}, \text{ let}$$
L_i be the line segment joining (α_i, β_i) to $(\alpha_{i+1}, \beta_{i+1})$. Next define $L(\beta)$ to be that continuous function whose graph is

$$\bigcup \{L_i : i \in \mathbf{P}_{n-1}\}.$$

Notice that for each $\beta \in \mathscr{B}$, $L(\beta) \in \mathscr{C}([a, b])$.
   (a) Prove that for each $\varepsilon > 0$ and $f \in \mathscr{C}([a, b])$, there exists a $\beta \in \mathscr{B}$ such that $\|f - L(\beta)\| < \varepsilon$. (Hint: Make use of the uniform continuity of f on $[a, b]$.)
   (b) Make use of (a) to prove that $\mathscr{C}([a, b])$ is separable.

5. Find a sequence (f_i) in $\mathscr{C}([a, b])$ such that $\|f_i\| = 1$ for each i and for $i \neq j$, $\|f_i - f_j\| = 1$.

6. Use the previous exercise or some other means to show that closed balls in $\mathscr{C}([a, b])$ are not compact.

7. A subset $\mathscr{S} \subset \mathscr{C}([a, b])$ is said to be uniformly equicontinuous provided that for each $\varepsilon > 0$ there is a $\delta > 0$ such that if x_1 and x_2 are in $[a, b]$ and $|x_1 - x_2| < \delta$, then $|f(x_1) - f(x_2)| < \varepsilon$ for all $f \in \mathscr{S}$.

From Exercise 6, we see that in $(\mathscr{C}([a, b]), \rho)$ subsets that are both bounded and closed need not be compact. However, if a subset $\mathscr{S} \subset \mathscr{C}([a, b])$ is bounded, closed, and equicontinuous, then it is compact. This fact is one form of an important theorem in analysis known as Arzela's (or Ascoli's) theorem. The purpose of this exercise is to present an outline of a proof of this theorem and to have the reader fill in the details that are omitted.

Let $\mathscr{S}$ be a bounded, closed, and equicontinuous set in $\mathscr{C}([a, b])$. Assuming the reader has already shown that $\mathscr{C}([a, b])$ is complete (Exercise 2), we need show only that every sequence (f_i) in $\mathscr{S}$ has a Cauchy subsequence. Then, because $\mathscr{S}$ is closed, the subsequence will converge to an $f \in \mathscr{S}$ and $\mathscr{S}$ is sequentially compact.

(a) Let (f_i) be a sequence in $\mathscr{S}$. Then there is a real number M such that $|f_i(x)| < M$ for each $x \in [a, b]$ and for $i \in \mathbf{P}$.

(b) Let $D = \{x_i : i \in \mathbf{P}\}$ be a dense subset of $[a, b]$. Our plan is to extract a subsequence of f_i that converges on D and to show, by making use of the equicontinuity of $\mathscr{S}$, that the subsequence is a Cauchy sequence in $\mathscr{S}$.

  (i) $(f_i(x_1))$ is a sequence of real numbers and the range of the sequence is bounded from part (a). Hence, there exists a subsequence (f_i^1) of (f_i) for which $(f_i^1(x_1))$ converges. Why?

  (ii) The sequence $(f_i^1(x_2))$ has a convergent subsequence. Hence, there is a subsequence (f_i^2) of (f_i^1) for which $(f_i^2(x_2))$ converges. Note that $(f_i^2(x_1))$ also converges.

  Next assume that sequences (f_i^1), (f_i^2), ..., (f_i^h) have been chosen so that (f_i^{j+1}) is a subsequence of (f_i^j) and such that (f_i^j) converges on $\{x_1, x_2, \ldots, x_j\}$. Then we may choose a subsequence (f_i^{h+1}) of (f_i^h) so that (f_i^{h+1}) converges on $\{x_1, x_2, \ldots, x_{h+1}\}$. Thus, by induction, there exists a sequence of sequences $((f_i^1), (f_i^2), \ldots, (f_i^n), \ldots)$ such that each is a subsequence of the preceding and, hence, of (f_i) and is such that for each n, $(f_i^n(x_j))_{i=1}^{\infty}$ converges for $1 \leq j \leq n$.

  (iii) Define $g_i = f_i^i$ and show that (g_i) is a subsequence of (f_i) and that $(g_i(x_j))_{i=1}^{\infty}$ converges for each $x_j \in D$.

  (iv) By making use of the facts that $(g_i(x_j))_{i=1}^{\infty}$ is a C.S. of real numbers for each $x_j \in D$ and that $\mathscr{S}$ is equicontinuous, show that (g_i) is a C.S. in $(\mathscr{C}([a, b]), \rho)$.

(c) We have been dealing in this exercise with the notion of *uniform equicontinuity*. A family of real-valued functions $\mathscr{F}$ defined on a metric space (X, d) is said to be *equicontinuous* provided that for each $x_0 \in X$ and $\varepsilon > 0$, there is a $\delta > 0$ such that if $d(x, x_0) < \delta$, then

$$|f(x) - f(x_0)| < \varepsilon$$

for all $f \in \mathscr{F}$. Show that if (X, d) is compact and $\mathscr{F}$ is an equicontinuous family of real-valued functions on X, then $\mathscr{F}$ is a uniformly equicontinuous family. Hence, the word "uniformly" can be dropped from the statement of Arzela's theorem as stated in this exercise.

(d) Prove that if $\mathscr{F}$ is a compact subset of the space $\mathscr{C}([a, b])$, then $\mathscr{F}$ is an equicontinuous family of functions. (Hint: Use the fact that a compact set is totally bounded.)

Note: For an alternate proof of Arzela's theorem that makes use of the fact that a set is compact if it is complete and totally bounded, see, for example, [15] or [21].

8. Let (X, d) be a compact metric space and let $\mathscr{C}(X)$ be the set of all real-valued continuous functions on X. Generalize the norm for $\mathscr{C}([a, b])$ to $\mathscr{C}(X)$ and prove that the resulting space is a Banach space. Prove further that if X and Y are homeomorphic compact metric spaces, then the spaces $\mathscr{C}(X)$ and $\mathscr{C}(Y)$ are also homeomorphic.

73. AN APPLICATION OF COMPLETENESS: CONTRACTION MAPPINGS

We have been studying the notion of complete metric spaces and have seen several examples that come up in analysis (e.g., $\mathbf{R}^n$, ℓ^2, $\mathscr{C}[a, b]$). We have already had occasion to use the property of completeness in proving that certain sequences converged. In this section we shall again make use of this property in an interesting and important application.

Our application is concerned with the solving of equations of the form $f(x) = x$, when f satisfies certain conditions and x takes on values in a complete metric space. (Incidentally, suppose we are dealing with a Banach space. Then considering a solution of an equation of the form $g(x) = \theta$, θ the zero element, is equivalent to considering a solution of an equation of the form $f(x) = x$, where $g(x) = x - f(x)$.) In mapping terminology we shall be considering a mapping $f:(X, d) \to (X, d)$, where (X, d) is complete and we shall be seeking a *fixed point* $x_0 \in X$ for f. A fixed point for f is a point x_0 such that $f(x_0) = x_0$.

We treat here a situation for which we can give a constructive proof of the existence of a unique solution to the type of equation just discussed.

73.1. Definition. Contraction mapping (or contractive mapping). *A mapping $f:(X, d) \to (X, d)$ is said to be a contraction mapping provided there exists a number* *k such that $0 \leq k < 1$ and for all x and y in X,*

73.1(a). $\qquad\qquad d(f(x), f(y)) \leq k\, d(x, y).$

Notice that if f is contractive, then it is uniformly continuous.

Certain terminology is useful in considering contraction mappings. We set $f^{(1)} = f$ and define $f^{(2)}$ to be $f \circ f$. Assuming $f^{(n-1)}$ has been defined, we define $f^{(n)}$ to be $f \circ f^{(n-1)}$. $f^{(n)}$ is called the nth iterate of f. Thus, if $f : X \to X$ is defined, we can define inductively the sequence $(f^{(n)})$ of iterates of f. If $x_0 \in X$, we speak of $f^{(n)}(x_0)$ as the nth iterate of x_0 under f.

73.2. Theorem. (Banach Fixed Point). *Let $f : (X, d) \to (X, d)$ be a contraction mapping from a complete metric space into itself. Let $x_0 \in X$. Then the sequence $(f^{(n)}(x_0))$ of the iterates of x_0 under f converges to a point $z_0 \in X$ and $f(z_0) = z_0$. Furthermore, z_0 is the unique fixed point for f.*

PROOF. (The proof consists of showing that $(f^{(n)}(x_0))$ is a C.S. in X and that its limit is the unique fixed point for f.)

By definition there is a k, $0 \leq k < 1$, such that 73.1(a) is satisfied. Let $x_i = f^{(i)}(x_0)$ for each $i \in \mathbf{P}$. Then

$$d(x_2, x_1) = d(f(x_1), f(x_0))$$
$$\leq k\, d(x_1, x_0).$$

Likewise,

$$d(x_3, x_2) \leq k\, d(x_2, x_1)$$
$$\leq k^2\, d(x_1, x_0).$$

Note that if we assume that

$$d(x_m, x_{m-1}) \leq k^{m-1}\, d(x_1, x_0),$$

then, as in the above, we obtain

$$d(x_{m+1}, x_m) = d(f(x_m), f(x_{m-1}))$$
$$\leq k\, d(x_m, x_{m-1})$$
$$\leq k^m\, d(x_1, x_0).$$

Thus, we have by induction that for each positive integer m

73.2(a). $\qquad\qquad d(x_{m+1}, x_m) \leq k^m\, d(x_1, x_0).$

By using first the triangle inequality and then 73.2(a), we get

$$d(x_{m+p}, x_m) \leq \sum_{i=1}^{p} d(x_{m+i}, x_{m+i-1})$$
$$\leq \sum_{i=1}^{p} k^{m+i-1}\, d(x_1, x_0)$$
$$= k^m \frac{1 - k^p}{1 - k}\, d(x_1, x_0)$$
$$< \frac{k^m}{1 - k}\, d(x_1, x_0).$$

Since $\lim (k^m) = 0$, if $\varepsilon > 0$ is given, then for m large enough

$$d(x_{m+p}, x_m) < \varepsilon \quad \text{for all} \quad p \in \mathbf{P}.$$

Hence, (x_m) is a C.S. in X and there is a $z_0 \in X$ such that $\lim (x_m) = z_0$.

We next show that $f(z_0) = z_0$. Since f is contractive, f is continuous. Hence,

$$\lim (f(x_m)) = f(\lim (x_m)) = f(z_0).$$

Then since $\lim (f(x_m)) = \lim (x_m)$,

$$z_0 = \lim (x_m) = f(z_0).$$

Furthermore, z_0 is the only fixed point for f. Suppose $f(z_1) = z_1$ and $f(z_0) = z_0$. Then $d(z_1, z_0) = d(f(z_1), f(z_0)) \leq k\, d(z_1, z_0)$ and

$$(1 - k)\, d(z_1, z_0) \leq 0.$$

Hence, since $(1 - k) > 0$, $d(z_1, z_0) \leq 0$ and, thus, $d(z_1, z_0) = 0$ and $z_1 = z_0$.

There is a wide variety of applications of the contractive principle just proved. Some of the more elementary applications are concerned with the solution of equations of the form $f(x) = x$, where f is real-valued and x takes on real values; other elementary applications have to do with the finding of solutions of systems of such equations. Some of the more sophisticated applications are in the field of differential and integral equations. Included in the next set of exercises will be some elementary applications. In the next section we shall discuss an application to the theory of first order differential equations. Nothing that will be discussed later in the book depends on that section and so the reader may omit or defer reading it without interfering with the continuity of study.

EXERCISES: CONTRACTION MAPPINGS

1. Prove that if $f: [a, b] \to [a, b]$ is differentiable and $|f'(x)| \leq K < 1$ for $x \in [a, b]$, then f is a contraction mapping on $[a, b]$.

2. Prove that if $f: (X, d) \to (X, d)$ is continuous, X is complete, and some iterate $f^{(k)}$ of f is a contraction mapping, then f has a unique fixed point x_0 and x_0 is the limit of the sequence $(f^{(n)}(y))$ of iterates of y under f, where y is an arbitrary point in X.

3. Show that $\cos: \mathbf{R} \to \mathbf{R}$ is not contractive but that one of the iterates of $\cos$ is a contraction. Illustrate the results of Exercise 2 by finding an approximate solution to the equation $\cos x = x$.

4. Let B be a Banach space and consider the mapping $f: B \to B$.
   (a) Show that if f is linear and contractive, then the zero vector is the only fixed point for f.
   (b) Show that if $f: B \to B$ is linear and $(I + f): B \to B$ is contractive, where I is the identity map on B, then f is one-to-one.

5. Suppose (X, d) is a complete metric space and (S, d) is a subspace of (X, d). Let f be a mapping from S onto X such that for an $\alpha > 1$,

$$d(f(x), f(y)) \geq \alpha\, d(x, y)$$

   for all x and y in S. Show that f has a unique fixed point in S.

6. Suppose that (X, d) is a compact metric space and $f: X \to X$ satisfies $d(f(x), f(y)) < d(x, y)$ whenever $x \neq y$.
    (a) Show by an example that the condition is not sufficient to force f to be a contraction mapping.
    (b) Prove that the conclusion of the Banach fixed point theorem holds for f, thus giving a stronger version of that theorem for compact spaces. (Hint: See Exercise 5, page 126).

7. Suppose that (X, d) is a compact metric space and (f_i) is a sequence of contraction mappings from X into X. Suppose that (f_i) converges uniformly to a mapping $f: X \to X$. Show that f has a fixed point and show by an example that it need not be unique.

74. FUNDAMENTAL EXISTENCE THEOREM FOR FIRST ORDER DIFFERENTIAL EQUATIONS—AN APPLICATION OF THE BANACH FIXED POINT THEOREM

In this section we use the Banach contraction principle to prove a classic theorem concerning the existence of a unique solution for certain first order differential equations. In the more classic approach (see, for example, page 74 in [17]), certain aspects of the proof of the Banach fixed point theorem can be seen, specialized however to the case under consideration. The content of the proof to be presented here depends on showing that the hypothesis of the Banach theorem holds.

74.1. Theorem. (Fundamental Existence—Picard). *Let f be a continuous real-valued function defined on an open subset U of $\mathbf{R}^2$. Suppose further that there exists a positive number M such that*

$$|f(x, y_1) - f(x, y_2)| \leq M |y_1 - y_2|$$

for all (x, y_1) and (x, y_2) in U. If $(x_0, y_0) \in U$, then there exists a real interval $I = [x_0 - a, x_0 + a]$ and a function $y: I \to \mathbf{R}$ such that

74.1(a). $\qquad\qquad\qquad y(x_0) = y_0$

and

74.1(b). $\qquad\qquad\qquad y'(x) = f(x, y(x)) \quad \text{for} \quad x \in I.$

Further, y is the only function that satisfies 74.1(a) and 74.1(b) on the interval I.

We make some preliminary observations before proceeding with the proof. In proving this theorem we shall consider the solution of the following integral equation for y.

74.1(c). $\qquad\qquad\qquad y(x) = y_0 + \int_{x_0}^{x} f(t, y(t)) \, dt.$

It is easy to check the validity of the following remark.

74.1(d). A function y satisfies 74.1(a) and 74.1(b) if and only if it satisfies 74.1(c).

Our strategy will be to find a complete space $\mathscr{C}$ of functions and a contractive mapping $\Psi:\mathscr{C} \to \mathscr{C}$ that will have the following property: If there is a solution y to 74.1(c), then $y \in \mathscr{C}$ and y is a solution if and only if $\Psi(y) = y$. To accomplish this, notice first that any possible solution y must be such that $(x, y(x)) \in U$ for any x in the domain of y so that $f(x, y(x))$ will be defined. Other considerations that lead to the choice of the space $\mathscr{C}$ will become clear to the reader only after the proof progresses.

PROOF OF 74.1. Since $(x_0, y_0) \in U$ and U is open, there is a rectangle

$$S = [x_0 - c, x_0 + c] \times [y_0 - b, y_0 + b] \subset U.$$

Also define

$$K = \text{l.u.b.} \{|f(x, y)| : (x, y) \in S\} + 1.$$

Let a be a number chosen so that

74.1(e).
$$0 < a < \min\left\{c, \frac{1}{M}, \frac{b}{K}\right\}.$$

Note that

$$[x_0 - a, x_0 + a] \times [y_0 - b, y_0 + b] \subset U.$$

Let $\mathscr{C}([x_0 - a, x_0 + a], \rho)$ be the metric space studied in 72. (This space just defined will not be exactly suitable for our purposes, but the following closed subset of it will be. The reader should check for himself that the set to be defined next is a closed subset of the previously defined space which is known to be complete.) Let $\mathscr{C}$ be the closed subset of $\mathscr{C}([x_0 - a, x_0 + a], \rho)$ consisting of all elements y of that space such that $y(x) \in [y_0 - b, y_0 + b]$ for all $x \in [x_0 - a, x_0 + a]$. Since $\mathscr{C}$ is a closed subset of a complete space it is complete. Notice that if $y \in \mathscr{C}$, then for all $x \in [x_0 - a, x_0 + a]$,

74.1(f).
$$|y(x) - y_0| \leq b$$
and
$$(x, y(x)) \in U.$$

We define Ψ on $\mathscr{C}$ as follows:
For each $g \in \mathscr{C}$, let $\Psi(g)$ be the function given by

74.1(g).
$$\Psi(g)(x) = y_0 + \int_{x_0}^{x} f(t, g(t)) \, dt.$$

We first show that if $g \in \mathscr{C}$, then $\Psi(g) \in \mathscr{C}$ so that $\Psi(\mathscr{C}) \subset \mathscr{C}$. From 74.1(f), if $g \in \mathscr{C}$, then $|g(x) - y_0| \leq b$ for $x \in [x_0 - a, x_0 + a]$ and so $f(x, g(x))$ is defined for $x \in [x_0 - a, x_0 + a]$. Then, for all x in $[x_0 - a, x_0 + a]$,

$$|\Psi(g)(x) - y_0| = \left| \int_{x_0}^{x} f(t, g(t)) \, dt \right|$$

$$\leq \left| \int_{x_0}^{x} |f(t, g(t))| \, dt \right|$$

$$\leq \left| \int_{x_0}^{x} \text{l.u.b.} \{|f(t, y)| : (t, y) \in S\} \, dt \right|$$

$$\leq \text{l.u.b.} \{|f(t, y)| : (t, y) \in S\}a.$$

From 74.1(e) and the definition of K, it follows that

$$|\Psi(g)(x) - y_0| < b \quad \text{for} \quad x \text{ in } [x_0 - a, x_0 + a].$$

The last inequality and the definition of the set $\mathscr{C}$ places $\Psi(g)$ in $\mathscr{C}$. Thus, we have shown that $\Psi[\mathscr{C}] \subset \mathscr{C}$.

We next show that $\Psi : \mathscr{C} \to \mathscr{C}$ is a contraction. Let $I = [x_0 - a, x_0 + a]$. Let g_1 and g_2 be elements of $\mathscr{C}$. Then

$$\rho(\Psi(g_1), \Psi(g_2)) = \text{l.u.b. } \{|\Psi(g_1)(x) - \Psi(g_2)(x)| : x \in I\}$$

$$= \text{l.u.b. } \left\{ \left| \int_{x_0}^{x} (f(t, g_1(t)) - f(t, g_2(t))) \, dt \right| : x \in I \right\}$$

$$\leq \text{l.u.b. } \left\{ \left| \int_{x_0}^{x} |f(t, g_1(t)) - f(t, g_2(t))| \, dt \right| : x \in I \right\}$$

$$\leq \text{l.u.b. } \left\{ \left| \int_{x_0}^{x} M \, |g_1(t) - g_2(t)| \, dt \right| : x \in I \right\}$$

$$\leq \text{l.u.b. } \left\{ \left| \int_{x_0}^{x} M \rho(g_1, g_2) \, dt \right| : x \in I \right\}$$

$$\leq \text{l.u.b. } \{ M \rho(g_1, g_2) \, |x - x_0| : x \in I \}$$

$$= M a \rho(g_1, g_2).$$

From 74.1(e), $Ma < 1$ and, hence, from the last inequality we see that $\Psi : \mathscr{C} \to \mathscr{C}$ is a contraction. Hence, from the Banach fixed point theorem there is a unique y in $\mathscr{C}$ such that $\Psi(y) = y$. Now y is a solution for 74.1(b) subject to the condition 74.1(a). If there were another solution y_1 for 74.1(b) satisfying 74.1(a) and which is defined on the interval I, then y_1 is not in $\mathscr{C}$. Hence for some x in I, $|y_1(x) - y_0| > b$. Then there would exist a first x in I for which $|y_1(x) - y_0| = b$. Hence for that particular x we would have

$$y_1(x) = y_0 + \int_{x_0}^{x} f(t, y_1(t)) \, dt.$$

But then

$$|y_1(x) - y_0| \leq \left| \int_{x_0}^{x} K \, dt \right|$$

$$\leq Ka.$$

But $Ka < b$ follows from 74.1(e). Thus we have a contradiction. Hence, $y_1 = y$. Thus, we have shown that there is an interval $I = [x_0 - a, x_0 + a]$ and a unique solution on I to 74.1(a) and 74.1(b).

It is to be noted that in a situation covered by the previous theorem we can find an approximate solution by first determining a suitable space $\mathscr{C}$ and then choosing any g_0 in $\mathscr{C}$ to generate the various Ψ-iterates of g_0.

In Chapter 5 of Reference [17], both a classic proof of the Picard theorem and the contraction mapping approach are given. Also included there are some exercises illustrating the method. The reader who wishes to pursue this topic further is referred to the reference just cited. For other references to this topic, see [5], [14].

6

General Topological Spaces and Mappings on Topological Spaces

We have seen that the notion from analytic geometry of distance can be generalized to distance between n-tuples of real numbers or points in $\mathbf{R}^n$ and also to the more general notion of a metric. Ultimately, all the concepts for metric spaces with which we have been dealing rest on the notion of distance between points. For a metric space (X, d) we used the phrase "topology generated by d" to refer to the collection $\mathcal{T}(d)$ of all open subsets of (X, d). Most, but not all, of the concepts that we have studied thus far depend only on the open subsets of the space. Recall that we referred to these properties as topological properties.

Notice that by means of the topology of a metric space (X, d) we can capture some of the qualitative aspects of "closeness." For example, consider the property possessed by a point $x \in X$ if x is "very close" to a set S in the sense that $x \in \text{cl}(S)$; this property can be characterized by requiring that every open subset U containing x intersect S. Likewise, we may roughly think of $\lim(x_i) = x$ as asserting that the x_i's get and stay arbitrarily close to x; this will hold if and only if for every open set U containing x, $x_i \in U$ for all but a finite number of i's. Also we may think of $f: (X, d) \to (Y, \rho)$ as being continuous at x_0 provided that points "close to" x_0 map onto points "close to" $f(x_0)$; in terms of open sets, f is continuous at x_0 provided that for each open set V containing $f(x_0)$, there is an open set U containing x_0 such that $f[U] \subset V$.

The considerations in the previous paragraph give rise to the following: Suppose we are studying a collection M of mathematical objects and mappings defined on this collection and we think it useful to consider such questions as *continuity*, *connectedness*, *compactness*, etc. Suppose on the other hand that we do not seem to be able to define a metric that appears to be useful or natural for the purpose at hand. Can we by-pass the defining of a metric and still define a collection of subsets of X that in some useful way behaves like the collection of open subsets of a metric space? In so doing we would wish to retain some of the qualitative aspects of closeness. Often this can be done, and the notion suggested here leads to the concept of a general topological space.

Our plan of attack will be as follows: Given a set X, we shall designate a certain collection $\mathcal{T}$ of subsets of X as the *open* sets or *topology* for X provided the collection satisfies certain conditions (to be discussed later) and we shall then call $(X, \mathcal{T})$ a topological space. The more properties $\mathcal{T}$ possesses that are similar to the properties enjoyed by the topology of a metric space, the more nearly will $(X, \mathcal{T})$ behave like a metric space.

Once we embark on the approach that we have outlined, another natural question arises. Suppose we have a topological space. Does there exist a metric d for X such that the collection $\mathcal{T}(d)$ of all open subsets of the metric space (X, d) is the same as the collection $\mathcal{T}$? This cannot always be done and when there is such a metric d, in general it is not unique. As we shall see, however, if $\mathcal{T}$ satisfies enough conditions there does exist a metric d such that the d-open sets are precisely the elements of $\mathcal{T}$. In such a case we shall say that the topological space is *metrizable*. On the other hand, if $\mathcal{T}$ fails to possess even one property possessed by the topologies of all metric spaces (for example, normality), then obviously $(X, \mathcal{T})$ could not be metrizable.

In this chapter we shall define the notion of topological space. We shall then be able to extend such notions as *closed set*, *limit point*, and *closure* of a set. Some of the special types of topological spaces to be studied in this chapter are T_1, *Hausdorff*, *regular*, *normal*, *separable*, *first countable*, and *second countable* *spaces*. We shall also extend the notion of *continuity* and *homeomorphism* to mappings from one topological space onto another.

75. TOPOLOGICAL SPACES

We define next what is meant by a topology $\mathcal{T}$ for a set X. As remarked previously, the definition is motivated by some of the properties possessed by the collection $\mathcal{T}(d)$ of open sets generated by a metric d for a set X. We also shall give other examples of topologies for various sets.

75.1. Definition. Topology for a set. *Let X be a set and let $\mathcal{T}$ be a collection of subsets of X. The collection $\mathcal{T}$ is called a topology for X provided $\mathcal{T}$ satisfies the following set of axioms:*

(a) $\varnothing$ *and X are elements of $\mathcal{T}$,*
(b) *if $U_1 \in \mathcal{T}$ and $U_2 \in \mathcal{T}$, then $U_1 \cap U_2 \in \mathcal{T}$,*
(c) *if $\mathcal{K}$ is an arbitrary subcollection of $\mathcal{T}$, then $\bigcup \mathcal{K} \in \mathcal{T}$.*

It is clear from (b) that if $\mathcal{K}$ is any nonempty *finite* subcollection of $\mathcal{T}$, then $\bigcap \mathcal{K} \in \mathcal{T}$.

75.2. Definitions. Topological space; open set. *If $\mathcal{T}$ is a topology for a set X, then $(X, \mathcal{T})$ is said to be a topological space. A subset U of X is said to be open in $(X, \mathcal{T})$ or an open subset of $(X, \mathcal{T})$ provided $U \in \mathcal{T}$.*

Analogous to the language used for metric spaces, we often speak of a set as being *open in X* rather than *open in* $(X, \mathcal{T})$. The expression $\mathcal{T}$*-open* is also used for *open in* $(X, \mathcal{T})$. We shall follow the same convention relative to other concepts to be introduced (e.g., closed set, closure of a set, limit point). In previous chapters we dealt with the notion of metric space. Suppose $\mathcal{T}(d)$ is the collection of open sets in a metric space (X, d). From Theorem 47.1, it follows that $\mathcal{T}(d)$ is a topology for X. Recall that we referred to $\mathcal{T}(d)$ as the *topology generated by d*. We see then that, in accordance with Definition 75.2, $(X, \mathcal{T}(d))$ is an example of a topological space. Thus, given a metric space (X, d) there is associated with it a topological space $(X, \mathcal{T}(d))$. Also we see that if d_1 and d_2 are equivalent metrics for a set X, then $(X, \mathcal{T}(d_1)) = (X, \mathcal{T}(d_2))$.

75.3. *Example.* Let X be a set. Let $\mathscr{P}(X)$ be the collection of all subsets of X (i.e., the power set of X). Then $\mathscr{P}(X)$ is a topology for X and is referred to as the *discrete* topology for X. Let $\mathcal{T} = \{\varnothing, X\}$. The collection $\mathcal{T}$ is known as the *trivial* topology for X.

75.4. *Example.* Let X be a nonempty set and let $a \in X$. Further, let $\mathcal{T} = \{\varnothing, \{a\}, X\}$. The collection $\mathcal{T}$ is a topology for X and, thus, $(X, \mathcal{T})$ is a topological space.

75.5. *Example.* Let X be a set consisting of four objects a, b, c, and d. Let $\mathcal{T} = \{\{a\}, \{a, b\}, \{a, c\}, \{a, b, c\}, X, \varnothing\}$. The collection $\mathcal{T}$ is a topology for X.

75.6. *Example.* Let X be the set of all real numbers. Let $\mathcal{T} = \{\varnothing\} \cup \{X - F : F$ is a finite subset of $X\}$. The collection $\mathcal{T}$ is a topology for X.

75.7. *Example.* Let X be the set of all real numbers. Let $\mathcal{T} = \{\varnothing\} \cup \{X - C : C$ is a countable subset of $X\}$. Then $\mathcal{T}$ is a topology for X.

For a topological space, as for a metric space, we shall call a property a *topological property* if it can be characterized by the open subsets or the topology of the space. In dealing with the topological properties of a metric space (X, d), we need be concerned only with the topology $\mathcal{T}(d)$ generated by d and, hence, with the topological space $(X, \mathcal{T}(d))$. Generally, in considering a metric space, we shall use the notation (X, d) for both the metric space (X, d) and for the topological space associated with it.

75.8. **Definition. Metrizable space.** *If $(X, \mathcal{T})$ is a topological space and there is a metric d for X such that the topology $\mathcal{T}(d)$ for X generated by d is the same as $\mathcal{T}$, then the topological space $(X, \mathcal{T})$ is called a metrizable space.*

In Example 75.3 $(X, \mathscr{P}(X))$ is a metrizable space since $\mathscr{P}(X)$ can be generated by the metric given in Example 45.4. On the other hand, the space $(X, \mathcal{T})$ in Example 75.5 is not metrizable. Suppose that it were. We know that in a metric space if x and y are distinct points in the space, then there exist disjoint open sets U and V such that $x \in U$ and $y \in V$. Clearly this cannot be done in this example.

EXERCISES: TOPOLOGICAL SPACES

1. In Examples 75.3 through 75.7, verify that the collection of subsets is a topology for the given set.

2. Show that the topological space $(X, \mathcal{T})$ of Example 75.6 is not metrizable. Do the same for the topological space of Example 75.7.

3. Let $X = \{1, 2, 3\}$.
    (a) Find all the topologies there are for the set X.
    (b) Find a collection $\mathcal{K}$ of subsets of X such that $\varnothing \in \mathcal{K}$, $X \in \mathcal{K}$, and which is such that $\mathcal{K}$ is not a topology for X.
    (c) Could one do (b), if X had fewer than three elements?

4. Let $\mathbf{P}$ be the set of all positive integers. For each $n \in \mathbf{P}$, let $\mathbf{P}_n$ denote the set consisting of the first n positive integers. Let $\mathcal{K} = \{\varnothing\} \cup \{\mathbf{P}\} \cup \{\mathbf{P}_n : n \in \mathbf{P}\}$. Is $\mathcal{K}$ a topology for $\mathbf{P}$?

5. Let X be the set of all real numbers. Let $\mathcal{K} = \{\varnothing\} \cup \{U : U \subset X - \{0\}$ or $U = X\}$. Is $\mathcal{K}$ a topology for X?

6. Let X be the set of all real numbers. For each $a \in X$, let $U_a = \{r : a < r\}$. Is the collection $\{\varnothing\} \cup \{X\} \cup \{U_a : a \in X\}$ a topology for X? For each a in X, let $F_a = \{r : a \leq r\}$. Is $\{\varnothing\} \cup \{X\} \cup \{F_a : a \in X\}$ a topology for X?

7. Let $\mathcal{C}$ be the set of all continuous real-valued functions defined on $\mathbf{R}$. For every compact subset K of $\mathbf{R}$, f in $\mathcal{C}$, and positive number $\varepsilon > 0$, let
    $$U(f, K, \varepsilon) = \{g : g \in \mathcal{C} \quad \text{and}$$
    $$|f(x) - g(x)| < \varepsilon \quad \text{for all} \quad x \in K\}.$$
    Let $\mathcal{K} = \{U(f, K, \varepsilon) : K$ is a compact subset of $\mathbf{R}$, $f \in \mathcal{C}$, $\varepsilon > 0\}$. Further, let $\mathcal{T} = \{\bigcup \mathcal{S} : \mathcal{S}$ is a subcollection of $\mathcal{K}\}$. Is $\mathcal{T}$ a topology for $\mathcal{C}$? ($\varnothing$ is in $\mathcal{T}$ since $\varnothing$ is the union of the empty subcollection of $\mathcal{K}$.)

76. BASE FOR A TOPOLOGY

Recall that we generated the open subsets of a metric space by means of the ε-neighborhoods. A subset U is open provided that for each $x \in U$ there is a neighborhood $N(x; \varepsilon)$ of x that is contained in U. This suggests the concept of a *base* for a topology, to be defined next.

76.1. Definition. Base for a topology. *Suppose that $(X, \mathcal{T})$ is a topological space. Then a subcollection $\mathcal{B}$ of $\mathcal{T}$ is said to be a base for $\mathcal{T}$ provided the following condition holds: for each $U \in \mathcal{T}$ and $x \in U$, there is a $W_x \in \mathcal{B}$ such that $x \in W_x \subset U$.*

We see from the definition that each element of a base $\mathcal{B}$ is an open set and that $\mathcal{B}$ is a covering of the space. However, a base is not an arbitrary open covering. A base $\mathcal{B}$ has the property that every open set in the space is the union of some subcollection of $\mathcal{B}$. The empty set, of course, is the union of the empty subcollection of $\mathcal{B}$. If U is a nonempty open subset of the space, then for each $x \in U$, we may choose a $W_x \in \mathscr{B}$ such that $x \in W_x \subset U$ and, hence, we may write $U = \bigcup \{W_x : x \in U\}$. It is also important to note that a topology may have different bases and that the topology is itself a base. For example, if d_1 and d_2 are two different but equivalent metrics for a set X, then the collections

$$\{N_{d_1}(x; \varepsilon) : x \in X, \varepsilon > 0\}$$

and

$$\{N_{d_2}(x; \varepsilon) : x \in X, \varepsilon > 0\}$$

are two different bases for the same topology for X. Thus, the collection of all "circular" neighborhoods and the collection of all "square" neighborhoods are two different bases for the same topology.

One of the reasons that makes the notion of a base useful is that the base can be given first and the topology subsequently generated. However, not every covering of a set can generate a topology for which it is a base. The following theorem gives a necessary and sufficient condition for a collection of subsets of a set to be a base for some topology for that set.

76.2. Theorem. *Let X be a set and let $\mathscr{B}$ be a collection of subsets of X such that $X = \bigcup \mathscr{B}$. Then $\mathscr{B}$ is a base for a topology for X if and only if the following condition holds:*

If $U \in \mathscr{B}$, $V \in \mathscr{B}$, and $x \in U \cap V$, then there is a $W \in \mathscr{B}$ such that $x \in W \subset U \cap V$.

PROOF. Suppose first that $\mathscr{B}$ is a base for a topology $\mathscr{T}$ for X. Assume that $U \in \mathscr{B}$, $V \in \mathscr{B}$, and $x \in U \cap V$. From the definition of base, U and V are elements of $\mathscr{T}$. Hence, $U \cap V \in \mathscr{T}$. Again from the definition of base, there is a $W \in \mathscr{B}$ such that $x \in W \subset U \cap V$. Therefore, the condition is satisfied. Conversely, suppose that the condition is satisfied. Let $\mathscr{T}(\mathscr{B})$ be the collection of all unions of subcollections of $\mathscr{B}$. That is, let

$$\mathscr{T}(\mathscr{B}) = \{\bigcup \mathscr{F} : \mathscr{F} \subset \mathscr{B}\}$$

Then $\mathscr{T}(\mathscr{B})$ is a topology for X and $\mathscr{B}$ is a base for $\mathscr{T}(\mathscr{B})$. The explanation of this is left as an exercise for the reader.

It is easy to verify that if $\mathscr{B}$ is a base for a topology then it is a base for only one topology. Hence, if a collection $\mathscr{B}$ of subsets of X satisfies the condition stated in the previous theorem, it is appropriate to speak of *the* topology $\mathscr{T}(\mathscr{B})$ for X *generated by $\mathscr{B}$.*

It is convenient to have a test for determining whether two collections $\mathscr{B}_1$ and $\mathscr{B}_2$ which are bases for topologies $\mathscr{T}(\mathscr{B}_1)$ and $\mathscr{T}(\mathscr{B}_2)$ for X, respectively, are *equivalent* in the sense of generating the same topologies for X (i.e., $\mathscr{T}(\mathscr{B}_1) = \mathscr{T}(\mathscr{B}_2)$). The answer is given by the following theorem which is an analog of Theorem 48.4.

76.3. Theorem. *Suppose that $\mathscr{T}_1$ and $\mathscr{T}_2$ are topologies for a set X. Suppose furthermore that $\mathscr{B}_1$ is a base for $\mathscr{T}_1$ and $\mathscr{B}_2$ is a base for $\mathscr{T}_2$. Then $\mathscr{T}_1 = \mathscr{T}_2$ if and only if the following condition is satisfied: For each $B_1 \in \mathscr{B}_1$ and $x \in B_1$, there is a $B_2 \in \mathscr{B}_2$ such that $x \in B_2 \subset B_1$; and for each $B_2 \in \mathscr{B}_2$ and $x \in B_2$ there is a $B_1 \in \mathscr{B}_1$ such that $x \in B_1 \subset B_2$.*

The proof of this theorem is left as an exercise. It is suggested that the theorem be proved by first proving the next statement. Theorem 76.3 will then follow as an immediate corollary.

76.3(a) Suppose that $\mathcal{T}_1$ and $\mathcal{T}_2$ are topologies for X. Suppose further that $\mathcal{B}_1$ is a base for $\mathcal{T}_1$ and $\mathcal{B}_2$ is a base for $\mathcal{T}_2$; then $\mathcal{T}_1 \subset \mathcal{T}_2$ if and only if the following condition is satisfied: for each $B_1 \in \mathcal{B}_1$ and $x \in B_1$, there is a $B_2 \in \mathcal{B}_2$ such that $x \in B_2 \subset B_1$.

Sometimes it is convenient to generate a base itself from what is called a *subbase*.

76.4. Definition. Subbase. *Let* $(X, \mathcal{T})$ *be a topological space. A subcollection* $\mathcal{S}$ *of* $\mathcal{T}$ *is said to be a subbase for* $\mathcal{T}$ *provided the family of intersections of nonempty finite subcollections of* $\mathcal{S}$ *is a base for* $\mathcal{T}$. *That is,*

$$\{\bigcap \mathcal{F} : \mathcal{F} \text{ is a nonempty finite subcollection of } \mathcal{S}\}$$

is a base for $\mathcal{T}$.

From this definition we see that an open covering $\mathcal{S}$ of $(X, \mathcal{T})$ is a subbase for $\mathcal{T}$ if and only if it has the following property: For each $U \in \mathcal{T}$ and $x \in U$, there is a subcollection $\{U_1, U_2, \ldots, U_n\}$ of $\mathcal{S}$ such that $x \in \bigcap \{U_i : i \in \mathbf{P}_n\} \subset U$.

76.5. Example. For each $a \in \mathbf{R}$, let $(-\infty, a)$ denote the infinite interval $\{x : x < a\}$. Similarly, (a, ∞) will denote the infinite interval $\{x : a < x\}$. The collection

$$\mathcal{S} = \{(-\infty, a) : a \in \mathbf{R}\} \cup \{(a, \infty) : a \in \mathbf{R}\}$$

is a subbase for the Euclidean topology for $\mathbf{R}$.

Suppose that $\mathcal{S}$ is a nonempty collection of subsets of X such that $\bigcup \mathcal{S} = X$. $\mathcal{S}$ need not be a base for a topology for X since $\mathcal{S}$ need not satisfy the conditions in Theorem 76.2. However, the situation is different with respect to a subbase, for if we let $\mathcal{B}$ be the collection of finite intersections of $\mathcal{S}$, then $\mathcal{B}$ is the base for a topology $\mathcal{T}(\mathcal{B})$ for X. This is seen at once by an application of Theorem 76.2. Suppose $x \in U \cap V$ where U and V are elements of $\mathcal{B}$. Since U and V are each the intersection of a finite subcollection of $\mathcal{S}$, so is $U \cap V$. Hence, $x \in W = U \cap V$ and the condition in 76.2 is satisfied. Thus, we can make the following definition.

76.6. Definition. Topology generated by a nonempty collection of sets. *Let* X *be a set and let* $\mathcal{S}$ *be a nonempty collection of subsets of* X *such that* $X = \bigcup \mathcal{S}$. *Let* $\mathcal{B}$ *be the collection of intersections of finite nonempty subcollections of* $\mathcal{S}$. *That is,*

$$\mathcal{B} = \{\bigcap \mathcal{F} : \mathcal{F} \text{ is a nonempty finite subcollection of } \mathcal{S}\}.$$

Then the topology $\mathcal{T}(\mathcal{B})$ *for* X *generated by* $\mathcal{B}$ *(see 76.2} is called the topology for* X *generated by* $\mathcal{S}$ *as a subbase.*

76.7. Example. Let X be the set consisting of four objects a, b, c, and d. Let $\mathcal{S} = \{\{a, b\}, \{a, c\}, \{a, b, c, d\}\}$. Notice that $\mathcal{S}$ is not a base for a topology for X. For $a \in \{a, b\} \cap \{a, c\}$ but for no $W \in \mathcal{S}$ is it true that $a \in W \subset \{a, b\} \cap \{a, c\}$. However, if we take $\mathcal{B}$ as in the previous definition, we can check directly that $\mathcal{B}$ is a base for a topology for X. For $\mathcal{B} = \{\{a\}, \{a, b\}, \{a, c\}, \{a, b, c, d\}\}$.

Notice that the collection of all arbitrary unions of subcollections (including the empty subcollection) of $\mathscr{B}$ is the collection of sets $\{\varnothing, \{a\}, \{a, b\}, \{a, c\}, \{a, b, c\}, \{a, b, c, d\}\}$. This collection is the same topology given for X in Example 75.5.

Let $\{(X_i, \mathscr{T}_i) : i \in \mathbf{P}_n\}$ be a finite collection of topological spaces. Making use of the notion of a subbase is a convenient way of defining a useful topology for $\mathsf{X}\{X_i : i \in \mathbf{P}_n\}$, called the product topology. In the special case that each $\mathscr{T}_i$ is the topology generated by a metric d_i, the product topology to be defined is the same as that generated by the product metric (see 51). In what follows, recall that if $\{X_i : i \in \mathbf{P}_n\}$ is a finite collection of sets, then by the Cartesian set $\mathsf{X}\{X_i : i \in \mathbf{P}_n\}$ is meant the collection of all finite-sequences or ordered n-tuples $(x_1, x_2, \ldots, x_n)$ such that $x_i \in X_i$ for each $i \in \mathbf{P}_n$. Recall also that for each $i \in \mathbf{P}_n$, the projection map $\pi_i : \mathsf{X}\{X_i : i \in \mathbf{P}_n\} \to X_i$ is defined by $\pi_i(x) = \pi_i((x_1, x_2, \ldots, x_n)) = x_i$.

76.8. Definition. Product topology for finite collections. *Let* $\{(X_i, \mathscr{T}_i) : i \in \mathbf{P}_n\}$ *be a finite collection of topological spaces. Then by the product topology* $\mathscr{T}$ *we shall mean the topology for* $X = \mathsf{X}\{X_i : i \in \mathbf{P}_n\}$ *generated by*

$$\mathscr{S} = \{U : U = \pi_i^{-1}[U_i] \quad \text{for an} \quad i \in \mathbf{P}_n \quad \text{and an open set} \quad U_i \subset X_i\}$$

as a subbase. The space $(X, \mathscr{T})$ *is called the product space.*

It is useful to notice, for example, that $\pi_1^{-1}[U_1] = U_1 \times X_2 \times X_3 \times \cdots \times X_n$. It is also useful to notice that the base $\mathscr{B}$ generated by $\mathscr{S}$ is the collection of all subsets of the form $U_1 \times U_2 \times \cdots \times U_n$ where each U_i is open in X_i. Thus, it follows from Theorem 51.2 that in the metric case the product topology is the same as that generated by the product metric.

76.9. *Example.* Let $X_1 = \{a, b, c\}$ and let $X_2 = \{1, 2, 3\}$. For topologies for X_1 and X_2, respectively, we take the collections $\mathscr{T}_1 = \{\varnothing, \{a\}, \{a, b\}, \{a, b, c\}\}$ and $\mathscr{T}_2 = \{\varnothing, \{1\}, \{1, 2\}, \{1, 2, 3\}\}$. Then the collection $\mathscr{S}$ as given next is a subbase for the product topology. $\mathscr{S} = \{\{a\} \times \{1, 2, 3\}, \{a, b\} \times \{1, 2, 3\}, \{a, b, c\} \times \{1, 2, 3\}, \{a, b, c\} \times \{1\}, \{a, b, c\} \times \{1, 2\}, \{a, b, c\} \times \{1, 2, 3\}, \varnothing\}$. The reader should write out the base that is generated by $\mathscr{S}$.

In the last chapter of the text we shall generalize the notion of product spaces to include infinite products. At that juncture we shall discuss product spaces in more detail. However, it will be instructive to give some consideration to finite products in the next set of exercises and in some of the subsequent sections. Some of our work on finite products will serve later as motivation for the more general considerations in the last chapter.

EXERCISES: BASE FOR A TOPOLOGY

1. Complete the proof of Theorem 76.2. Prove Theorem 76.3.

2. Show that the following collection $\mathscr{S}$ of subsets of $\mathbf{R}^2$ is a subbase for the topology generated by the Euclidean metric.

   $\mathscr{S}$ is the collection of all sets that have one of the following forms:

   $$\{(x, y) : y < a\}, \{(x, y) : y > b\}, \{(x, y) : x < c\}, \{(x, y) : x > d\}.$$

3. Let $\mathscr{B}$ be the collection $\{N(p;r):p \in \mathbf{R}^2$ and both coordinates of p are rational, r is rational$\}$. Show that $\mathscr{B}$ is a base for the topology for $\mathbf{R}^2$ generated by the Euclidean metric.

4. Let $\mathscr{C}$ be the collection of all nondegenerate closed intervals in $\mathbf{R}$. Is $\mathscr{C}$ a base for a topology for the set $\mathbf{R}$?

5. Let $\mathscr{C}$ be the collection of all subsets of $\mathbf{R}$ of the form $\{x:\alpha \le x < \beta\}$. Is $\mathscr{C}$ a base for a topology for the set $\mathbf{R}$?

6. Let Q consist of the points of $\mathbf{R}^2$ with nonnegative y coordinates. Let $\mathscr{B}$ be the collection of all subsets S of Q such that S is either a Euclidean open neighborhood if the circular boundary of the neighborhood does not intersect the x-axis, or S is a Euclidean open neighborhood, together with the point of tangency if its boundary is tangent to the x-axis. Show that $\mathscr{B}$ is a base for a topology for Q.

7. Let X be a set. Let $\mathscr{B} = \{\{x\}:x \in X\}$. Show that $\mathscr{B}$ is a base for the discrete topology for X. (See Example 75.3.)

8. For each real number r, define the following subsets of the set $\mathbf{R}^2$. $F_r = \{(x,y):x+y \ge r\}$ and $U_r = \{(x,y):x+y > r\}$. Let $\mathscr{F} = \{F_r:r \in \mathbf{R}\}$ and let $\mathscr{U} = \{U_r:r \in \mathbf{R}\}$. Is the collection $\mathscr{F}$ a base for a topology for the set $\mathbf{R}^2$? Is $\mathscr{U}$ a base for a topology for the set $\mathbf{R}^2$?

9. For each $(a,b) \in \mathbf{R}^2$ and $\varepsilon > 0$, define the following subset of $\mathbf{R}^2$:

    $U(a,b,\varepsilon) = \{(x,y):a - \varepsilon < x \le a + \varepsilon$ and

    $$b - \varepsilon < y \le b + \varepsilon\}.$$

    Let $\mathscr{U} = \{U(a,b,\varepsilon):(a,b) \in \mathbf{R}^2,\ \varepsilon > 0\}$. Is $\mathscr{U}$ a base for a topology for $\mathbf{R}^2$?

10. Let $(X, \le)$ be a totally ordered set, with X consisting of more than one point (see 22.1 and 22.5). We shall use the notation $a < b$ to mean that $a \le b$ and $a \ne b$. Thus, for each two different elements of X exactly one of the following holds: $a < b$ or $b < a$. Let $\mathscr{S}$ be the collection of all subsets $U \subset X$ such that U has the form $\{x:x < a\}$ or $\{x:a < x\}$. Show that $\mathscr{S}$ is a subbase for a topology for X. Describe the base $\mathscr{B}$ that is generated by $\mathscr{S}$. The topology generated in this manner is called the *order topology* for the totally ordered system $(X, \le)$. Notice that for the special case of the real number system with the usual ordering, the order topology is the same as the Euclidean topology for $\mathbf{R}$ (see 76.5).

11. Let $(X, \mathscr{T})$ be the topological space in Example 75.4. Determine the product topology for $X \times X$.

12. Let $X_1 = \{a, b, c\}$ and $X_2 = \{a, b, c\}$. Let

$$\mathscr{T}_1 = \{\varnothing, \{a\}, X_1\} \quad \text{and let} \quad \mathscr{T}_2 = \{\varnothing, \{a\}, \{a, b\}, X_2\}.$$

Determine the product topology for the collection

$$\{(X_i, \mathscr{T}_i) : i = 1, 2\}.$$

77. SOME BASIC DEFINITIONS

In dealing with metric spaces in previous sections the reader should have become thoroughly familiar with certain concepts completely dependent on the notion of openness. In this section we list a number of such definitions extended to the setting of general topological spaces. A few are introduced for the first time in the text here but, of course, would have applied equally well for metric spaces. In the next section we shall consider fundamental theorems involving these concepts.

In each of the following definitions $(X, \mathscr{T})$ is a topological space.

77.1. Definition. Limit point. *Suppose $S \subset X$ and $p \in X$. Then p is a limit point of S (with respect to $(X, \mathscr{T})$) provided that for each open set U containing p, U intersects S in at least one point distinct from p.*

(In general, we shall omit "with respect to $(X, \mathscr{T})$" in the previous and following terms unless there is a chance for confusion, as, for example, if two different topologies on X are competing for our attention.)

77.2. Definition. Closed set. *A set $S \subset X$ is said to be closed provided it contains all its limit points.*

77.3. Definition. Closure of a set. *The union of a subset S of X and the set of all its limit points is called the closure of S and will be denoted by* cl (S).

77.4. Definition. Interior point and interior of a set. *If $S \subset X$, then x is said to be an interior point of S provided there exists an open subset U such that $x \in U \subset S$. The set of all interior points of a set S is called the interior of S and abbreviated* int (S).

It should be noted that a set U is open if and only if $U =$ int (U). Also a set F is closed if and only if cl $(F) = F$. Furthermore, it is obvious from the definitions that int $(A) \subset A \subset$ cl (A) for all subsets A of X.

77.5. Definition. Exterior of a set. *The interior of the complement of a set S is called the exterior of S and is abbreviated* ext (S).

Observe that each point of the exterior of a set S is neither a point of S nor a limit point of S. Thus, if a set is closed, its exterior is simply its complement.

77.6. Example. Let $\mathbf{R}^2$ be endowed with the Euclidean topology and let $S = \{(x, y) : (x, y) \in \mathbf{R}^2 \text{ and } 0 < x^2 + y^2 < 1\}$. Then cl $(S) = \{(x, y) : x^2 + y^2 \leq 1\}$, int $(S) = S$, and ext $(S) = \{(x, y) : x^2 + y^2 > 1\}$.

A point is in the closure of a set if and only if it is either a point of the set or a limit point of the set. Hence, one thinks of a point in the closure of a set as being "close" to the set. We next define the notion of *boundary point* of a set. We shall want to think of a boundary point of a set as being a point that is both "close" to the set and to the complement of the set.

77.7. Definition. Boundary point; boundary of a set. *If $S \subset X$ and $p \in X$, then p is said to be a boundary or frontier point of S if every open set containing p intersects both S and the complement of S. The set of all boundary points of S is called the boundary or frontier of S and is abbreviated* Fr (S).

It is easy to see that Fr $(S) = $ cl $(S) \cap$ cl $(X - S)$. For the ball $B(\Theta; r)$ and for the neighborhood $N(\Theta; r)$ in $\mathbf{R}^2$, the boundary is the circle

$$\{(x, y) : x^2 + y^2 = r^2\}.$$

The boundary of the set $\{(x, y) : 1 < x^2 + y^2 < 4\}$ is the union of the circles given by $x^2 + y^2 = 1$ and $x^2 + y^2 = 4$. However, boundaries of sets need not look like what we might ordinarily think of as "boundary curves." For example, the boundary of the set $\{(x, y) : x$ and y are rational$\} \subset \mathbf{R}^2$ is $\mathbf{R}^2$ itself.

77.8. Definition. Dense set. *A subset D of a topological space X is said to be dense in X provided* cl $(D) = X$.

We saw that, for a metric space, D is dense in the space if and only if every nonempty open subset of the space has a nonempty intersection with D. It is easy to see that this characterization also holds for topological spaces in general. A point x in a space is called an *isolated point* provided $\{x\}$ is an open subset of the space. Observe that if D is a dense subset of a space, then it must contain all the isolated points in the space. Thus, if a space has the discrete topology (Example 75.3), no proper subset of the space can be a dense subset of the space. On the other hand, if a set has the trivial topology (75.3), every nonempty subset of the space is a dense subset.

We next extend the definition of *convergent sequence* to topological spaces. Recall from 50.5 that in a metric space a subset S is closed if and only if every convergent sequence in S converges to a point in S. Thus, since open subsets can be characterized in terms of closed sets (a set is open if and only if its complement is closed), any notion in a metric space that can be characterized in terms of open sets can also be characterized in terms of convergent sequences. It is useful to do this in, for example, the cases of compactness and continuity (see 52.4 and 62). We shall give an example to show that for general topological spaces, convergent sequences are *not* adequate to characterize closedness. However, in subsequent sections we shall study a class of topological spaces for which convergent sequences are adequate for this purpose.

77.9. Definition. Convergent sequence. *A sequence (x_i) in X is said to converge to a point $x \in X$ provided the following condition is satisfied: For each open subset U of X that contains x, there is a positive integer N such that $x_i \in U$ for all $i \geq N$.*

Recall that if a sequence (x_n) in a metric space converges, then the unique point x to which it converges is denoted by lim (x_n). Thus, for metric spaces we are justified in speaking of *the limit* of a convergent sequence. However, for general topological spaces a convergent sequence might converge to more than one point. We will reserve the notation $\lim (x_n) = x$ for cases in which the sequence converges to a unique point.

77.10. Example. Let X be the set of all real numbers and let $\mathscr{T}$ be the topology for X given in Example 75.6. Consider the sequence (x_n) given by $x_n = n$ for each $n \in \mathbf{P}$. A nonempty set U is open if and only if it is the complement in X of a finite set. Now let $x \in X$. Let U be an open subset of X that contains x. Then $x_n \in U$ for all but a finite number of n's. Hence, the sequence (x_n) converges to x. Thus, we see that (x_n) converges to each point in the space.

77.11. Example. Let X be the set of all reals and let $\mathscr{T} = \{\varnothing\} \cup \{X - C : C$ is a countable subset of $X\}$ (see Example 75.7). Let $a < b$ and consider the set $S = \{x : a < x < b\}$. First observe that every point of X is a limit point of S. Hence, S is not a closed set. On the other hand, no sequence in S can converge to a point in $X - S$. Suppose that (x_i) is a sequence in S and $x \in X - S$. Consider the set $X - \bigcup \{x_i : i \in \mathbf{P}\}$. This set is open but does not intersect any x_i. Hence, (x_i) does not converge to x. Thus, no sequence in S can converge to a point in $X - S$. Yet S is not closed. We see, then, that the condition stated in 50.5(b) does not hold for all spaces.

EXERCISES: SOME BASIC DEFINITIONS

1. List all the closed sets for the space $(X, \mathscr{T})$ of Example 75.4. By reviewing some of the properties possessed by the topologies of all metric spaces, show that $(X, \mathscr{T})$ is not metrizable if X has more than one point.

2. Let $\mathbf{R}$ be endowed with the Euclidean topology. Let $\mathbf{Q}$ be the set of all rational numbers. Find each of the following:
    (a) cl $(\mathbf{Q})$.
    (b) int $(\mathbf{Q})$.
    (c) ext $(\mathbf{Q})$.
    (d) Fr $(\mathbf{Q})$.

3. Let the set $\mathbf{R}$ be endowed with the topology given to it in Example 75.7. For the set $\mathbf{Q}$ of all rational numbers find each of the following: cl $(\mathbf{Q})$, int $(\mathbf{Q})$, ext $(\mathbf{Q})$, Fr $(\mathbf{Q})$.

4. Repeat the previous exercise; this time, however, let $\mathbf{R}$ be endowed with the topology given to it in Example 75.6. Also let $I = \{x : 0 \leq x \leq 1\}$ and find: cl (I), int (I), ext (I), Fr (I).

5. Prove that if (X, d) is a metric space and S is a subset of X, then the set of all limit points of S is closed. Show by an example that the result is not true for general topological spaces.

6. (a) A subset S in a topological space is said to be nowhere dense provided int $(\mathrm{cl}\,(S)) = \varnothing$. Show that if S is an open set then the Fr (S) is a nowhere dense set.

(b) Is the statement in (a) still true if S is a closed set rather than an open set?

(c) Show by an example that the statement in (a) is not valid for arbitrary subsets of a topological space.

(d) Prove the following proposition or show with an example that it is not correct: A set S is nowhere dense if and only if the complement of its closure is dense.

78. SOME BASIC THEOREMS FOR TOPOLOGICAL SPACES

In this section we shall include elementary theorems for general topological spaces, some of which have been given previously for metric spaces. In each case the proof is left as an exercise for the reader.

78.1. Theorem. *A subset of a topological space is closed (open) if and only if its complement is open (closed).*

78.2. Theorem. *Let X be a topological space. The empty set and X are closed subsets of X. The union of each finite collection of closed subsets is closed. Furthermore, if $\mathscr{H}$ is an arbitrary nonempty collection of closed subsets of X, then $\bigcap \mathscr{H}$ is closed.*

78.3. Example. Let $(X, \mathscr{T})$ be the topological space given in Example 75.5. Let $\mathscr{F}$ be the collection of all closed subsets of X. Then $\mathscr{F} = \{\{b, c, d\}, \{c, d\}, \{b, d\}, \{d\}, \varnothing, X\}$.

78.4. Example. Let $(X, \mathscr{T})$ be the space given in Example 75.6. Note that a set U is closed if and only if $U = \varnothing$, $U = X$, or U is a finite subset of X.

78.5. Theorem. *A subset S of a topological space is closed if and only if* cl $(S) = S$. *A subset S is open if and only if* int $(S) = S$. *Furthermore, the closure of each set is a closed set and the interior of each set is open.*

78.6. Theorem. *The closure of a set S is the intersection of all closed subsets that contain S. The interior of a set S is the union of all open sets that are contained in S.*

Notice that the closed ball $B(p; r)$ in $\mathbf{R}^n$ contains its boundary $\{x : |x - p| = r\}$, whereas an open neighborhood $N(p; r)$ does not contain any of its boundary $\{x : |x - p| = r\}$. This suggests what happens in general as seen in the following theorem.

78.7. Theorem. *A set is closed if and only if it contains its boundary. A set is open if and only if it does not contain any of its boundary points.*

78.8. Theorem. *A set S is dense in a topological space X if and only if every nonempty open subset of X intersects S.*

78.9. Example. Let X be the set of all real numbers. Let $\mathscr{T} = \{\varnothing, X, \{0\}\}$ (see 75.4). Notice that each nonempty open subset of X contains the point 0. Hence, the set $\{0\}$ is a dense subset of X. Furthermore, the set $X - \{0\}$ is not dense since 0 is an isolated point of the space.

78.10. Let $(X, \mathcal{T})$ be the topological space given in 75.6. A subset of this space is dense if and only if it is an infinite set. For the space in Example 75.7, a set is dense if and only if it is an uncountable set.

78.11. **Theorem.** *The closure operator* cl *possesses the following properties. Let A and B be subsets of X. Then,*

78.11(a).   cl $(\varnothing) = \varnothing$.

78.11(b). $A \subset$ cl (A).

78.11(c).   cl (cl (A)) = cl (A).

78.11(d).   cl $(A \cup B) =$ cl $(A) \cup$ cl (B).

78.12. **Theorem.** *For each subset* $S \subset X$, Fr $(S) =$ cl $(S) \cap$ cl $(\sim S)$.

78.13. **Theorem.** *The interior operator* int *possesses the following properties: Let A and B be subsets of X. Then*

78.13(a).   int $(X) = X$.

78.13(b).   int $(A) \subset A$.

78.13(c).   int (int A) = int (A).

78.13(d).   int $(A \cap B) =$ int $(A) \cap$ int (B).

EXERCISES: SOME BASIC THEOREMS FOR TOPOLOGICAL SPACES

1. Prove each of the following theorems:
(a) 78.1	(f) 78.8
(b) 78.2	(g) 78.11
(c) 78.5	(h) 78.12
(d) 78.6	(i) 78.13
(e) 78.7	

2. Definitions 77.1, 77.7, and 77.9 are concerned with conditions that must be met by every open set containing a point. Prove that it would be equivalent to assume only that the requirements hold for elements of a base. More explicitly, prove the following:

    Let $(X, \mathcal{T})$ be a topological space and let $\mathcal{B}$ be a base for $\mathcal{T}$. Then each of the following holds:
    (a) A point p in X is a limit point of a set $S \subset X$ if and only if for each $U \in \mathcal{B}$ that contains p, U intersects S in at least one point distinct from p.
    (b) Suppose $S \subset X$ and $p \in X$. Then p is a boundary point of S if and only if for every $U \in \mathcal{B}$ that contains p, U intersects both S and the complement of S.
    (c) A sequence (x_i) in X converges to a point $x \in X$ provided that for each $U \in \mathcal{B}$ that contains x, there is a positive integer N such that $x_i \in U$ for all $i \geq N$. (In connection with this exercise, recall that in the case of a metric space

we originally defined the concept of limit point in terms of the base of ε-neighborhoods.)

3. Give an example of a topological space for which it is not true that a set consisting of exactly one point is closed.

4. Give an example of a topological space for which it is true that sets consisting of a single point are closed and for which there exist at least two points x and y for which there do not exist disjoint open sets U_x and U_y such that $x \in U_x$ and $y \in U_y$. Note that such a space could not be metrizable.

5. Are there sequences in Example 78.9 that converge to more than one point?

6. Give an example of a space consisting of more than one point in which every sequence converges.

7. The following properties show how closed sets can characterize the topology of a space:

    Let X be a set and suppose $\mathscr{F}$ is a collection of subsets of X that satisfies the following (see Theorem 78.2):
    (a) $\varnothing \in \mathscr{F}$ and $X \in \mathscr{F}$.
    (b) The union of each finite subcollection of $\mathscr{F}$ is an element of $\mathscr{F}$.
    (c) The intersection of each nonempty subcollection of $\mathscr{F}$ is an element of $\mathscr{F}$.

    Let $\mathscr{T} = \{U : U = X - W \text{ for some } W \in \mathscr{F}\}$.
    Show that $\mathscr{T}$ is a topology for X and that K is closed in $(X, \mathscr{T})$ if and only if $K \in \mathscr{F}$.

8. Let X be a set and let $\kappa : \mathscr{P}(X) \to \mathscr{P}(X)$ be a map where $\mathscr{P}(X)$ is the power set for X. Suppose κ satisfies the following properties (called the Kuratowski closure axioms):
    (a) $\kappa(\varnothing) = \varnothing$.
    (b) $A \subset \kappa(A)$ for $A \in \mathscr{P}(X)$.
    (c) $\kappa(\kappa(A)) = \kappa(A)$ for all A in $\mathscr{P}(X)$.
    (d) $\kappa(A \cup B) = \kappa(A) \cup \kappa(B)$ for all A and B in $\mathscr{P}(X)$.

    Let $\mathscr{F} = \{F : \kappa(F) = F, F \in \mathscr{P}(X)\}$. Show that $\mathscr{F}$ satisfies the properties in Exercise 7 and, hence, determines a topological space $(X, \mathscr{T})$, where $\mathscr{T}$ is the collection of complements in X of elements of $\mathscr{F}$. Show further that in the space $(X, \mathscr{T})$ cl $(S) = \kappa(S)$ for each $S \subset X$.

9. Using Theorem 78.13 as a guide, set up conditions for a mapping

$$I : \mathscr{P}(X) \to \mathscr{P}(X)$$

which will guarantee that

$$\mathscr{T} = \{U : I(U) = U, U \in \mathscr{P}(X)\}$$

is a topology for X and $I(U) = \text{int } (U)$ for each $U \in \mathscr{P}(X)$.

10. Let S be a subset of a topological space X. In each of the following, tell whether the statement is necessarily true. Justify your answer by a proof or a counterexample.
   (a) Fr $(S) = $ cl $(S) - (X - S)$.
   (b) Fr (S) is closed.
   (c) int $(S) = S - $ Fr (S).
   (d) Fr $(S) = \emptyset$ implies that S is both open and closed.

11. Let X be the following subset of $\mathbf{R}^2$:

$$X = \left\{ \left(\frac{1}{n}, 0 \right) : n \in \mathbf{P} \right\} \cup \{(0, 1), (0, -1)\}.$$

For each $n \in \mathbf{P}$, let $A_n^+ = \left\{ \left(\frac{1}{j}, 0 \right) : j \geq n \right\} \cup \{(0, 1)\}$ and $A_n^- = \left\{ \left(\frac{1}{j}, 0 \right) : j \geq n \right\} \cup \{(0, -1)\}$. Let $\mathscr{B}$ be the collection of all subsets of X that are of the form A_n^+, A_n^-, or $\left\{ \left(\frac{1}{j}, 0 \right) \right\}$ for some $j \in \mathbf{P}$.

   (a) Show that $\mathscr{B}$ is a base for a topology $\mathscr{T}(\mathscr{B})$ for X.
   (b) Show that convergent sequences in $(X, \mathscr{T}(\mathscr{B}))$ do not necessarily have unique limits.
   (c) Show that sets consisting of single points are closed subsets of this space.
   (d) Point out how you know that this space is not metrizable.

12. Let X_1 and X_2 be topological spaces and let $X_1 \times X_2$ denote the product space. In each of the following, determine if the statement is true:
   (a) If $A_1 \subset X_1$ and $A_2 \subset X_2$, then cl $(A_1 \times A_2) = $ cl $(A_1) \times$ cl (A_2).
   (b) If A_1 is a closed subset of X_1 and A_2 is a closed subset of X_2, then $A_1 \times A_2$ is a closed subset of $X_1 \times X_2$.
   (c) If $A_1 \subset X_1$ and $A_2 \subset X_2$, then int $(A_1 \times A_2) = $ int $(A_1) \times$ int (A_2).
   (d) If $\mathscr{B}_1$ is a base for the topology for X_1 and $\mathscr{B}_2$ is a base for the topology for X_2, then $\mathscr{B} = \{U_1 \times U_2 : U_1 \in \mathscr{B}_1$ and $U_2 \in \mathscr{B}_2\}$ is a base for the product topology for $X_1 \times X_2$.

79. NEIGHBORHOODS AND NEIGHBORHOOD SYSTEMS

We define next the notion of neighborhood of a point. For metric spaces it will turn out that the sets of the form $N(p; \varepsilon)$ and $B(p; \varepsilon)$ will both be examples of a more general type of set called *neighborhood* of p.

79.1. Definition. Neighborhood of a point. *Let X be a topological space. By a neighborhood of a point x, we shall mean a set N for which there exists an open set U such that $x \in U \subset N$.*

It is important to note that a neighborhood of a point p need not be an open set, but it is necessary that $p \in $ int (N) in order for N to be a neighborhood of p.

Note also that an open set U is necessarily a neighborhood of each point $x \in U$. Thus, if we wish to say that U is an open set such that $x \in U$, it is equivalent to say that U is an open neighborhood of the point x.

79.2. Example. Let $N = \{(x, y): x^2 + y^2 \leq 1$ or $x = 0$ or $y = 0\}$. As a subset in the space $\mathbf{R}^2$, N is a neighborhood of each of the points in $\{(x, y): x^2 + y^2 < 1\}$ but is not a neighborhood of any other point in N.

The following statement, whose easy proof is left as an exercise, gives a characterization of open sets in terms of neighborhoods.

79.3. Theorem. *A set U in a topological space is open if and only if for each $x \in U$, U contains a neighborhood of x.*

79.4. Definition. Neighborhood system of a point. *If X is a topological space and $x \in X$, then the collection $\mathcal{N}_x$ of all neighborhoods of x is called the neighborhood system of x. A base for the neighborhood system $\mathcal{N}_x$ of a point x is a subcollection $\mathcal{B}_x$ of $\mathcal{N}_x$ such that for each $N_x \in \mathcal{N}_x$, there is a $U_x \in \mathcal{B}_x$ such that $x \in U_x \subset N_x$.*

The elements of a base for a neighborhood system need not be open but often are taken that way.

Note that if $\mathcal{B}$ is a base for the topology $\mathcal{T}$ of a space $(X, \mathcal{T})$ and $x \in X$, then $\{U: U \in \mathcal{B}$ and $x \in U\}$ is a base for the neighborhood system of x.

EXERCISES: NEIGHBORHOODS AND NEIGHBORHOOD SYSTEMS

1. Prove Theorem 79.3.

2. Let X be a topological space. For each x let $\mathcal{N}_x$ be the neighborhood system of x. Prove:
   (a) If $U_x \in \mathcal{N}_x$ and $V_x \in \mathcal{N}_x$, then $U_x \cap V_x \in \mathcal{N}_x$.
   (b) If $U_x \in \mathcal{N}_x$ and $Q \supset U_x$, then $Q \in \mathcal{N}_x$.
   (c) For each $x \in X$, if $U \in \mathcal{N}_x$, there is a $V \in \mathcal{N}_x$ such that for each $y \in V$, $U \in \mathcal{N}_y$.

3. Suppose that X is a set, and for each $x \in X$ there is a nonempty collection $\mathcal{N}_x$ of subsets of X such that $x \in U$ for each $U \in \mathcal{N}_x$. Suppose further that the $\mathcal{N}_x$'s satisfy properties (a), (b), and (c) of the previous exercise.
   Let $\mathcal{T} = \{U: U \subset X$ and $U \in \mathcal{N}_x$ for each $x \in U\}$.
   Prove that $\mathcal{T}$ is a topology for X and that for each x, $\mathcal{N}_x$ is the neighborhood system for x in the space $(X, \mathcal{T})$.

4. Characterize the notion of limit point in terms of neighborhoods.

80. SUBSPACES

Suppose that Y is a subset of a metric space (X, d). Recall from Definition 49.1 that if we let $d^* = d \,|\, Y \times Y$, then (Y, d^*) is called a subspace of (X, d). On the basis of this definition, we were able to prove the following:

Let Y be a subspace of a metric space (X, d). Then $S \subset Y$ is open in Y if and only if there is an open subset U of X such that $S = U \cap Y$.

This characterization of open sets for subspaces of a metric space suggests a generalization of the notion of subspace.

80.1. Definition. Subspace. *Let $(X, \mathscr{T})$ be a topological space and suppose $Y \subset X$. Let $\mathscr{T} \mid Y$ be the collection $\{U \cap Y : U \in \mathscr{T}\}$. It is easy to verify that $\mathscr{T} \mid Y$ is a topology for Y. This collection $\mathscr{T} \mid Y$ is called the relative topology for Y induced by $\mathscr{T}$. Furthermore, $(Y, \mathscr{T} \mid Y)$ is called a subspace of $(X, \mathscr{T})$.*

80.2. Theorem. *Let $(X, \mathscr{T})$ be a topological space and let $(Y, \mathscr{T} \mid Y)$ be a subspace of $(X, \mathscr{T})$. Then:*

80.2(a). *A subset S of Y is closed in $(Y, \mathscr{T} \mid Y)$ if and only if $S = F \cap Y$ for some closed subset F of X.*

80.2(b). *If $\mathscr{B}$ is a base (subbase) for $\mathscr{T}$, then*

$$\mathscr{B} \mid Y = \{U \cap Y : U \in \mathscr{B}\} \text{ is a base (subbase) for } \mathscr{T} \mid Y.$$

PROOF OF (a). Suppose first that $S \subset Y$ is closed in Y. Then $Y - S$ is open in Y. Hence, from Definition 80.1, there is an open subset U of X such that $Y - S = U \cap Y$. Hence, $S = Y - U \cap Y = Y \cap (X - U)$. Thus, as was to have been shown, S is the intersection of Y with a closed subset of X. Next suppose that $S = F \cap Y$, where F is a closed subset of X. Then $Y - S = Y - F \cap Y = Y \cap (X - F)$. But since $X - F$ is open in X, $Y - S$ is open in Y. Hence, S is closed in Y. This completes the proof of (a).

The proof of (b) is left as an exercise.

80.3. *Example.* Let S be a circle and L a line in $\mathbf{R}^2$. The collection of all open arcs contained in S forms a base for the relative topology for S. Similarly, the collection of all open line intervals contained in L is a base for the relative topology for L.

80.4. Theorem. *Let X be a topological space and let Y be an open (a closed) subspace of X. Then $W \subset Y$ is open (closed) in Y if and only if W is open (closed) in X.*

The proof of this theorem is left as an exercise.

EXERCISES: SUBSPACES

1. Show that $\mathscr{T} \mid Y$ in Definition 80.1 is a topology for Y.

2. Prove Theorem 80.2 (b).

3. Prove Theorem 80.4.

4. Let P be a plane and S^2 a spherical surface in $\mathbf{R}^3$. By making use of Theorem 80.2(b), give a base for the relative topology for P. Also give a base for the relative topology for S^2. In each case answer the question by giving a description of the elements of the base.

81. CONTINUOUS AND TOPOLOGICAL MAPPINGS

With the notion of topological spaces we can extend the definitions of continuous and topological mappings from the framework of metric spaces to that of topological spaces.

81.1. Definition. Continuous mapping. *Let $(X, \mathcal{T}(X))$ and $(Y, \mathcal{T}(Y))$ be topological spaces and let f be a mapping from X into Y. Then f is said to be continuous at $x_0 \in X$ provided that for each open neighborhood V of $f(x_0)$ in Y, there is an open neighborhood U of x_0 in X such that $f[U] \subset V$. If f is continuous at each point in X, then f is said to be a continuous mapping from $(X, \mathcal{T}(X))$ into $(Y, \mathcal{T}(Y))$.*

In previous chapters we have studied examples of continuous functions from one metric space into another. We next consider an example in the more general setting of topological spaces. Other examples will be considered in the next set of exercises.

81.2. *Example.* Let X and Y each denote the set of real numbers. Let $\mathcal{T}(X)$ be the topology for X generated by the Euclidean metric and let $\mathcal{T}(Y)$ be the topology for Y as in Example 75.6. Thus, $U \in \mathcal{T}(Y)$ if and only if $U = \varnothing$ or $U = Y - F$ for some finite set $F \subset Y$. We consider that mapping $f: X \to Y$ given by $f(x) = x$ for all $x \in X$ and show that f is a continuous mapping from $(X, \mathcal{T}(X))$ into $(Y, \mathcal{T}(Y))$. To see this let $x_0 \in X$ and let V be an open neighborhood of $f(x_0)$. Then $V = Y - F$ for some finite subset F of Y. Now $f(x_0) \notin F$ so that $x_0 \notin F$. There exists an open interval $U \subset X$ such that $x_0 \in U \subset X - F$. But then $f[U] \subset Y - F = V$. Upon noticing that U is an open subset of $(X, \mathcal{T}(X))$, we see that f is continuous at x_0.

If we wish to say that f is a continuous mapping from $(X, \mathcal{T}(X))$ into $(Y, \mathcal{T}(Y))$, we shall abbreviate this by saying that $f: (X, \mathcal{T}(X)) \to (Y, \mathcal{T}(Y))$ is continuous. Likewise, in other situations if we are considering a map $f: X \to Y$ and the topologies $\mathcal{T}(X)$ and $\mathcal{T}(Y)$ on X and Y, respectively, are relevant to our discussion, we shall use the notation $f: (X, \mathcal{T}(X)) \to (Y, \mathcal{T}(Y))$ for the map. Analogous with the metric situation, we shall feel free to revert to the notation $f: X \to Y$ if it is clear from the context which topologies are involved.

81.3. Definition. Topological mapping. *If $f: (X, \mathcal{T}(X)) \to (Y, \mathcal{T}(Y))$ is a bijection such that f and f^{-1} are continuous, then f is said to be a topological mapping or a homeomorphism. Two topological spaces are said to be homeomorphic or topologically equivalent provided there exists a homeomorphism that carries one of the spaces onto the other.*

EXERCISES: CONTINUOUS AND TOPOLOGICAL MAPPINGS

1. Let X and Y each be the set of all real numbers. Let $\mathcal{T}(X)$ be the topology for X as given in Example 75.6 and let $\mathcal{T}(Y)$ be the topology as given in Example 75.7. Further, let f be given by $f(x) = x$ for all x in X.
   (a) Is $f: (X, \mathcal{T}(X)) \to (Y, \mathcal{T}(Y))$ continuous?
   (b) Is $f^{-1}: (Y, \mathcal{T}(Y)) \to (X, \mathcal{T}(X))$ continuous?

2. Let $(X, \mathcal{T})$ be the topological space as given in Example 75.5. Let $\mathcal{T}_1$ be the following topology for X: $\mathcal{T}_1 = \{\{d\}, \{d, c\}, \{d, b\}, \{d, b, c\}, X, \varnothing\}$. Is the space $(X, \mathcal{T})$ homeomorphic to the space $(X, \mathcal{T}_1)$?

3. Suppose that $f:(X, \mathcal{T}(X)) \to (Y, \mathcal{T}(Y))$ is a surjection. In each of the following determine whether the statement is necessarily true.
   (a) If $\mathcal{T}(Y)$ is the trivial topology for Y, then f is continuous.
   (b) If $\mathcal{T}(X)$ is the discrete topology for X, then f is continuous.
   (c) If f is continuous and $\mathcal{T}(Y)$ is the discrete topology for Y, then $\mathcal{T}(X)$ is the discrete topology for X.

4. Suppose that $f:(X_1, \mathcal{T}_1) \to (X_2, \mathcal{T}_2)$ is a mapping. Suppose also that $(S, \mathcal{T}_2 \,|\, S)$ is a subspace of $(X_2, \mathcal{T}_2)$ such that $f[X_1] \subset S$. Prove that $f:(X_1, \mathcal{T}_1) \to (X_2, \mathcal{T}_2)$ is continuous if and only if $f:(X_1, \mathcal{T}_1) \to (S, \mathcal{T}_2 \,|\, S)$ is continuous.

5. Let $(X, \mathcal{T})$ be a topological space and let $(S, \mathcal{T} \,|\, S)$ be a subspace of $(X, \mathcal{T})$. Let $i:(S, \mathcal{T} \,|\, S) \to (X, \mathcal{T})$ be the inclusion map given by $i(x) = x$ for all x in S. Show that i is continuous.

6. Let $\mathcal{T}(d)$ be the topology for $\mathbf{R}$, generated by the Euclidean metric d. Let $\mathcal{T}$ be the topology for the set of real numbers $\mathbf{R}$ as given in Example 75.7. Hence, $U \in \mathcal{T}$ if and only if $U = \varnothing$ or if $U = \mathbf{R} - C$, where C is a countable set. Let i be the identity function on $\mathbf{R}$.
   (a) Show that $i:(\mathbf{R}, \mathcal{T}) \to (\mathbf{R}, \mathcal{T}(d))$ is not continuous at any point.
   (b) Describe the convergent sequences in $(\mathbf{R}, \mathcal{T})$.
   (c) Show that for each sequence (x_i) that converges to a point x_0 in the space $(\mathbf{R}, \mathcal{T})$, it is true that $\lim (i(x_i)) = i(x_0)$. Compare this with the situation for metric spaces as given in Theorem 52.4(b).

82. SOME BASIC THEOREMS CONCERNING MAPPINGS

In Exercise 6, above , an example was given to show that not all characterizations of continuity in metric spaces hold in the more general setting of topological spaces. Included among the following theorems are some useful characterizations that do hold. If a proof is not provided, it is left to the reader as an exercise.

82.1. Theorem. *Let $f:(X, \mathcal{T}(X)) \to (Y, \mathcal{T}(Y))$ be a mapping. Then f is continuous if and only if any one of the following conditions holds.*

82.1(a). *If $U \in \mathcal{T}(Y)$, then $f^{-1}[U] \in \mathcal{T}(X)$.*

82.1(b). *If F is a closed subset of Y, then $f^{-1}[F]$ is a closed subset of X.*

82.1(c). *For each subset $A \subset X$, $f[\mathrm{cl}\,(A)] \subset \mathrm{cl}\,(f[A])$.*

82.1(d). *For each $x \in X$, if V is a neighborhood of $f(x)$ in the space $(Y, \mathcal{T}(Y))$, then there is a neighborhood U of x in $(X, \mathcal{T}(X))$ such that $f[U] \subset V$.*

Proof that (b) implies (c): Let $A \subset X$. Then, since $f[A] \subset \mathrm{cl}\,(f[A])$, $A \subset f^{-1}[\mathrm{cl}\,(f[A])]$. By (b), $f^{-1}[\mathrm{cl}\,(f[A])]$ is a closed set. Then since $\mathrm{cl}\,(A)$ is a minimal closed set that contains A, we have $A \subset \mathrm{cl}\,(A) \subset f^{-1}[\mathrm{cl}\,(f[A])]$. From this we then have $f[\mathrm{cl}\,(A)] \subset f[f^{-1}[\mathrm{cl}\,(f[A])]] \subset \mathrm{cl}\,(f[A])$.

Actually in applying 82.1(a) to verify the continuity of a function

$$f: (X, \mathcal{T}(X)) \to (Y, \mathcal{T}(Y)),$$

it is necessary only to check the condition for a subbase for $\mathcal{T}(Y)$. Suppose that $\mathcal{S}$ is a subbase for $\mathcal{T}(Y)$ and $f^{-1}[S]$ is open for every S in $\mathcal{S}$. Let $U \in \mathcal{T}(Y)$. Then for each $y \in U$, there exists a finite subcollection $\{S_i : i \in \mathbf{P}_n\}$ of $\mathcal{S}$ such that $y \in \bigcap \{S_i : i \in \mathbf{P}_n\} \subset U$. Let $U_y = \bigcap \{S_i : i \in \mathbf{P}_n\}$. Then because the inverse of a function preserves intersections, we have

$$f^{-1}[U_y] = f^{-1}[\bigcap \{S_i : i \in \mathbf{P}_n\}] = \bigcap \{f^{-1}[S_i] : i \in \mathbf{P}_n\}.$$

Since each $f^{-1}[S_i]$ is open, so is $f^{-1}[U_y]$. Now we may choose one such U_y for each $y \in U$. Hence, $U = \bigcup \{U_y : y \in U\}$. Thus, since the inverse of a function preserves unions, $f^{-1}[U] = f^{-1}[\bigcup \{U_y : y \in U\}] = \bigcup \{f^{-1}[U_y] : y \in U\}$. Then since each $f^{-1}[U_y]$ is open, so is $f^{-1}[U]$. Thus, we have shown that f satisfies 82.1a. Conversely, if f satisfies 82.1a, for each $S \in \mathcal{S}$, $f^{-1}[S]$ is open since each S in $\mathcal{S}$ is open in Y. We therefore have the following alternate form of 82.1(a):

82.1(e). *Let $f: (X, \mathcal{T}(X)) \to (Y, \mathcal{T}(Y))$ be a mapping and suppose that $\mathcal{S}$ a subbase for $\mathcal{T}(Y)$. Then f is continuous if and only if for each $S \in \mathcal{S}$, $f^{-1}[S]$ is open in X.*

It should be clear that since each base for a topology is also a subbase, the proposition just stated remains valid if the word *base* is substituted for the word *subbase*.

We next define two additional important types of mappings.

82.2. Definitions. Open mappings; closed mappings. *Let $f: (X, \mathcal{T}(X)) \to (Y, \mathcal{T}(Y))$ be a mapping. Then f is said to be an open mapping provided $f[U]$ is open in Y if U is open in X. The mapping f is said to be a closed mapping provided $f[F]$ is closed in Y if F is closed in X.*

Notice that if $f: (X, \mathcal{T}(X)) \to (Y, \mathcal{T}(Y))$ is continuous, then the invariant action on open (closed) sets is "backward going," i.e., f^{-1} carries open (closed) sets in Y onto open (closed) sets in X. For open mappings, the action is "forward going," i.e., f carries open sets in X onto open sets in Y. Similarly, the action of a closed mapping on closed sets is forward. We may picture these remarks schematically as follows: In the diagrams, $\mathcal{F}(X)$ will denote the collection of *closed* subsets of $(X, \mathcal{T}(X))$ and, similarly, $\mathcal{F}(Y)$ will denote the collection of all closed subsets of $(Y, \mathcal{T}(Y))$.

$$\mathcal{T}(X) \xrightarrow[f,\ \text{open}]{f} \mathcal{T}(Y)$$

$$\mathcal{T}(X) \xleftarrow[f,\ \text{continuous}]{f^{-1}} \mathcal{T}(Y)$$

$$\mathcal{F}(X) \xleftarrow[f,\ \text{continuous}]{f^{-1}} \mathcal{F}(Y)$$

$$\mathcal{F}(X) \xrightarrow[f,\ \text{closed}]{f} \mathcal{F}(Y)$$

This diagram suggests the following important proposition that follows at once from 82.1(a) and (b).

82.3.   Theorem. *Suppose that* $f:(X, \mathcal{T}(X)) \to (Y, \mathcal{T}(Y))$ *is a continuous bijection. Then f is a homeomorphism if and only if f is an open (closed) mapping.*

We shall call a property that is preserved under a homeomorphism a *topological invariant*. If f is a homeomorphism, then f is a bijection and both f and f^{-1} preserve openness. Thus if a property can be characterized in terms of the topology of the space (i.e. a topological property), then it is a topological invariant.

82.4.   Theorem. *Suppose X, Y, and Z are topological spaces and $f: X \to Y$ and $g: Y \to Z$ are mappings. If f and g are continuous (open) ((closed)), so is the mapping $g \circ f: X \to Z$. If f and g are homeomorphisms, then so is $g \circ f$.*

The proof of the first part is the same as the proof given in 52.13. The remainder of the proof is left as an exercise.

The following theorem, previously stated for sequences of mappings defined on metric spaces, holds in the more general setting indicated next. The proof of Theorem 67.4 carries over directly (see Definition 67.3).

82.5.   Theorem. *Let* $(f_n:(X, \mathcal{T}) \to (Y, \rho))$ *be a sequence of continuous mappings that converges uniformly to $f:(X, \mathcal{T}) \to (Y, \rho)$, where (Y, ρ) is a metric space. Then f is continuous.*

In the next set of exercises, the reader will be asked to prove the following characterization of open mappings.

82.6.   Theorem. *Suppose that* $f:(X, \mathcal{T}(X)) \to (Y, \mathcal{T}(Y))$ *is a mapping and $\mathcal{B}$ is a base for $\mathcal{T}(X)$. Then f is an open mapping if and only if $f[U]$ is open for each $U \in \mathcal{B}$.*

We shall use the theorem just stated in obtaining the following useful fact about projection mappings defined on product spaces.

82.7.   Theorem. *Let* $\{(X_i, \mathcal{T}_i): i \in \mathbf{P}_n\}$ *be a finite collection of topological spaces and let $(X, \mathcal{T})$ denote the product space $\times \{(X_i, \mathcal{T}_i): i \in \mathbf{P}_n\}$. Then for each $i \in \mathbf{P}_n$, the projection mapping $\pi_i:(X, \mathcal{T}) \to (X_i, \mathcal{T}_i)$ is an open continuous surjection.*

PROOF. Let $i \in \mathbf{P}_n$. Since $\pi_i(X) = X_i$, it follows that π_i is a surjection. Furthermore, it follows from the definition of the product topology that $\pi_i^{-1}[U_i]$ is open for each open set U_i in X_i. Hence, π_i is a continuous mapping. That π_i is an open mapping can be seen from the following argument: Let $\mathcal{B}$ be the base for $\mathcal{T}$ generated by the subbase $\mathcal{S}$ as defined in 76.6. Let $B \in \mathcal{B}$. Then $B = U_1 \times U_2 \times \cdots \times U_n$ where each U_j is open in X_j. But then $\pi_i[B] = U_i$. Hence, $\pi_i[B]$ is open in X_i and from 82.6 it follows that π_i is an open mapping.

EXERCISES: SOME BASIC THEOREMS CONCERNING MAPPINGS

1.   Complete the proof of 82.1.

2.   Complete the proof of 82.4. Prove Theorem 82.6.

3. Suppose that $f:(X, \mathcal{T}(X)) \to (Y, \mathcal{T}(Y))$ is an open (closed) mapping and S is an open (closed) subset of X. Prove that $f\,|\,S:S \to Y$ is an open (closed) mapping, where S is given the relative topology.

4. Suppose $f:X \to Y$ is an open (a closed) surjection. Suppose $S \subset X$ and $f^{-1}[f[S]] = S$. Show that the mapping $f\,|\,S:S \to f[S]$ is open (closed) where the relative topologies are put on S and $f[S]$. Give an example to show that the conclusions are not necessarily true if S is not an inverse set (an inverse set S for f is one such that $f^{-1}[f[S]] = S$).

5. Prove the following form of the Weierstrass M test:

    Suppose for each $i \in \mathbf{P}$, $g_i:X \to \mathbf{R}^n$ is a continuous mapping from a topological space X into $\mathbf{R}^n$. Suppose furthermore $|g_i(x)| \leq M_i$ for $x \in X$, where $\sum\limits_{i=1}^{\infty} M_i$ is a convergent sequence of positive terms. Then the sequence of partial sums $\left(\sum\limits_{i=1}^{n} g_i \right)_{n=1}^{\infty}$ converges uniformly to a continuous mapping $f:X \to \mathbf{R}^n$.

6. Let $\{(X_i, \mathcal{T}_i): i \in \mathbf{P}_n\}$ be a collection of topological spaces and let $(X, \mathcal{T})$ be the product space for this collection. Let $i \in \mathbf{P}_n$ and $c \in X$. In this exercise we shall define a subset X_i^c that contains the point c. Show that X_i^c is a homeomorphic copy of the coordinate space X_i:

$$X_i^c = \{x: c_j = x_j \text{ for } j \neq i\}.$$

(Note that we may write $X_i^c = \{c_1\} \times \{c_2\} \times \cdots \times \{c_{i-1}\} \times X_i \times \{c_{i+1}\} \times \cdots \times \{c_n\}$.)

83. SEPARATION PROPERTIES FOR TOPOLOGICAL SPACES

In 54, we proved that metric spaces enjoyed certain separation properties. For example, the property called normality was one of the stronger properties studied. Some of the examples of topological spaces that we have considered in this chapter do not even satisfy some of the weakest of the separation properties. For example, in Example 75.5 there is no neighborhood of the point d that does not contain the point a. However, there are many spaces that do satisfy separation properties of varying strengths, and it has been fruitful to study the effect on spaces of assuming various ones or combinations of them. We shall list these separation properties here and consider them in connection with other topics in subsequent sections. In each case the label attached to the space possessing the given property is given first and then the definition of the property is given.

83.1. Definition. T_1-**space or space satisfying the weak separation axiom.** *For each pair of distinct points x and y in the space, there is an open neighborhood U of x that does not contain y and an open neighborhood V of y that does not contain x.*

The T_1 property is equivalent to the property that sets consisting of a single point are closed subsets of the space. The reader will be asked to prove this in the next set of exercises.

83.2. Definition. Hausdorff (T_2) space or space satisfying the strong separation axiom.   *For each pair of distinct points x and y in X, there exists a pair of disjoint open neighborhoods U_x and U_y of x and y, respectively.*

83.3. Definition. Regular space.   *If L is a closed set and x is a point not in L, then there exists a pair of disjoint open sets U_L and U_x containing L and x, respectively.*

83.4. Definition. T_3-space.   *A T_3-space is a space that is both T_1 and regular.*

83.5. Definition. Normal space.   *If H and K are disjoint closed subsets, then there exists a pair of disjoint open sets U_H and U_K that contain H and K, respectively.*

83.6. Definition. T_4-space.   *A space that is both T_1 and normal is called a T_4 space.*

83.7. *Example.*   Let X be the set of all real numbers and let $\mathcal{T}_1$ be the discrete topology. The space $(X, \mathcal{T}_1)$ is metrizable and, from our work in metric spaces, we know that this space possesses all the separation properties listed in this section. On the other hand, let $\mathcal{T}_2$ be the trivial topology for X. It is easy to see

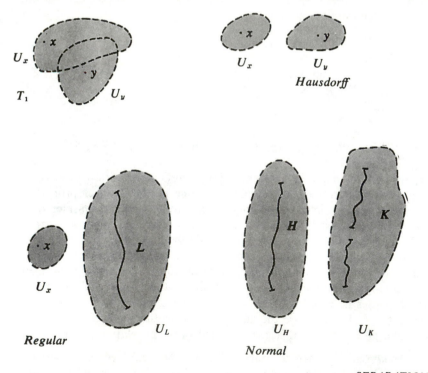

T_1

Hausdorff

Regular

Normal

Figure 19

that $(X, \mathcal{T}_2)$ is not a T_1-space. On the other hand, since $(X, \mathcal{T}_2)$ has no closed subsets other than the empty set and X itself, it follows that it is normal and regular in a vacuous way.

83.8. *Example.* Let $(X, \mathcal{T})$ be the space in Example 75.6. It is easy to verify that this space is a T_1-space. However, each pair of nonempty open subsets intersect. Hence, the space is not a Hausdorff space.

It should be pointed out to the reader that in some books the terms *regular* and *normal* are used to designate the stronger properties T_3 and T_4 respectively. In this text we will generally say "T_1 and normal" rather than use the label T_4, and similarly for T_3-spaces. With the numerical labeling as defined here, it follows immediately that if a space is a T_i space then it is a T_{i-1} space for $i \in \{2, 3, 4\}$.

EXERCISES: SEPARATION PROPERTIES FOR TOPOLOGICAL SPACES

1. In each of the following, list the separation properties possessed by the given space:
   (a) The space in Example 75.4.
   (b) The space in Example 75.5.
   (c) The space in Example 75.6.
   (d) The space in Example 75.7.
   (e) The space $(X, \mathcal{T})$, where $X = \{1, 2, 3, 4\}$ and

   $$\mathcal{T} = \{\varnothing, \{1\}, \{1, 2\}, \{1, 2, 3\}, \{1, 2, 3, 4\}\}.$$

2. Does there exist a finite T_1-space that is not a T_2-space? If there is no such example, prove why there is not.

3. Suppose X is a Hausdorff space. Show that if a sequence in X converges, then it converges to only one point. Give an example to show that this property does not necessarily hold for T_1-spaces.
   In Exercises 4 through 13, prove the proposition stated in the exercise.

4. (a) $(X, \mathcal{T})$ is a T_1-space if and only if sets consisting of a single point are closed sets in $(X, \mathcal{T})$.
   (b) $(X, \mathcal{T})$ is a T_1-space if and only if the following holds: If $S \subset X$, then $p \in X$ is a limit point of S if and only if every open neighborhood of p contains an infinite subset of S.

5. A topological space is regular if and only if the following condition is satisfied: For each point x in the space and open set U containing the point, there is an open set V such that $x \in V \subset \operatorname{cl}(V) \subset U$.

6. A topological space is normal if and only if for each closed subset A and open set U containing the set A, there is an open set V such that $A \subset V \subset \operatorname{cl}(V) \subset U$.

7. $(X, \mathcal{T})$ is a T_1-space if and only if finite subsets of X are closed.

8. If $(X, \mathcal{T})$ is a Hausdorff space and $x_1, x_2, \ldots, x_n$ are distinct elements of X, then there exist pairwise disjoint open sets $U_1, U_2, \ldots, U_n$ such that $x_i \in U_i$.

9. If a topological space X has the following property, then it is regular. If $x \in X$ and A is a nonempty closed subset that does not contain x, then there is a continuous mapping from X into the real closed interval $[0, 1]$ such that $f(x) = 0$ and $f[A] = \{1\}$.

10. If a topological space X has the following property, then it is normal. If A and B are disjoint nonempty closed subsets, then there exists a continuous mapping f from X into the real closed interval $[0, 1]$ such that $f[A] = \{0\}$ and $f[B] = \{1\}$.

11. Suppose X is normal and $\{A_1, A_2, \ldots, A_n\}$ is a pairwise disjoint finite collection of nonempty closed subsets of X. Then there exists a collection of open subsets $\{U_1, U_2, \ldots, U_n\}$ such that $A_i \subseteq U_i$ and cl $(U_i) \cap$ cl $(U_j) = \varnothing$ for $i \neq j$.

12. Each of the following properties is inherited by every subspace of the space if the space has the property: T_1, T_2, regular.

13. Every *closed* subspace of a normal space is normal.

14. In this exercise we define a separation property that is even weaker than the property referred to as the *weak separation property* defined in 83.1. A topological space is called a T_0-space provided that for each pair of distinct points x and y in the space, there exists an open neighborhood U_x of x that does not contain y *or* an open neighborhood U_y of y that does not contain x.
    (a) Give an example of a T_0-space that is not a T_1-space.
    (b) Determine whether the following statement is true: A topological space is a T_0-space if and only if for each pair of distinct points x and y in the space, x is not a limit point of $\{y\}$ or y is not a limit point of $\{x\}$.
    (c) Determine whether the following statement is true: A topological space is a T_1-space if and only if for each pair of distinct points x and y in the space, x is not a limit point of $\{y\}$ and y is not a limit point of $\{x\}$.
    (d) Suppose that $f : X \to Y$ is a continuous bijection. If X is a T_i-space, $i = 0, 1,$ or 2, is Y necessarily a T_i-space? If Y is a T_i-space, $i = 0, 1,$ or 2, is X necessarily a T_i-space?

15. Suppose that X is a topological space and Y is a Hausdorff space. Let $f : X \to Y$ and $g : X \to Y$ be continuous. Show that $\{x : x \in X \text{ and } f(x) = g(x)\}$ is a closed subset of X. Give an example to show that the conclusion does not necessarily hold if Y is not a Hausdorff space.

16. Suppose that f and g are continuous mappings from a topological space X into a Hausdorff space Y. Show that if $f \mid D = g \mid D$ for some dense subset D of X, then $f = g$.

17. Suppose that f is a continuous closed surjection from a normal space X onto a topological space Y. Show that Y is normal.

18. Let $\{(X_i, \mathcal{T}_i) : i \in \mathbf{P}_n\}$ be a finite collection of topological spaces and let $(X, \mathcal{T})$ be the product space for this collection. Prove that X is a Hausdorff space if and only if each of the coordinate spaces X_i is a Hausdorff space.

84. A CHARACTERIZATION OF NORMALITY

Recall from Theorem 54.6 that each metric space X has the following property.

84.1. PROPERTY. *If A and B are two nonempty disjoint closed subsets of X, then there exists a continuous mapping $f : X \to [0, 1]$ such that $f[A] = \{0\}$ and $f[B] = \{1\}$.*

It was remarked in 54.7 that this property can be used to prove that metric spaces are normal. In fact, it is easy to show that any topological space possessing the property is normal (see Exercise 10, page 186). We shall show in this section that if X is a normal space, then it possesses the property stated in 84.1. Thus, the statement gives a characterization of normality. The statement that normality implies 84.1 is a key step in a theorem of Urysohn that characterizes separable metric spaces.

84.2. Urysohn's lemma. *Suppose X is a normal space. For each pair of nonempty disjoint closed subsets A and B of X there exists a continuous mapping f from X into the closed real interval $[0, 1]$ such that $f[A] = \{0\}$ and $f[B] = \{1\}$.*

(The proof makes use of the fact that the set D of rationals of the form $\frac{m}{2^n}$, n is a positive integer and $1 \leq m \leq 2^n$, is dense in $[0, 1]$. The plan is to define a nested collection $\{U_r : r \in D\}$ of open subsets containing A such that for r_1 and r_2 in D, if $r_1 < r_2$, then

$$\text{cl}\,(U_{r_1}) \subset U_{r_2} \subset U_1 = X - B.$$

The required function will be defined in terms of the indices, the functional value at each point in some sense denoting in which of the U_r's the point is located. With this much of a hint, it is suggested that the reader try to construct a proof before reading the proof below.)

PROOF. Let D be the set of all diadics $\frac{m}{2^n}$ in $(0, 1]$. Let A and B be nonempty disjoint closed subsets. Let U_1 be the open subset $X - B$. Since X is normal, by Exercise 6, page 185, there is an open subset $U_{\frac{1}{2}}$ such that

$$A \subset U_{\frac{1}{2}} \subset \text{cl}\,(U_{\frac{1}{2}}) \subset U_1.$$

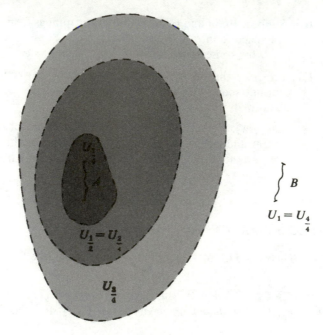

Figure 20

Using normality again we obtain an open subset $U_{\frac{1}{4}}$ such that

$$A \subset U_{\frac{1}{4}} \subset \text{cl }(U_{\frac{1}{4}}) \subset U_{\frac{1}{2}}.$$

Also since cl $(U_{\frac{1}{2}})$ is closed, we can find an open subset $U_{\frac{3}{4}}$ such that

$$\text{cl }(U_{\frac{1}{2}}) \subset U_{\frac{3}{4}} \subset \text{cl }(U_{\frac{3}{4}}) \subset U_1.$$

Notice that we now have a collection of open sets $\mathscr{U}_2 = \{U_r : r = \frac{1}{4}, \frac{2}{4}, \frac{3}{4}, \frac{4}{4}\}$, it being understood that any equivalent representative of a diadic may be used once a subscript is chosen; e.g., $U_{\frac{1}{2}} = U_{\frac{2}{4}}$. Note that for r_1 and r_2 in $\{\frac{1}{4}, \frac{2}{4}, \frac{3}{4}\}$, if $r_1 < r_2$, then

$$A \subset U_{r_1} \subset \text{cl }(U_{r_1}) \subset U_{r_2} \subset \text{cl }(U_{r_2}) \subset U_1.$$

Next assume that for $n = 2, 3, \ldots, h$ a collection of open sets

$$\mathscr{U}_n = \left\{ U_{\frac{m}{2^n}} : m = 1, 2, \ldots, 2^n \right\}$$

has been chosen so that for r_1 and r_2 in $\left\{ \frac{m}{2^n} : m = 1, 2, \ldots, 2^n - 1 \right\}$ and $r_1 < r_2$.

84.2(a). $$A \subset U_{r_1} \subset \text{cl }(U_{r_1}) \subset U_{r_2} \subset \text{cl }(U_{r_2}) \subset U_1.$$

We next show that we can choose $\mathscr{U}_n$ for $n = h + 1$ so that the property in 84.2(a) holds.

For m even, $U_{\frac{m}{2^{h+1}}} = U_{m/2} \in \mathscr{U}_h$ so that $U_{\frac{m}{2^{h+1}}}$ has already been chosen.

For m odd, we choose $U_{\frac{m}{2^{h+1}}}$ as follows: Choose $U_{\frac{1}{2^{h+1}}}$ so that

$$A \subset U_{\frac{1}{2^{h+1}}} \subset \text{cl }\left(U_{\frac{1}{2^{h+1}}} \right) \subset U_{\frac{2}{2^{h+1}}}.$$

For m odd and $1 < m \leq 2^{\lambda+1} - 1$, choose $U_{\frac{m}{2^{\lambda+1}}}$ so that

$$\mathrm{cl}\left(U_{\frac{m-1}{2^{\lambda+1}}}\right) \subset U_{\frac{m}{2^{\lambda+1}}} \subset \mathrm{cl}\left(U_{\frac{m}{2^{\lambda+1}}}\right) \subset U_{\frac{m+1}{2^{\lambda+1}}}.$$

Then, by induction, for each positive integer n we can choose a collection of open sets $\mathscr{U}_n$ that satisfies the conditions listed in 84.2(a). Let

$$\mathscr{U} = \bigcup \{\mathscr{U}_n : n \in \mathbf{P}\}.$$

Define $f: X \to [0, 1]$ as follows:

$$f(x) = 1 \quad \text{for } x \in B$$
$$f(x) = \text{g.l.b. } \{r : x \in U_r \quad \text{and} \quad U_r \in \mathscr{U}\}$$

Note next that $f[A] = \{0\}, f[B] = \{1\}$ and $f[X] \subset [0, 1]$. We complete the proof below by showing that f is continuous.

Let $x \in X$ and $\varepsilon > 0$. Let n be chosen so that $\dfrac{1}{2^n} < \dfrac{\varepsilon}{2}$. It is easy to see that

$$X = U_{\frac{2}{2^n}} \cup \left[U_{\frac{3}{2^n}} - \mathrm{cl}\left(U_{\frac{1}{2^n}}\right)\right] \cup \left[U_{\frac{4}{2^n}} - \mathrm{cl}\left(U_{\frac{2}{2^n}}\right)\right] \cup \cdots$$

$$\cup \left[U_{\frac{2^n}{2^n}} - \mathrm{cl}\left(U_{\frac{2^n-2}{2^n}}\right)\right] \cup \left[X - \mathrm{cl}\left(U_{\frac{2^n-1}{2^n}}\right)\right].$$

We consider several cases corresponding to the terms in this union. Suppose $x \in U_{\frac{2}{2^n}}$. Then for any y in the open set $U_{\frac{2}{2^n}}$, $0 \leq f(y) \leq \dfrac{2}{2^n}$ so that

$$|f(x) - f(y)| \leq \frac{2}{2^n} < \varepsilon.$$

Next, if for $2 < m \leq 2^n$, $x \in \left[U_{\frac{m}{2^n}} - \mathrm{cl}\left(U_{\frac{m-2}{2^n}}\right)\right]$, then for any

$$y \in \left[U_{\frac{m}{2^n}} - \mathrm{cl}\left(U_{\frac{m-2}{2^n}}\right)\right], \quad \frac{m-2}{2^n} \leq f(y) \leq \frac{m}{2^n}$$

and

$$|f(x) - f(y)| \leq \frac{2}{2^n} < \varepsilon.$$

Finally, consider the case $x \in X - \mathrm{cl}\left(U_{\frac{2^n-1}{2^n}}\right)$. For such an x, $\dfrac{2^n - 1}{2^n} \leq f(x) \leq 1$

and for any y in the open set $X - \mathrm{cl}\left(U_{\frac{2^n-1}{2^n}}\right)$, $f(y)$ is also in $\left[\dfrac{2^n - 1}{2^n}, 1\right]$. Hence,

$$|f(x) - f(y)| \leq \frac{1}{2^n} < \frac{\varepsilon}{2} < \varepsilon.$$

Thus, we have shown that f is continuous on X.

With Urysohn's lemma it is easy to prove the following useful variation of the lemma.

84.3. Urysohn's lemma (alternate form). *Suppose X is a normal space, A and B are nonempty disjoint closed subsets of X, and $[a, b]$ is a closed real interval. Then there is a continuous mapping $f: X \to [a, b]$ such that $f[A] = \{a\}$ and $f[B] = \{b\}$.*

Suppose that A is a subset of a topological space X and $f: A \to Y$ is a continuous mapping. A continuous mapping $g: X \to Y$ is called a *continuous extension* of $f: A \to Y$ provided $g \mid A = f$ (i.e., $g(x) = f(x)$ for all x in A). We have already met the notion of continuous extension for metric spaces (See 63.6 and 66.15), and the notion is also important in the study of general topological spaces. It should be clear to the reader that not every continuous mapping has a continuous extension. For example, suppose that $A = \{x : x \in \mathbf{R}$ and $x \neq 0\}$. The mapping $f: A \to \mathbf{R}$ given by $f(x) = \dfrac{1}{x}$ for $x \in A$ has no continuous extension that takes $\mathbf{R}$ into $\mathbf{R}$.

In terms of extension, Urysohn's lemma can be stated as follows:

Suppose X is a normal space and A and B are disjoint closed nonempty subsets of X. Suppose f is the real-valued function defined on $A \cup B$ such that $f[A] = \{0\}$ and $f[B] = \{1\}$. Then there is a continuous extension $g: X \to [0, 1]$ of $f: A \cup B \to [0, 1]$.

In the next theorem, we shall use Urysohn's lemma to prove an important extension theorem that says that if X is a normal space and f is a continuous mapping from a *closed* subset L of X into the real interval $[-M, M]$, then there is a continuous extension g of f that takes all of X into $[-M, M]$. Notice that there are two aspects of the conclusion. First of all, f is extended to all of X so that the extension is continuous. Secondly, $|f(x)| \leq M$ for all $x \in L$, and this bound is preserved for the extension.

84.4. Tietze's extension theorem. *Suppose X is a normal space, L is a nonempty closed subset of X, and $[-M, M]$ is a closed real interval. If $f: L \to [-M, M]$ is continuous, then there exists a continuous extension of f that takes X into $[-M, M]$.*

PROOF. The conclusion is obvious if $M = 0$. We assume $M > 0$ and let

$$A = \left\{ x : x \in L \text{ and } f(x) \geq \frac{M}{3} \right\}$$

$$B = \left\{ x : x \in L \text{ and } f(x) \leq -\frac{M}{3} \right\}.$$

A and B are closed subsets of the closed subset L and, hence, are also closed in X. Suppose first that A and B are nonempty. Note that they are disjoint. By 84.3, there exists a continuous mapping $g_1 : X \to \left[-\dfrac{M}{3}, \dfrac{M}{3} \right]$ such that

$$g_1[A] = \{ \tfrac{1}{3}M \}, \quad g_1[B] = \{ -\tfrac{1}{3}M \}$$

and, thus,

$$|f(x) - g_1(x)| \leq \tfrac{2}{3}M \quad \text{for } x \in L.$$

If A or B is empty an appropriate constant function g_1 can be chosen such that the last inequality holds. This detail is left to the reader. Next let $f_1 = f - g_1$ on

L and $M_1 = \frac{2}{3}M$. By repeating the same argument with f_1 replacing the f and M_1 replacing the M, we obtain a continuous mapping

$$g_2 : X \to \left[-\frac{M_1}{3}, \frac{M_1}{3} \right]$$

such that for $x \in L$,

$$|f_1(x) - g_2(x)| \leq \tfrac{2}{3}M_1 = (\tfrac{2}{3})(\tfrac{2}{3})M$$

or

$$|f(x) - (g_1(x) + g_2(x))| \leq (\tfrac{2}{3})^2 M$$

and

$$|g_2(x)| \leq (\tfrac{1}{3})M_1 = \tfrac{1}{3}(\tfrac{2}{3})M.$$

We next make the inductive hypothesis that a finite sequence $g_1, g_2, \ldots, g_h$ of continuous mappings defined on X has been chosen so that

$$|g_i(x)| \leq \tfrac{1}{3}(\tfrac{2}{3})^{i-1}M$$

and such that

$$\left| f(x) - \sum_{i=1}^{h} g_i(x) \right| \leq (\tfrac{2}{3})^h M \quad \text{for} \quad x \in L.$$

Again by letting $f_h(x) = f(x) - \sum_{i=1}^{h} g_i(x)$ for $x \in L$, $M_h = (\tfrac{2}{3})^h M$ and letting M_h replace M and f_h replace f in the argument of the first paragraph, we obtain a continuous mapping g_{h+1} defined on X such that

$$|g_{h+1}(x)| \leq \tfrac{1}{3}(\tfrac{2}{3})^h M$$

$$\left| f(x) - \sum_{i=1}^{h+1} g_i(x) \right| \leq (\tfrac{2}{3})^{h+1} M \quad \text{for} \quad x \in L.$$

Thus, we have by induction a sequence $g_1, g_2, \ldots, g_n, \ldots$ of mappings satisfying

84.4(a)

$$|g_n(x)| \leq (\tfrac{1}{3})(\tfrac{2}{3})^{n-1}M \quad \text{for} \quad x \in X.$$

84.4(b)

$$\left| f(x) - \sum_{i=1}^{n+1} g_i(x) \right| \leq (\tfrac{2}{3})^{n+1} M \quad \text{for} \quad x \in L.$$

From 84.4(b) we see that $F(x) = \sum_{i=1}^{\infty} g_i(x)$ converges pointwise to f on L. From 84.4(a) and the Weierstrass M-test (Exercise 5, page 183), we see that F is continuous on X and that

$$|F(x)| \leq \sum_{i=1}^{\infty} (\tfrac{2}{3})^{i-1}\left(\frac{M}{3}\right) = \frac{M}{3}\frac{1}{1 - \frac{2}{3}} = M.$$

Thus, F is a continuous extension of f to X and $F[X] \subset [-M, M]$.

It is easy to prove that if a space satisfies the conclusion of Tietze's extension theorem, then it also satisfies the conclusion in Urysohn's lemma. To see this suppose that X has the extension property as in 84.4. Let A and B be a pair of disjoint nonempty closed subsets of X. Let g be defined on $A \cup B$ by $g(x) = 0$ for every x in A and $g(x) = 1$ for every x in B. We observe that g is continuous on $A \cup B$. By the condition assumed for X, there is a continuous extension $g^*: X \to [-1, 1]$ of g. Then the mapping f given by $f(x) = |g^*(x)|$ for all x in X is also a continuous extension of g. Furthermore, $f[X] \subset [0, 1]$ and f satisfies the properties in the conclusion of Urysohn's lemma.

The equivalence of the statements in Urysohn's lemma and in Tietze's extension theorem, together with Exercises 6 and 10, page 185, give the following characterization of normality.

84.5. Theorem. Let X be a topological space. Then the normality of X is equivalent to each of the following:

(a) If A is a closed subset of X and U is an open subset of X such that $A \subset U$, then there is an open subset V such that $A \subset V \subset \mathrm{cl}\,(V) \subset U$.

(b) For each pair of disjoint nonempty closed subsets A and B of X, there is a continuous function $f: X \to [0, 1]$ such that $f[A] = \{0\}$ and $f[B] = \{1\}$.

(c) If A is a closed subset of X and $f: A \to [-M, M]$ is continuous, then there is a continuous extension $g: X \to [-M, M]$ of f.

Suppose that a topological space X has the following property: If A is a nonempty closed subset of X and $x \in X - A$, then there is a continuous mapping $f: X \to [0, 1]$ such that $f(x) = 0$ and $f[A] = \{1\}$. It is easy to prove that this property implies that the space is regular (see Exercise 9, page 186). Because of 84.5, one might be tempted to guess that this property is equivalent to regularity. However, this is *not* the case and we have the following definition.

84.6. Definition. Completely regular space. *A topological space X is said to be completely regular provided that for each nonempty closed subset A and $x \in X - A$, there is a continuous mapping $f: X \to [0, 1]$ such that $f(x) = 0$ and $f[A] = \{1\}$. A Tychonoff space is a space that is both T_1 and completely regular.*

There exist examples of regular topological spaces that are not completely regular. An example of such a space is found in an article entitled *A Regular Space, not Completely Regular*, by John Thomas [31].

EXERCISES: A CHARACTERIZATION OF NORMALITY

1. Prove 84.3.

2. Prove that complete regularity is a topological invariant. Point out why the other separation properties are obviously topological invariants.

3. Prove that every subspace of a completely regular space is completely regular.

4. Prove that every T_1 normal space is a Tychonoff space.

5. Let $Y = \mathbf{R} - \{0\}$ with the relative topology induced by $\mathbf{R}$. Let $A = [-1, -\frac{1}{2}] \cup [\frac{1}{2}, 1]$ and $f: A \to \mathbf{R} - \{0\}$ be given by $f(x) = x$ for $x \in A$. Note that f is continuous. Show that there is no continuous extension of f to all of $\mathbf{R}$ that carries $\mathbf{R}$ into Y. (This simple example points out not only the importance of the

domain space of the contemplated extension but also the space into which we are seeking to map.)

6. Let $(X, \mathcal{T})$ be the topological space in Exercise 6, page 169. This space (of Niemytzki—see page 88 in [9]) is an example of a Tychonoff space that is not normal.

    (a) Show that X is a Hausdorff and, hence, a T_1-space.

    (b) Show that this space is a completely regular space. (See 3K, page 50 in [12] for a hint if you cannot get this part and for further comments about this example.)

    (c) Show that $(X, \mathcal{T})$ is not normal. The following is an outline of one approach to this problem. Assume that X is normal.

        Let $S = \{(r, 0) : r \in \mathbf{R}\}$. For $A \subset S$ show that there exists a continuous function $f_A : (X, \mathcal{T}) \to [0, 1]$ such that $f_A(x) = 0$ for x in A and $f_A(x) = 1$ for x in $S - A$. For each $A \subset S$, choose one such f_A and let $\mathcal{F} = \{f_A : A \subset S\}$. Point out why it is that $\mathcal{F}$ has larger cardinality than does the set $\mathbf{R}$ (see Theorem 28.2). Let $C = \{(a, b) : a$ is rational and b is positive and rational$\}$. Note that C is a countable dense subset of $(X, \mathcal{T})$. Observe that if $f_A \,|\, C = f_B \,|\, C$, where $f_A \in \mathcal{F}$ and $f_B \in \mathcal{F}$, then $f_A = f_B$ (see Exercise 16, page 187). Use a cardinality argument to get a contradiction by considering the cardinality of the collection of real-valued functions defined on C. (For further discussion of this space see page 49 in [12].)

85. SEPARABILITY AXIOM

We have been introduced to the separability property that holds for some metric spaces, for example, $\mathbf{R}^n$, ℓ^2, and $\mathcal{C}([a, b])$, and not for certain others. The concept is also important in the study of more general topological spaces.

85.1. Definition. Separable space. *A topological space X is said to be separable provided there exists a countable set D that is dense in X.*

For metric spaces we showed that every subspace of a separable space is separable. This is not so for topological spaces in general, and this fact will be brought out in the exercises. However, it is easy to establish the following.

85.2. Theorem. *Every open subspace of a separable topological space is separable.*

Not only is separability a topological property but we have the following stronger result.

85.3. Theorem. *If $f : X \to Y$ is a continuous surjection and X is separable, then so is Y.*

EXERCISES: SEPARABILITY AXIOM

1. Show that the topological space in Exercise 6, page 169, is separable but has a subspace that is not separable.

2. Verify Theorems 85.2 and 85.3.

3. Suppose $(X, \mathcal{T})$ has the discrete topology and is known to be separable. Does this imply anything about its cardinality?

4. Consider a product space $\bigtimes \{(X_i, \mathcal{T}_i) : i \in \mathbf{P}_n\}$. Suppose that this product space is separable. Is each coordinate space X_i necessarily separable? Suppose for each $i \in \mathbf{P}_n$, X_i is separable. Is the product space necessarily separable?

5. Suppose that X is the countable union of separable subspaces. Is X necessarily separable?

86. SECOND COUNTABLE SPACES

86.1. Definition. Second countable space. *A topological space $(X, \mathcal{T})$ is said to be a second countable space or a perfectly separable space provided there exists a countable base $\mathcal{B}$ for the topology $\mathcal{T}$.*

Recall that we showed in Theorem 58.7 that, for metric spaces, the properties of separability and second countability are equivalent. This is not so for the general situation, second countability being the stronger of the two properties. The reader should verify the following.

86.2. Theorem. *If X is a second countable space, then it is a separable space.*

For metric spaces we found that a subspace of a separable metric space is separable (Theorem 58.10). This results from the fact that, for metric spaces, second countability and separability are equivalent. For topological spaces in general, we cannot prove that subspaces of separable spaces are necessarily separable. However, we do have the following theorem. The proof furnished is probably similar to the reader's proof for Theorem 58.10.

86.3. Theorem. *Every subspace of a second countable space is second countable.*

PROOF. Let S be a subspace of a second countable space X. Suppose $\{U_i : i \in \mathbf{P}\}$ is a countable base for the topology for X. Define $\mathcal{B} = \{W_i : W_i = U_i \cap S\}$. The claim is made that $\mathcal{B}$ is a base for the relative topology for S. To see this let $x \in S$ and U be an open S-neighborhood of x. Then there is a V, open in X, such that $x \in U = S \cap V$. For some $i \in \mathbf{P}$, $x \in U_i \subset V$. But then $x \in U_i \cap S \subset V \cap S$. Since $(U_i \cap S) \in \mathcal{B}$, the proof is complete.

It is an interesting and useful fact that second countability together with regularity strengthens the separation property as seen in the next theorem.

86.4.   Theorem. *If $(X, \mathcal{T})$ is regular and second countable, then it is normal.*

PROOF.   Let A and B be two nonempty disjoint closed subsets of X. Since X is second countable there exists a countable base $\mathcal{B}$ for $\mathcal{T}$. Let $\mathcal{B}_A$ be the subcollection of all elements of $\mathcal{B}$ that intersect A and whose closures do not intersect B. Similarly, let $\mathcal{B}_B$ be the subcollection of all elements of $\mathcal{B}$ that intersect B and have closures disjoint from A. From the fact that $\mathcal{B}$ is a base and that X is regular, it follows easily that $\bigcup \mathcal{B}_A \supset A$, and $\bigcup \mathcal{B}_B \supset B$. However, we have no assurance that these two open sets are disjoint so we modify them as follows: Index $\mathcal{B}_A$ and $\mathcal{B}_B$ with the positive integers and write

$$\mathcal{B}_A = \{U_i : i \in \mathbf{P}\}$$
$$\mathcal{B}_B = \{V_i : i \in \mathbf{P}\}.$$

Let $R_1 = U_1 - \mathrm{cl}\,(V_1)$, $S_1 = V_1 - \mathrm{cl}\,(U_1)$ and, in general,

$$R_n = U_n - \bigcup \{\mathrm{cl}\,(V_i) : i \in \mathbf{P}_n\}$$

and

$$S_n = V_n - \bigcup \{\mathrm{cl}\,(U_i) : i \in \mathbf{P}_n\}$$

Let

$$R = \bigcup \{R_n : n \in \mathbf{P}\}$$

and

$$S = \bigcup \{S_n : n \in \mathbf{P}\}.$$

It is clear that the open sets R and S contain A and B, respectively. We complete the proof by showing that $R \cap S = \varnothing$.

Suppose $x \in R \cap S$. Then there exist positive integers m and n such that $x \in R_m \cap S_n$ and either $m \geq n$ or $n \geq m$. We consider the case $m \geq n$, and the other case is analogous. Since $x \in R_m = U_m - \bigcup \{\mathrm{cl}\,(V_k) : k \in \mathbf{P}_m)\}$, $x \notin \mathrm{cl}\,(V_k)$ for $k \leq m$. Thus, $x \notin V_n$ and $x \notin S_n$ and we have a contradiction.

Once again, as for metric spaces, we have another extension of the Lindelöf theorem.

86.5.   Lindelöf theorem. *Let X be a second countable space. If S is a subset of X and $\mathcal{H}$ is a collection of open subsets of X such that $\bigcup \mathcal{H} \supset S$, then some countable subcollection of $\mathcal{H}$ also covers S.*

The proof of Theorem 58.9 carries over without any change to prove Theorem 86.5.

86.6.   Definition. Lindelöf space. *If every open covering of a space X has a countable subcovering, then X is called a Lindelöf space.*

If X is a metric space and is a Lindelöf space, then for every $\varepsilon > 0$, X can be covered by a countable collection of open ε-spheres. By Exercise 6, page 118 , X is separable and, hence, since X is metric, X is second countable. Thus, a metric space X is second countable if and only if X is a Lindelöf space. In the next set of exercises, the reader will be asked to verify that this is not correct for topological spaces.

A slight modification of the first part of the proof of Theorem 86.4, gives the following.

86.7.   Theorem. *If X is a regular Lindelöf space, then it is normal.*

EXERCISES: SECOND COUNTABLE SPACES

1. Prove Theorem 86.2.

2. Show that the topological space in Exercise 6, page 169, is not second countable. Recall that it is separable.

3. Is the space in Example 75.6 second countable? Is it a Lindelöf space?

4. Give an example to show that second countability is not invariant under continuous mappings.

5. Show that second countability is invariant under open continuous surjections.

6. Show that the space in Example 75.7 is not second countable but that it is Lindelöf.

7. Give an example of a Lindelöf space that is not separable.

8. Point out the modification in the proof of 86.4 that must be made in order to prove 86.7.

87. FIRST COUNTABLE SPACES

If (X, d) is a metric space, then for each point $x \in X$, the collection of all $\frac{1}{i}$-neighborhoods of x forms a countable base for the neighborhood system of x. As will be seen from the next set of exercises, there are spaces that do not have this property of having a countable base for the neighborhood system of each point.

87.1. Definition. First countable space. *If X is a topological space and for each $x \in X$ the neighborhood system of x has a countable base, then X is said to be a first countable space.*

Note that every second countable space is also first countable.

87.2. Theorem. *If X is a first countable space, then for each $x \in X$ the neighborhood system of x has a countable base $\mathscr{B}_x = \{B_x^i : i \in \mathbf{P}\}$ such that each B_x^i is open and $B_x^i \supset B_x^{i+1}$. Furthermore, if X is T_1, then $\bigcap \mathscr{B}_x = \{x\}$.*

PROOF. Let $x \in X$ and suppose $\mathscr{B}_x^*$ is a countable base for the neighborhood system of x. For each $C_i^* \in \mathscr{B}_x^*$, choose an open set C_i such that $x \in C_i \subset C_i^*$. Next, let $B_x^1 = C_1$, $B_x^n = \bigcap \{C_i : i \in \mathbf{P}_n\}$. Note that each B_x^i is open and $B_x^i \supset B_x^{i+1}$. Let $\mathscr{B}_x = \{B_x^i : i \in \mathbf{P}\}$. That $\mathscr{B}_x$ is a base for the neighborhood system of x is seen as follows: Let Q be a neighborhood of x. Then there is a $C_i^* \in \mathscr{B}_x^*$ such that $x \in C_i^* \subset Q$. But then $x \in B_x^i \subset C_i^* \subset Q$. So $\mathscr{B}_x$ is a base for the neighborhood system of x.

If, further, X is T_1, $\bigcap \mathscr{B}_x = \{x\}$. Suppose $y \neq x$. Then there is an open neighborhood V of x such that $y \notin V$. But then there is a $j \in \mathbf{P}$ such that $x \in B_x^j \subset V$. Since $y \notin B_x^j$, $y \notin \bigcap \mathscr{B}_x$.

Following are some generalizations of theorems previously given for metric spaces.

87.3. Theorem. *Let X be a first countable space. Then x is a limit point of a set S if and only if there is a sequence in $S - \{x\}$ that converges to x.*

87.4. Theorem. *Suppose that f is a mapping from a first countable space X into a topological space Y. Then f is continuous if and only if for each sequence (x_i) in X and $x \in X$, (x_i) converges to x implies that $(f(x_i))$ converges to $f(x)$.*

EXERCISES: FIRST COUNTABLE SPACES

1. Prove Theorem 87.3.

2. Prove Theorem 87.4.

3. Give an example of a T_1-space that is not first countable.

4. Characterize closure in terms of sequences for first countable spaces.

5. Suppose $f: X \to Y$ is a continuous open surjection. If X is first countable, is Y necessarily first countable? If Y is first countable, is X necessarily first countable?

6. Refer to the proof of Theorem 58.7(a). Note that the proof essentially made use of the first countability of metric spaces to show that separable metric spaces are second countable. Give an example of a separable first countable space that is not second countable.

7. Is every subspace of a first countable space first countable?

8. Prove the following proposition:
   Let X be a T_1 first countable space. Then x is a limit point of a set S if and only if there is a sequence of distinct points in S that converges to x. (Compare this to Theorem 87.3.)

9. Prove that a first countable space is Hausdorff if and only if each convergent sequence has a unique limit.

88. COMPARISON OF TOPOLOGIES

Suppose $\mathcal{T}_1$ and $\mathcal{T}_2$ are topologies for a set X such that $\mathcal{T}_1 \subset \mathcal{T}_2$. We then say that $\mathcal{T}_1$ is a *smaller* topology for X than $\mathcal{T}_2$ and $\mathcal{T}_2$ is a *larger* topology than $\mathcal{T}_1$. It should be emphasized that $\mathcal{T}_1 \subset \mathcal{T}_2$ means inclusion between these collections and *not* inclusion between members of the collections. In fact, the larger collection $\mathcal{T}_2$ has smaller members in the following sense: For every $T_1 \in \mathcal{T}_1$ and $T_2 \in \mathcal{T}_2$, $T_1 \cap T_2 \in \mathcal{T}_2$ and $T_1 \cap T_2 \subset T_1$. (This observation is reflected in the following alternate terminology: The phrase $\mathcal{T}_2$ is *finer* than $\mathcal{T}_1$ is sometimes used to indicate that $\mathcal{T}_2$ is *larger* than $\mathcal{T}_1$; or the phrase $\mathcal{T}_1$ is *coarser* than $\mathcal{T}_2$ may be used to express the same relationship.)

The collection of all topologies for a set X is partially ordered (see Definition 22.1) by inclusion. However, the collection of topologies for X is not totally ordered by inclusion (Definition 22.5). Note that the discrete topology is the largest topology that a set can have and that the trivial topology is the smallest topology that a set can have.

88.1. *Example.* Let $X = \{1, 2, 3, 4\}$. Consider the topologies

$$\mathcal{T}_1 = \{\varnothing, \{1\}, \{1, 2\}, \{1, 2, 3\}, \{1, 2, 3, 4\}\}$$

and

$$\mathcal{T}_2 = \{\varnothing, \{3\}, \{2, 3\}, \{1, 2, 3\}, \{1, 2, 3, 4\}\}.$$

It is interesting to note that $\mathcal{T}_1 \cap \mathcal{T}_2$ is a topology for X but $\mathcal{T}_1 \cup \mathcal{T}_2$ is not a topology. The fact that in this case $\mathcal{T}_1 \cap \mathcal{T}_2$ is a topology raises an interesting question that will be pursued in the next set of exercises.

In the following set of exercises, questions will be asked about various topologies for a set. In answering these questions the reader will also be reviewing various other concepts studied in this chapter.

EXERCISES: COMPARISON OF TOPOLOGIES

1. Suppose $\{\mathcal{T}_\alpha : \alpha \in A\}$ is a collection of topologies for a set X. Is $\bigcap \{\mathcal{T}_\alpha : \alpha \in A\}$ a topology for X?

2. Suppose $\mathcal{T}_1$ and $\mathcal{T}_2$ are topologies for X. Suppose $\mathcal{T}_1$ is smaller than $\mathcal{T}_2$. Which of the following properties possessed by $(X, \mathcal{T}_1)$ is necessarily possessed by $(X, \mathcal{T}_2)$?
   (a) T_1.   (d) Normal.
   (b) T_2.   (e) Separable.
   (c) Regular.   (f) Second countable.
   Repeat the question under the assumption that $\mathcal{T}_2 \subset \mathcal{T}_1$.

3. Assume that $f : (X, \mathcal{T}) \to (Y, \mathcal{T}_1)$ is continuous.
   (a) Suppose $\mathcal{T}_2$ is a topology for Y such that $\mathcal{T}_2$ is larger than $\mathcal{T}_1$. Is $f : (X, \mathcal{T}) \to (Y, \mathcal{T}_2)$ necessarily continuous?
   (b) Suppose $\mathcal{T}_3$ is a topology for Y such that $\mathcal{T}_3$ is smaller than $\mathcal{T}_1$. Is $f : (X, \mathcal{T}) \to (Y, \mathcal{T}_3)$ necessarily continuous?
   (c) Suppose $\mathcal{T}^*$ is a smaller topology for X than is $\mathcal{T}$. Is $f : (X, \mathcal{T}^*) \to (Y, \mathcal{T}_1)$ necessarily continuous?
   (d) Suppose $\mathcal{T}^{**}$ is a larger topology for X than is $\mathcal{T}$. Is $f : (X, \mathcal{T}^{**}) \to (Y, \mathcal{T}_1)$ necessarily continuous?

4. Suppose X is a set and $\mathcal{T}_1$ and $\mathcal{T}_2$ are topologies for X such that $\mathcal{T}_1 \subset \mathcal{T}_2$. Suppose $p \in X$ and p is a limit point of a subset S in the space $(X, \mathcal{T}_1)$. Is p necessarily a limit point in the space $(X, \mathcal{T}_2)$? Repeat the question, this time under the assumption that $\mathcal{T}_2 \subset \mathcal{T}_1$.

89. URYSOHN'S METRIZATION THEOREM

It has been an important problem in topology to determine sufficient conditions for a topological space to be metrizable. In this section we state and prove a classic theorem that gives a sufficient but by no means necessary condition for metrizability.

One method of showing that a topological space X is metrizable is to show that there is a topological mapping from X onto a metric space. It will then follow that X is metrizable because of the following theorem.

89.1.   Theorem.   *If $h:(X, \mathcal{T}(X)) \to (Y, \mathcal{T}(Y))$ is a topological mapping from X onto Y, then $(X, \mathcal{T}(X))$ is metrizable if and only if $(Y, \mathcal{T}(Y))$ is metrizable. (Thus, metrizability is a topological invariant.)*

PROOF.   Let h be a homeomorphism from $(X, \mathcal{T}(X))$ onto $(Y, \mathcal{T}(Y))$. Notice the symmetry of the hypothesis; that is, h^{-1} is also a homeomorphism. Therefore, it will be sufficient to show that if $(Y, \mathcal{T}(Y))$ is metrizable so is $(X, \mathcal{T}(X))$. Let d be a metric that generates $\mathcal{T}(Y)$. We define a metric d^* for X as follows:

$$d^*(a, b) = d(h(a), h(b)).$$

From the fact that h is one-to-one it follows easily that $d^*(a, b) = 0$ if and only if $a = b$. It is likewise easy to show that d^* inherits the triangle inequality and symmetric properties from d. We show next that d^* generates $\mathcal{T}(X)$. In order to do this let $\mathcal{T}(d^*)$ be the topology generated by d^*. We show first that $\mathcal{T}(X) \subset \mathcal{T}(d^*)$. Let $U \in \mathcal{T}(X)$. We need consider only the case $U \neq \emptyset$. Let $x \in U$. We shall show that there is a d^*-neighborhood of x that is contained in U. Since $x \in U, h(x) \in h[U]$. Since h is a homeomorphism, it is an open mapping so that $h[U] \in \mathcal{T}(Y)$. Then there is an $\varepsilon > 0$ such that $N_d(h(x); \varepsilon) \subset h[U]$. But

$$h^{-1}[N_d(h(x); \varepsilon)] = N_{d^*}(x; \varepsilon)$$

from the way in which d^* was defined. Upon noticing that $x \in N_{d^*}(x; \varepsilon) \subset U$, we see that x is an interior point of U in the space $(X, \mathcal{T}(d^*))$. Thus, we have shown that $U \in \mathcal{T}(d^*)$. We next show that $\mathcal{T}(d^*) \subset \mathcal{T}(X)$. Let $U \in \mathcal{T}(d^*)$. If $x \in U$, then there is an $\varepsilon > 0$ such that $x \in N_{d^*}(x; \varepsilon) \subset U$. Since $N_d(h(x); \varepsilon) \in \mathcal{T}(Y)$ and $h:(X, \mathcal{T}(X)) \to (Y, \mathcal{T}(Y))$ is continuous, $h^{-1}[N_d(h(x); \varepsilon)] \in \mathcal{T}(X)$. But since $N_{d^*}(x; \varepsilon) = h^{-1}[N_d(h(x); \varepsilon)]$, it follows that $N_{d^*}(x; \varepsilon) \in \mathcal{T}(X)$. Thus, we have shown that U is open in $(X, \mathcal{T}(X))$ and $U \in \mathcal{T}(X)$. We have now shown that $\mathcal{T}(d^*) \subset \mathcal{T}(X)$ and, together with the first part of the proof, we have shown that $\mathcal{T}(d^*) = \mathcal{T}(X)$. Hence, d^* is a metric for X that generates $\mathcal{T}(X)$ and, by definition, $(X, \mathcal{T}(X))$ is metrizable.

In the next theorem we shall prove that every T_1 regular second countable space is metrizable. This will be done by showing that such a space can be mapped topologically onto a subset of the Hilbert cube I^∞ (see 71) and, hence, by the previous theorem is metrizable. The homeomorphism h to be defined will be defined coordinatewise. Each coordinate function h_i will be defined by making use of Urysohn's lemma (84.3). To apply Urysohn's lemma it will be necessary to keep in mind that regular second countable spaces are normal.

89.2.   Urysohn's metrization theorem.   *Let $(X, \mathcal{T})$ be a T_1 regular second countable space. Then X is homeomorphic to a subset of the Hilbert cube I^∞ and, hence, is metrizable.*

PROOF.   Let $\mathcal{B} = \{B_i : i \in \mathbf{P}\}$ be a countable base for the topology $\mathcal{T}$. Also let $\mathcal{R}$ be the collection of all pairs $(B_i, B_j) \in \mathcal{B} \times \mathcal{B}$ such that $\text{cl}\,(B_i) \subset B_j$. (Since

X is regular, there will be an adequate supply of such pairs for our purposes.) Index $\mathscr{R}$ with the set $\mathbf{P}$ of all positive integers so that we may write $\mathscr{R} = \{R_i : i \in \mathbf{P}\}$. Let $i \in \mathbf{P}$. Then R_i is a pair (B_{n_i}, B_{m_i}) such that cl $(B_{n_i}) \subset B_{m_i}$. Since X is regular and second countable (by theorem 86.4), it is normal. Hence, by Urysohn's lemma (84.3), there is a continuous mapping $h_i : X \rightarrow \left[0, \dfrac{1}{i}\right]$ such that $h_i[\mathrm{cl}\,(B_{n_i})] = \{0\}$ and $h_i[X - B_{m_i}] = \left\{\dfrac{1}{i}\right\}$. We find such an h_i for each $i \in \mathbf{P}$ and define $h : X \rightarrow I^\infty$ as follows: For each $x \in X$, let

$$h(x) = (h_1(x), h_2(x), \ldots, h_n(x), \ldots).$$

Note that since $|h_i(x)| \le \dfrac{1}{i}$, $h(x) \in I^\infty$. We shall show that h is a homeomorphism in the following steps: (a) h is continuous; (b) h is one-to-one from X onto $h[X]$; and (c) h maps open subsets of X onto subsets of I^∞ that are open relative to $h[X]$. The proof will then have been completed.

(a) Since by hypothesis X is second countable, we may use the sequential characterization of continuity (87.4). Let (x_i) be a sequence in X that converges to a point x. For each $j \in \mathbf{P}$, the sequence $(h_j(x_i))_{i=1}^\infty$ converges to $h_j(x)$ since h_j is continuous. Then from 71.1, the sequence $(h(x_i))$ converges to $h(x) = (h_1(x), h_2(x), \ldots, h_n(x), \ldots)$. Thus, h is continuous.

(b) To show that h is one-to-one, suppose x and y are in X and $x \ne y$. Since X is a T_1-space, there is an open neighborhood U of x that does not contain y. Then there is a B' in $\mathscr{B}$ such that $x \in B' \subset U$. By regularity we can find a B'' in $\mathscr{B}$ such that $x \in B'' \subset \mathrm{cl}\,(B'') \subset B'$. For some $i \in \mathbf{P}$, $R_i = (B'', B') \in \mathscr{R}$. But $y \in X - B'$ and $x \in \mathrm{cl}\,(B'')$ so that $h_i(x) = 0$ and $h_i(y) = \dfrac{1}{i}$. Thus, $h_i(x) \ne h_i(y)$ and, consequently, $h(x) \ne h(y)$.

(c) Let U be open in X. We wish to show that $h[U]$ is open relative to the subspace $h[X]$ of (I^∞, d). Let $z \in h[U]$. We will show that $h[U]$ is open relative to $h[X]$ by showing that z is not a limit point of $h[X] - h[U]$. Let $x = h^{-1}(z)$. Then there is a positive integer i such that $R_i = (B_{n_i}, B_{m_i})$, $x \in B_{n_i} \subset \mathrm{cl}\,(B_{n_i}) \subset B_{m_i} \subset U$. Recall that $h_i(x) = 0$ and $h_i(q) = \dfrac{1}{i}$ for $q \in X - B_{m_i}$. Thus, $h_i(q) = \dfrac{1}{i}$ for $q \in X - U$. Hence, for each $q \in X - U$, $d(h(x), h(q)) \ge \dfrac{1}{i}$. From this we see that z could not be a limit point of $h[X] - h[U]$, since any point of $h[X] - h[U]$ is at a distance of at least $\dfrac{1}{i}$ from z.

The following interesting proposition follows from the previous theorem and from the fact that all separable metric spaces are T_1, regular, and second countable.

89.3. Theorem. *The Hilbert cube I^∞ contains a topological copy of each separable metric space.*

The conditions given in the hypothesis of Theorem 89.2 are not necessary for metrizability. However, they are necessary for a space to be metrizable and separable, as indicated by the next theorem. The proof is left as an exercise for the reader.

89.4. Theorem. *A topological space is separable and metrizable if and only if it is a T_1 regular second countable space.*

EXERCISES: URYSOHN'S METRIZATION THEOREM

1. Suppose that (X, d) is a separable metric space and $f : (X, d) \to (Y, \mathscr{T})$ is a continuous surjection that is both open and closed. Prove that $(Y, \mathscr{T})$ is metrizable.

2. Prove Theorem 89.4.

3. In Exercise 3, page 121, the reader was asked to prove that if a metric space (X, d) is not separable, then for no metric d^* equivalent to d is it true that (X, d^*) is totally bounded. Prove the converse of this statement; that is, prove that if (X, d) is a separable metric space, then there is a metric d^* that is equivalent to d and which is such that (X, d^*) is totally bounded. Hint: Use 89.2 and recall that I^∞ is compact and, hence, totally bounded.

7

Compactness and Related Properties

In Sections 60, 61, and 62 we studied the concepts of compactness, sequential compactness, and Bolzano-Weierstrass properties for metric spaces. In this chapter we shall study these properties as well as an additional compactness property called countable compactness in the setting of topological spaces. Although all these compactness properties are equivalent for metric spaces, they are not equivalent in general; each merits study on its own. We shall show, for example, that for all spaces the Bolzano-Weierstrass property is implied by each of the other types of compactness and that they are all equivalent for T_1 second countable spaces. Sequential compactness does not lend itself well to spaces that are not first countable. It is, however, a valuable tool and is equivalent to the Bolzano-Weierstrass property for spaces that are both T_1 and first countable. We shall also study the notion of local compactness in a general setting.

As for metric spaces, compactness is preserved by continuous surjections. Moreover, if f is a continuous mapping from a compact space onto a Hausdorff space, f has some rather strong and useful properties. For example, such a mapping is closed and, furthermore, $f^{-1}[K]$ is compact for every compact subset K of the range space. Many of the other properties enjoyed by a continuous surjection f from a compact space onto a Hausdorff space result from the fact that such a mapping is closed and has the so-called compact point inverse property; that is, $f^{-1}[y]$ is compact for each y in the range of f. Continuous surjections with these two properties are called *perfect* mappings. The fact that perfect mappings are closed implies that normality is preserved under such mappings. We shall show further that the Hausdorff property, regularity, second countability, and metrizability are each preserved by perfect mappings.

90. DEFINITIONS OF VARIOUS COMPACTNESS PROPERTIES

In this section we list definitions of various compactness properties for general topological spaces. The reader is already familiar with most of these concepts from our previous work with metric spaces. By examining theorems and their proofs for metric spaces, we shall be able to see appropriate generalizations of some of the previous theorems. For example, for topological spaces, compactness and sequential compactness are not equivalent. However, a close examination of our previous work suggests that they are equivalent under conditions somewhat weaker than metrizability.

90.1. Definition. Compact space. *A topological space X is said to be compact provided every open covering of X contains a finite subcollection that covers X.*

A collection $\mathscr{C}$ of sets is said to have the *finite intersection property* provided $\mathscr{C}$ is nonempty and for every nonempty finite subcollection $\mathscr{F}$ of $\mathscr{C}$, $\bigcap \mathscr{F}$ is nonempty. By making use of De Morgan's laws, we can characterize compactness in terms of the finite intersection property.

90.2. Theorem. *A topological space is compact if and only if every collection of closed subsets of X with the finite intersection property has a nonempty intersection.*

The details of the proof of this theorem are left as an exercise.

90.3. Definition. Countably compact space. *A topological space X is said to be countably compact provided every countable open covering of X contains a finite subcollection that covers X.*

90.4. Definition. Sequentially compact space. *A topological space X is called sequentially compact provided every sequence in X has a subsequence that converges to a point in X.*

90.5. Definition. The Bolzano-Weierstrass property. *A space X is said to have the Bolzano-Weierstrass property provided every infinite subset of X has a limit point in X. If X has this property, we shall say that it is B.W. compact (the "B.W." for Bolzano-Weierstrass).*

90.6. *Example.* Let $\mathbf{Z}$ denote the set of all integers. It is easy to verify that $\mathscr{B} = \{\{-n, n\} : n \text{ is a nonnegative integer}\}$ is a base for a topology $\mathscr{T}$ for $\mathbf{Z}$. The space $(\mathbf{Z}, \mathscr{T})$ is not T_1 and, hence, is not metrizable. The collection $\mathscr{B}$ is itself a countable open covering of $\mathbf{Z}$ that does not contain a finite subcovering of $\mathbf{Z}$. Thus, $(\mathbf{Z}, \mathscr{T})$ is not countably compact (and, hence, not compact). Furthermore, if we let $z_i = i$ for each $i \in \mathbf{P}$, we see that $(\mathbf{Z}, \mathscr{T})$ is not sequentially compact, because the sequence (z_i) has no convergent subsequence. On the other hand, $(\mathbf{Z}, \mathscr{T})$ is B.W. compact. To see this, let S be an infinite subset of $\mathbf{Z}$. Then there is a $z \in S$ such that $z \neq 0$. The point $-z$ is a limit point of the set $\{z\}$ and, hence, also of S. Thus, we have shown that $(\mathbf{Z}, \mathscr{T})$ is B.W. compact. It is of interest to note that $(\mathbf{Z}, \mathscr{T})$ is second countable since $\mathscr{B}$ is countable. This example also points out that compact subsets of a space need not be closed. For example, the subset $\{1\}$ is a compact subset of $(\mathbf{Z}, \mathscr{T})$ but it is not a closed set.

EXERCISES: DEFINITIONS OF VARIOUS COMPACTNESS PROPERTIES

1. Previously we proved that, for metric spaces, the concepts of B.W. compactness, compactness, and sequential compactness are equivalent. In this section we added the notion of countable compactness. Show that it is also equivalent to compactness for metric spaces.

2. (a) Show that every finite subset of a topological space has each of the four types of compactness defined in this section.

    (b) In Theorem 60.2, we saw that sequentially compact subsets of a metric space are closed; hence, for metric spaces compact subsets are closed. Use part (a) of this exercise to construct an example that shows that in the setting of a general topological space, none of the types of compactness defined in this section implies closedness.

    (c) Is a compact subset of a Hausdorff space necessarily closed?

3. Which of the compactness properties does the topological space in Example 75.6 possess?

4. Let $\mathbf{Z}$ be the set of all integers. For each $n \in \mathbf{Z}$, let B_n be the half-open real interval $[n - 1, n)$. Let $\mathscr{B} = \{B_n : n \in \mathbf{Z}\}$.

    (a) Show that $\mathscr{B}$ is a base for a topology for the set of all real numbers $\mathbf{R}$. Let $\mathscr{T}(\mathscr{B})$ be the topology generated by $\mathscr{B}$.

    (b) Which of the separation properties does $(\mathbf{R}, \mathscr{T}(\mathscr{B}))$ possess?

    (c) Is $(\mathbf{R}, \mathscr{T}(\mathscr{B}))$ second countable?

    (d) Which, if any, of the compactness properties does this space possess?

5. Let $\mathscr{B}$ be the collection of all subsets of $\mathbf{R}$ of the form $[a, b)$. Show that $\mathscr{B}$ is a base for a topology $\mathscr{T}(\mathscr{B})$ for the set $\mathbf{R}$.

    Repeat questions (b), (c), and (d) of the previous exercise for this space.

6. Suppose X is a nonempty set and $\mathscr{T}$ is the trivial topology for X. Is $(X, \mathscr{T})$ necessarily a compact space?

7. Suppose X is a nonempty set and $\mathscr{T}$ is the discrete topology for X. Suppose it is known that $(X, \mathscr{T})$ is compact. Does this fact give you any information about the cardinality of X?

8. Prove Theorem 90.2.

9. Let $g : \mathbf{R} \to \mathbf{R}$ be given by $g(x) = -x$. Let $\mathscr{T}$ be the collection of all subsets of $\mathbf{R}$ of the form $U \cup g[U]$ where U is open in $\mathbf{R}$ with the Euclidean metric.

    (a) Show that $(\mathbf{R}, \mathscr{T})$ is a topological space.

    (b) Show that there exist compact subsets of $(\mathbf{R}, \mathscr{T})$ that are not closed.

(c) Is $(\mathbf{R}, \mathscr{T})$ a T_1-space? a T_2-space?

(d) Show that the intersection of compact subsets of $(\mathbf{R}, \mathscr{T})$ need not be compact.

10. Suppose $\mathscr{T}_1$ and $\mathscr{T}_2$ are two topologies for X and $\mathscr{T}_1 \subset \mathscr{T}_2$. Does the compactness of $(X, \mathscr{T}_1)$ imply the compactness of $(X, \mathscr{T}_2)$? Does the compactness of $(X, \mathscr{T}_2)$ imply the compactness of $(X, \mathscr{T}_1)$?

91. SOME CONSEQUENCES OF COMPACTNESS

Recall that a closed subset of a compact metric space is compact. This is true for topological spaces in general.

91.1. Theorem. *Suppose that X is a compact space and F is a closed subset of X. Then F is compact.*

The proof of this theorem is left as an exercise.

In the previous section we saw that compact subsets of a topological space need not be closed (see Example 90.6). However, if the space is a Hausdorff space, such examples no longer exist.

91.2. Theorem. *Let K be a compact subset of a Hausdorff space X. Then K is a closed subset of X.*

We shall prove Theorem 91.2 by proving the following more general statement.

91.3. Theorem. *Let K be a compact subset of a Hausdorff space X and let $x \in X - K$. Then there exists a pair of disjoint open subsets U and V such that $K \subset U$ and $x \in V$. (Since $x \in V \subset X - K$, this shows in particular that $X - K$ is open and, hence, that K is closed.)*

PROOF. We assume that $K \neq \varnothing$, for otherwise the conclusion is obvious. Suppose $x \in X - K$. For each $y \in K$ there exists a pair of disjoint open neighborhoods $U(y)$ and $V^x(y)$ of y and x, respectively. The collection $\{U(y) : y \in K\}$ is an open covering of K. Since K is compact, there is a finite subcovering

$$\{U(y_1), U(y_2), \ldots, U(y_n)\}.$$

Let $U = \bigcup \{U(y_i) : i \in \mathbf{P}_n\}$ and $V = \bigcap \{V^x(y_i) : i \in \mathbf{P}_n\}$. U and V are a pair of disjoint open subsets. Moreover, $K \subset U$ and $x \in V$.

By making use of Theorem 91.3, we can prove the following more general result. The proof is left as an exercise.

91.4. Theorem. *Suppose X is a Hausdorff space and H and K are a pair of disjoint compact subsets. Then there exists a pair of disjoint open subsets U and V such that $H \subset U$ and $K \subset V$.*

The previous theorem suggests the following important result that follows from it and Theorem 91.1.

91.5.   **Theorem.**   *If a space is Hausdorff and compact, then it is normal.*

Recall that a compact metric space is second countable. This need not be the case for topological spaces. The space in Example 75.6 is compact but not second countable. However, if a compact Hausdorff space happens to be second countable, then from Theorem 91.5 and Urysohn's metrization theorem, we have the following.

91.6.   **Theorem.**   *A compact Hausdorff second countable space is metrizable.*

The following two important theorems are easy to prove and their proofs are left as exercises.

91.7.   **Theorem.**   *If $f: X \to Y$ is continuous and S is a compact subset of X, then $f[S]$ is a compact subset of Y.*

91.8.   **Theorem.**   *Suppose X is compact and Y is Hausdorff. If $f: X \to Y$ is a continuous surjection, then f is a closed mapping. If, in addition, f is one-to-one, then f is a homeomorphism.*

Recall from Exercise 5, page 196, that second countability is invariant under continuous open surjections. The following theorem gives another condition under which second countability is invariant.

91.9.   **Theorem.**   *Suppose that X is compact, Y is Hausdorff and $f: X \to Y$ is a continuous surjection. Then, if X is second countable, so is Y.*

Proof.   Let $\mathscr{B} = \{B_i : i \in \mathbf{P}\}$ be a countable base for X. Let $\mathscr{B}^*$ be the collection of all finite unions of elements of $\mathscr{B}$. Since $\mathscr{B}^*$ is countable, we may write $\mathscr{B}^* = \{B_i^* : i \in \mathbf{P}\}$. (Note that the B_i^*'s have the property that if K is compact and U is an open subset of X containing K, then for some j, $K \subset B_j^* \subset U$.) From the previous theorem, f is a closed mapping. Hence, $U_j = Y - f[X - B_j^*]$ is open for each $j \in \mathbf{P}$. We will show that $\{U_j : j \in \mathbf{P}\}$ is a base for Y. To see this let $y \in U$, where U is open in Y. Then $f^{-1}[y]$ is closed and, hence, compact. Furthermore, $f^{-1}[y] \subset f^{-1}[U]$. Thus, there is a B_j^* such that

$$f^{-1}[y] \subset B_j^* \subset f^{-1}[U].$$

By taking complements, we get from the last set of inclusions

$$X - f^{-1}[U] \subset X - B_j^* \subset X - f^{-1}[y].$$

Thus,

$$Y - U \subset f[X - B_j^*] \subset Y - \{y\}.$$

By taking complements again, we get

$$\{y\} \subset Y - f[X - B_j^*] \subset U$$

so that

$$y \in U_j \subset U.$$

By making use of the previous theorem, we can prove the following important result.

91.10.   **Theorem.**   *Suppose X is compact and metrizable and Y is Hausdorff. If there exists a continuous mapping from X onto Y, then Y is metrizable.*

EXERCISES: SOME CONSEQUENCES OF COMPACTNESS

1. Prove each of the following theorems
   (a) 91.1.   (d) 91.6.
   (b) 91.4.   (e) 91.7.
   (c) 91.5.   (f) 91.8.

2. Prove that if A and B are compact subsets of a Hausdorff space, then $A \cap B$ is compact.

3. Is the intersection of a closed subset of a topological space and a compact subset of the space necessarily compact?

4. Suppose X and Y are compact Hausdorff spaces and $f: X \to Y$ is a mapping. Is the following statement true?
   f is continuous if and only if $f^{-1}[K]$ is compact for each compact subset $K \subset Y$.

5. Prove that if $(X, \mathcal{T})$ is compact and $\mathcal{T}_1$ is another topology for X such that $\mathcal{T} \supset \mathcal{T}_1$, then $(X, \mathcal{T}_1)$ is also compact.

6. Give an example of a compact Hausdorff space that is not first countable and, hence, not metrizable. (If you cannot work this out, see page 79 in [29].)

7. Prove that a compact Hausdorff space is metrizable if and only if it is second countable.

8. Prove Theorem 91.10.

9. Suppose that $(X, \mathcal{T})$ is a compact Hausdorff space. Prove that if $\mathcal{T}_1$ is another topology for X that is properly contained in $\mathcal{T}$, then $(X, \mathcal{T}_1)$ is not Hausdorff.

10. Prove that if a space is compact and regular then it is normal.

11. Prove that if X is regular and K is a compact subset of X, then cl (K) is compact. Is this proposition true without any assumptions on X?

92. RELATIONS BETWEEN VARIOUS TYPES OF COMPACTNESS

In this section we shall study relations between various compactness properties. However, before doing so it will be useful to make some remarks about limit points. For a point p to be a limit point of a set S, we require only that every neighborhood of p contain at least one point of S distinct from p. However, for T_1-spaces, limit points have a stronger property, as seen in the following theorem.

92.1. Theorem. *A topological space X is a T_1-space if and only if it has the following property: A point $p \in X$ is a limit point of a set S if and only if every neighborhood of p contains an infinite subset of S.*

PROOF. (The reader who did Exercise 4, page 185, has already proved this theorem.) Suppose X is a T_1 space. Assume p is a limit point of a set S and for some neighborhood U of p, the set $U \cap S$ is a finite set. Then since finite subsets of a T_1-space are closed, the set $F = U \cap S - \{p\}$ is closed. Consequently, $U - F$ is an open neighborhood of p. But since p is a limit point of S, $(U - F) \cap S - \{p\}$ is nonempty and we have a contradiction. We have shown, therefore, that if X is a T_1-space and p is a limit point of a set S, then every open neighborhood of p contains an infinite subset of S. Suppose next that X is not a T_1-space. Then there must exist two points x and y such that every open neighborhood of x contains $\{y\}$. Hence, x is a limit point of $\{y\}$. Thus, we have found a limit point x that does not have the property in the statement of the theorem.

Suppose S is a subset of a space X and $p \in X$. Kelley in [26] uses the terminology p is an *ω-accumulation point* of S to indicate that every neighborhood of p contains an infinite subset of S. Since we have been using the term *limit point* rather than its synonym *accumulation point*, we shall call this strong type of limit point an *ω-limit point*. (In some texts, the name accumulation point is reserved for what we are calling ω-limit point.) We see that if a space is a T_1-space, then a point is an ω-limit point of a set if and only if it is a limit point of that set. Hence, for T_1-spaces we need not have a different name for this stronger type of limit point. Another concept that we shall relate in an interesting way to ω-limit point is the concept of a *cluster point of a sequence*. A point p will be called a cluster point of a sequence (x_i) provided each neighborhood of the point p intersects the set $\{x_i : i \in \mathbf{P}\}$ for an infinite number of i's. Obviously, if a sequence (x_i) converges to a point x, then x is a cluster point of the sequence. Furthermore, we have the following, the proof of which is left as an exercise.

92.2. *Remark.* If (x_i) is a sequence in a first countable space, then (x_i) has a cluster point if and only if (x_i) has a convergent subsequence.

We saw in Example 90.6 that a space can have the Bolzano-Weierstrass property without being countably compact. However, if we strengthen the Bolzano-Weierstrass property to assert that every infinite set has an ω-limit point, this can no longer happen and we have the following theorem.

92.3. **Theorem.** *A topological space is countably compact if and only if either one of the following conditions holds:*

92.3(a). *Every infinite set has an ω-limit point.*

92.3(b). *Every infinite sequence has a cluster point.*

PROOF. We shall prove first that if X is countably compact, then it has property (a). Suppose that S is an infinite set of points that does not have an ω-limit point. There will be no loss in generality if we assume that S is countably infinite so that we may write $S = \{x_i : i \in \mathbf{P}\}$. Then for each $x \in X$, we may choose an open set $U(x)$ containing x such that $U(x) \cap S$ is a finite set. For each $n \in \mathbf{P}$, let

$$X_n = \bigcup \{U(x) : U(x) \cap S \subset \{x_1, x_2, \ldots, x_n\}\}.$$

Observe that each X_n is open and that the collection $\{X_n : n \in \mathbf{P}\}$ covers X. Since X is countably compact, some finite subcollection $\{X_{n_1}, X_{n_2}, \ldots, X_{n_j}\}$ also covers X. However,

$$S = S \cap X = S \cap (\bigcup \{X_{n_i} : i \in \mathbf{P}_j\}) = \bigcup \{X_{n_i} \cap S : i \in \mathbf{P}_j\}$$

is a finite set and we have a contradiction. Thus, we have shown that if X is countably compact, then it has property (a).

Next suppose that every infinite subset has an ω-limit point and that there exists a countably open covering $\{U_i : i \in \mathbf{P}\}$ of X that contains no finite subcovering of X. Then, for each $n \in \mathbf{P}$, there is a point $x_n \in X - \bigcup \{U_i : i \in \mathbf{P}_n\}$. It is easy to see that the set $\{x_n : n \in \mathbf{P}\}$ is an infinite set from the way in which the x_n's were chosen. Hence, this set must have at least one ω-limit point p and it must be an element of some U_k. Hence, an infinite number of x_n's must also be in U_k, contrary to the way in which the x_n's were chosen. Thus, we have shown that some finite subcollection of $\{U_n : n \in \mathbf{P}\}$ covers X. This completes the proof that property (a) is equivalent to countable compactness. It is left as an exercise to show that countable compactness is equivalent to property (b).

92.4. Theorem. *Every countably compact space is a B.W. compact space, and for T_1-spaces, countable compactness is equivalent to B.W. compactness.*

PROOF. If X is countably compact then it has property (a) of Theorem 92.3. Hence, it is B.W. compact. Suppose next that X is B.W. compact and T_1. Because of Theorem 92.1, it has property (a) of Theorem 92.3 and is therefore countably compact.

By making use of 92.2 and 92.3, we can prove the following theorem. The details are left as an exercise for the reader.

92.5. Theorem. *If X is first countable and countably compact, then it is sequentially compact.*

The following theorem is easy to prove. The proof is left as an exercise.

92.6. Theorem. *If a Lindelöf space is countably compact, then it is compact. In particular, if a second countable space is countably compact, then it is compact.*

Figure 21 summarizes the relations obtained between various types of compactness properties. Any special hypothesis that is needed is mentioned by the appropriate implication arrow. Also, reference numbers are given to relevant theorems except when implications follow directly from the definitions.

Note that if a space is second countable, then it is also a Lindelöf and first countable space. Hence, we see from the chart that for T_1 second countable spaces all four compactness properties are equivalent. In Sections 61 and 62 we proved that for metric spaces, compactness, B.W. compactness, and sequential compactness are equivalent properties. All metric spaces are T_1 and first countable. Hence, we see from the chart that for metric spaces, sequential compactness, B.W. compactness, and countable compactness are equivalent. However, metric spaces are not necessarily Lindelöf, and in view of Figure 21 one might wonder why we are able to prove the equivalence of the four properties for metric spaces. Recall that we accomplished this in the following way: We proved that sequentially compact metric spaces are totally bounded and, hence, separable (Exercise 4, page 123). Then, since for metric spaces separability implies the Lindelöf property, we were able to prove that sequentially compact metric spaces are compact (see proof of Theorem 62.2).

In the next theorem we summarize implications concerning compactness

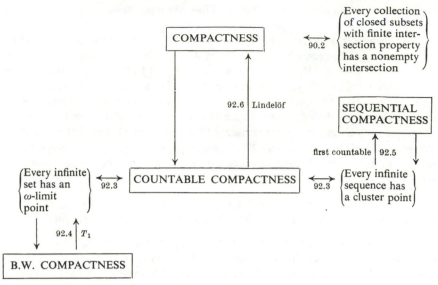

Figure 21. Relations between various types of compactness.

properties that hold for all spaces. We also include equivalences that hold under various conditions.

92.7. Theorem

92.7(a). *For all spaces, either compactness or sequential compactness implies countable compactness. Also countable compactness implies B.W. compactness.*

92.7(b). *For first countable spaces, countable compactness and sequential compactness are equivalent.*

92.7(c). *For T_1-spaces, B.W. compactness and countable compactness are equivalent. Hence, by (b), for T_1 first countable spaces, sequential compactness, countable compactness, and B.W. compactness are equivalent.*

92.7(d). *For spaces that are T_1 second countable spaces or metrizable spaces, compactness, countable compactness, sequential compactness, and B.W. compactness are equivalent.*

EXERCISES: RELATIONS BETWEEN VARIOUS TYPES OF COMPACTNESS

1. (a) Prove Remark 92.2.
    (b) Complete the proof of Theorem 92.3.
    (c) Prove Theorem 92.5.
    (d) Prove Theorem 92.6.

2. Show that the union of a finite collection of compact sets is compact. Is the same true for each of the other types of compactness?

3. Let f be a continuous mapping defined on a compact space X onto a Hausdorff space Y. Suppose $y \in Y$ and U is an open

subset of X that contains $f^{-1}[y]$. Show that there exists an open neighborhood V of y such that $f^{-1}[V] \subset U$.

4. Let z be an object that is not in **R** and $X = \mathbf{R} \cup \{z\}$. Define $\mathcal{T} = \{U : U$ is open in $\mathbf{R}\} \cup \{U : U = V \cup \{z\}$ for some subset V that is the complement of a compact subset of $\mathbf{R}\}$. Show that $\mathcal{T}$ is a topology for X and that $(X, \mathcal{T})$ is a compact space of which **R** is a dense subset. Show that $(X, \mathcal{T})$ is homeomorphic to a circle in $\mathbf{R}^2$.

93. LOCAL COMPACTNESS

Recall that although $\mathbf{R}^n$ is not compact, each point of $\mathbf{R}^n$ has a compact neighborhood. There are metric spaces that do not have this property; for example, ℓ^2 and $\mathscr{C}([a, b])$. As we shall see in this section, when a space does have the property, it also possesses certain other useful properties that hold for compact spaces but not necessarily for all topological spaces.

93.1. Definition. Locally compact spaces. *If X is a topological space and $x \in X$ has a compact neighborhood, then X is said to be locally compact at x. If X is locally compact at each of its points, then X is said to be a locally compact space. A subset S of a space is said to be locally compact when it is locally compact with respect to the relative topology.*

93.2. Remark. Suppose that $(X, \mathcal{T})$ is a topological space and $S \subset X$. Then S is locally compact at $x \in S$ if and only if there exists a $\mathcal{T}$-neighborhood N of x such that $N \cap S$ is compact.

It should be observed that subsets of locally compact spaces need not be locally compact. For example, the subset **Q** of rationals is not locally compact although **R** is locally compact. The following theorem gives some important special cases in which subsets inherit local compactness from the containing space. The proof is left as an exercise.

93.3. Theorem. *If X is a locally compact topological space and S is a closed subset of X, then S is locally compact. If X is locally compact and regular, then every open subspace of X is locally compact and regular.*

For Hausdorff spaces, it follows from 91.2 that compact neighborhoods are closed. Thus, if N is a compact neighborhood of a point in a Hausdorff space, then cl (int N) $\subset N$. This observation suggests the following characterization of local compactness for T_2-spaces. The proof is left as an exercise.

93.4. Theorem. *If X is a Hausdorff space, then X is locally compact if and only if each point has an open neighborhood whose closure is compact.*

It is easy to show that if (X, d) is a locally compact metric space, then for each $\varepsilon > 0$ and $x \in X$, there is a compact neighborhood K of x such that $K \subset N(x; \varepsilon)$. Thus, at each point there exist arbitrarily small compact neighborhoods. This observation suggests the following proposition.

93.5. Theorem. *Suppose X is a locally compact Hausdorff space. Then for each x ∈ X, the collection of all compact (and, hence, closed compact) neighborhoods of x is a base for the neighborhood system of x.*

Proof. Suppose $x \in X$ and N is a neighborhood of x. Since X is Hausdorff, there exists an open neighborhood W of x that has a compact closure. Hence, $x \in (\text{int } N) \cap W \subset \text{cl } (W)$. Note that cl (W) is a compact Hausdorff space and is, therefore, regular by 91.5. We let cl $(W) = S$ and, accordingly, we shall use the notation cl_S to denote the closure with respect to the subspace S. Since S is regular and (int N) $\cap$ W is open in S, there is a set V that is open in S and is such that $x \in V \subset \text{cl}_S (V) \subset (\text{int}(N)) \cap W \subset N$. Notice, however, that V is also open in X and that $\text{cl}_S (V) = \text{cl } (V)$. Thus, cl (V) is a compact neighborhood of x that is contained in N, and the proof is complete.

We get the following similar result for regular spaces, the proof of which is left as an exercise.

93.6. Theorem. *If X is a locally compact regular space, then for each x ∈ X, the collection of all closed compact neighborhoods of x is a base for the neighborhood system of x.*

Note that from Theorem 93.5 it follows immediately that locally compact Hausdorff spaces are regular. Actually there is a stronger result that can be obtained as seen in the statement of the following theorem, whose proof is left as an exercise.

93.7. Theorem. *If X is a locally compact Hausdorff space or if X is a locally compact regular space, then X is completely regular.*

EXERCISES: LOCAL COMPACTNESS

1. Let $\mathscr{B}$ be the collection of all subsets of **R** of the form $\{0, a\}$, $a \in \mathbf{R}$. Note that $\{0\} \in \mathscr{B}$ and that $\mathscr{B}$ is a base for a topology for **R**. Let $\mathscr{T}$ be the topology generated by $\mathscr{B}$. Show that $(\mathbf{R}, \mathscr{T})$ is locally compact. Show that there is no open neighborhood of 0 whose closure is compact.

2. Prove the remark stated in 93.2.

3. Prove each of the following:
    (a) Theorem 93.3.
    (b) Theorem 93.4.
    (c) Theorem 93.6.
    (d) Theorem 93.7.

4. Prove that if X is a locally compact Hausdorff space and K is a compact subset of X, then there is an open subset U containing K such that cl (U) is compact.

5. Prove that in a locally compact Hausdorff space, the intersection of a countable collection of open dense sets is dense. (This is another version of a theorem of Baire. See Theorem 64.2.)

6. Suppose f is a continuous mapping from a locally compact space X onto a space Y. Is Y necessarily locally compact? If the answer is no, answer the question assuming that f is a continuous open surjection.

7. Prove that if X is a locally compact second countable Hausdorff space, then there exists an increasing sequence of compact sets (X_i) such that $X = \bigcup \{\text{int } (X_i) : i \in \mathbf{P}\}$.

94. THE ONE-POINT COMPACTIFICATION

In Exercise 4, page 211, it was shown that $\mathbf{R}$ can be extended to a compact set by the addition of a single point. In this section we shall generalize the results of the example to which we just referred. In particular, we shall show that for any topological space $(X, \mathcal{T})$, we can adjoin a single element z to X and define a topology $\mathcal{T}_z$ for $X_z = X \cup \{z\}$ such that $(X_z, \mathcal{T}_z)$ is compact and $(X, \mathcal{T})$ is a subspace of $(X_z, \mathcal{T}_z)$.

Let $(X, \mathcal{T})$ be a topological space and let z be an object that is not in X. Let $X_z = X \cup \{z\}$. We define $\mathcal{T}_z$ as the collection $\{U : U$ is an open subset of $(X, \mathcal{T})$ or U is the union of $\{z\}$ and the complement of a closed compact subset of $(X, \mathcal{T})\}$. At this point the reader should verify the following facts.

94.1. Theorem. *The collection $\mathcal{T}_z$ as defined in the previous paragraph is a topology for X_z and the space $(X_z, \mathcal{T}_z)$ is compact.*

94.2. Definition. The one-point compactification. *Let $(X, \mathcal{T})$ and $(X_z, \mathcal{T}_z)$ be as in the previous theorem. The compact space $(X_z, \mathcal{T}_z)$ is called the one-point compactification of $(X, \mathcal{T})$.*

The one-point compactification of a space is sometimes called the Alexandroff compactification. If there is no chance of confusion, we shall write X_z instead of $(X_z, \mathcal{T}_z)$.

The next theorem lists a number of facts about the one-point compactification. The proof of each part is left as an exercise for the reader. Working through the proofs will provide a review of a number of concepts.

94.3. Theorem. *Let X_z be the one-point compactification of the topological space X.*

94.3(a). *X is compact if and only if the point z is an isolated point of X_z.*

94.3(b). *X is not compact if and only if X is a dense subset of X_z.*

94.3(c). *X is a T_1-space if and only if X_z is a T_1-space.*

94.3(d). *X_z is a Hausdorff space if and only if X is a locally compact Hausdorff space.*

94.3(e). *If X is a locally compact Hausdorff space, then X_z is second countable if and only if X is second countable.*

94.3(f). *If X is a locally compact separable metrizable space, then X_z is a separable metrizable space.*

As an example of how the one-point compactification can sometimes be a useful tool, we make use of it in giving a proof of the first part of Theorem 93.7. That is, we wish to show that if X is a locally compact Hausdorff space, then it is completely regular. By 94.3(d), a one-point compactification of X is Hausdorff and, of course, it is compact. However, a compact Hausdorff space is normal and, hence, is completely regular. But X is a subspace of X_z and, hence, it must also be completely regular.

A more general concept of compactification is one that makes use of the notion of topological embeddings. We shall say that $f:(X, \mathcal{T}(X)) \to (Y, \mathcal{T}(Y))$ is a *topological embedding* of $(X, \mathcal{T}(X))$ in $(Y, \mathcal{T}(Y))$ provided f is a homeomorphism of X onto $f[X]$. We have already seen a special case of this when we dealt with the notion of isometric embedding in 66.1.

94.4. Definition. Compactification. *Suppose that X is a topological space and Y is a compact space. If $h: X \to Y$ is a topological embedding such that $h[X]$ is a dense subset of Y, then the pair (h, Y) is called a compactification of X.*

Suppose that X is a noncompact space and i is the inclusion mapping from X into its one-point compactification X_z. Then the pair (i, X_z) is a compactification of X in the sense of Definition 94.4.

EXERCISES: THE ONE-POINT COMPACTIFICATION

1. Prove Theorem 94.1.

2. Prove Theorem 94.3.

3. By making use of Theorem 94.3(e) construct an example of a compact Hausdorff space that is not second countable (see Exercise 6, page 207).

95. SOME GENERALIZATIONS OF MAPPINGS DEFINED ON COMPACT SPACES

We have seen in previous sections that continuous mappings defined on compact spaces enjoy various useful properties, particularly when the domain and range spaces are Hausdorff. We review briefly some of the facts about such mappings that we have encountered in previous sections.

95.1. Theorem. *Assume that $f: X \to Y$ is a continuous surjection, that X is a compact space, and that Y is Hausdorff.*

95.1(a). *Under continuous mappings, compactness is preserved (Theorem 91.7). Then since compact Hausdorff spaces are automatically regular and normal, these properties are inherited by a Hausdorff range space if the domain is Hausdorff.*

95.1(b). *If X is second countable, so is Y (Theorem 91.9).*

95.1(c). *If X is metrizable, then Y is metrizable (Theorem 91.10).*

95.1(d). *If $y \in Y$ and U is an open subset of X that contains $f^{-1}[y]$, then there is an open neighborhood V of y such that $f^{-1}[V] \subset U$ (see Exercise 3, page 210).*

95.1(e). *$f^{-1}[K]$ is compact for each compact subset $K \subset Y$ (see Exercise 4, page 207).*

95.1(f). *f is a closed mapping and $f^{-1}[y]$ is compact for each $y \in Y$. (This follows from Theorem 91.8 and 95.1(e).)*

In the study of continuous mappings, much time has been devoted to investigation of situations in which the domain space is compact with various additional restrictions. (For example, see G. T. Whyburn's *Analytic Topology* [33].) It was found that some of the results could be extended to continuous mappings that were closed and still others to continuous mappings that satisfied the properties stated in 95.1(e) or (f) even though their domains were not compact. This observation led to investigations of continuous surjections of the types defined next.

95.2. Definitions. Perfect mappings; compact mappings. *A surjection $f: X \to Y$ is said to have the compact point inverse property provided that for each $y \in Y$, $f^{-1}[y]$ is a compact set. A closed continuous surjection with the compact point inverse property is called a perfect mapping. A surjection $f: X \to Y$ such that $f^{-1}[K]$ is compact for each compact set $K \subset Y$ is called a compact mapping.*

It is useful that the property stated in 95.1(d) is equivalent to closedness for continuous surjections. This is the content of the next theorem.

95.3. Theorem. *Suppose $f: X \to Y$ is a continuous surjection. Then the following property is equivalent to the closedness of f.*

For each $y \in Y$, if U is an open subset of X that contains $f^{-1}[y]$, then there is an open neighborhood W of y such that $f^{-1}[W] \subset U$.

PROOF. (We prove that the condition is implied by the closedness of f and leave the converse as an exercise.)

Suppose f is closed, $y \in Y$ and U is an open set that contains $f^{-1}[y]$. Then $X - U$ is closed and, hence, so is $f[X - U]$. Now $y \in Y - f[X - U]$, and we complete the proof by showing that this open set has the property required in the statement of the theorem. To see this, notice that

$$f^{-1}[Y - f[X - U]] = f^{-1}[Y] - f^{-1}[f[X - U]] \subset X - (X - U) = U.$$

This theorem can be used to prove the following characterization of closedness for continuous surjections. (The proof is left as an exercise.)

95.4. Theorem. *If $f: X \to Y$ is a continuous surjection, then the following property is equivalent to the closedness of f.*

For each subset A of Y, and open subset U containing $f^{-1}[A]$, there is an open subset W containing A such that $f^{-1}[W] \subset U$.

As an illustration of the usefulness of this theorem we prove the following important invariance theorem previously given as an exercise (Exercise 17, page 187).

95.5.   Theorem.   *Normality is invariant under closed continuous surjections.*

PROOF.   Suppose $f: X \to Y$ is a continuous closed surjection and X is normal. We wish to show that Y is also normal. Let A and B be two disjoint nonempty closed subsets of Y. Then $C = f^{-1}[A]$ and $D = f^{-1}[B]$ are closed disjoint subsets of X. Since X is normal, there are disjoint open sets U and V such that $C \subset U$ and $D \subset V$. From 95.4, there are open subsets Q and R in Y containing A and B, respectively, such that $C = f^{-1}[A] \subset f^{-1}[Q] \subset U$ and $D = f^{-1}[B] \subset f^{-1}[R] \subset V$. It now follows that Q and R are disjoint, for otherwise $f^{-1}[Q]$ and $f^{-1}[R]$ would not be disjoint. Thus, we have found disjoint open sets Q and R that contain A and B, respectively, and we have shown that Y is normal.

A somewhat similar proof can be used to prove each part of the next invariance theorem. The proof is left as an exercise.

95.6.   Theorem.   *Suppose $f: X \to Y$ is a continuous closed surjection with the compact point inverse property (i.e., f is a perfect mapping). Then if X is Hausdorff, so is Y. If X is regular, Y is regular.*

95.7.   Theorem.   *Suppose $f: X \to Y$ is a perfect mapping. Then compactness is preserved under f^{-1}; that is, f is a compact mapping.*

PROOF.   Suppose K is a compact subset of Y. We show that $f^{-1}[K]$ is compact. Let $\mathscr{C}$ be an open covering of $f^{-1}[K]$. For each $y \in K$ a finite subcollection $\mathscr{F}_y$ of $\mathscr{C}$ covers $f^{-1}[y]$, since $f^{-1}[y]$ is compact. Let $U_y = \bigcup \mathscr{F}_y$. By Theorem 95.4, there is an open neighborhood W_y of y such that $f^{-1}[W_y] \subset U_y$. Now the collection $\mathscr{W}$ of all W_y so chosen is an open covering of K. Some finite subcovering $\{W_{y_1}, W_{y_2}, \ldots, W_{y_n}\}$ also covers K. But then $\{f^{-1}[W_{y_i}] : i \in \mathbf{P}_n\}$ covers $f^{-1}[K]$. Since each $f^{-1}[W_{y_i}] \subset U_{y_i}$, the collection $\{U_{y_i} : i \in \mathbf{P}_n\}$ covers $f^{-1}[K]$. Since each $U_{y_i} = \bigcup \mathscr{F}_{y_i}$, $\mathscr{F} = \mathscr{F}_{y_1} \cup \mathscr{F}_{y_2} \cup \cdots \cup \mathscr{F}_{y_n}$ covers $f^{-1}[K]$. Also, each $\mathscr{F}_{y_i}$ consisted of only a finite number of elements of the original covering $\mathscr{C}$ and so $\mathscr{F}$ itself is a finite subcollection of $\mathscr{C}$ that covers $f^{-1}[K]$. Hence, $f^{-1}[K]$ is compact.

It is known that compact mappings are not always closed. However, the following theorem gives one of the situations in which they are.

95.8.   Theorem.   *Suppose that X and Y are first countable Hausdorff spaces and $f: X \to Y$ is a continuous compact surjection. Then f is a closed mapping.*

The proof is left as an exercise. (Hint: Review 87.3 and 87.4. Note also that if a sequence (y_i) converges to a point y, then the set $\{y_i : i \in \mathbf{P}\} \cup \{y\}$ is a compact set.)

By making use of this theorem and Theorem 95.7, we obtain the following.

95.9.   Theorem.   *Suppose that X and Y are Hausdorff first countable spaces and $f: X \to Y$ is a continuous surjection. Then f is a perfect mapping if and only if it is a compact mapping. (See Theorem 8.2 in [36].)*

To the reader who is familiar with the theory of functions of a complex variable, it should be interesting to note that complex polynomials are compact mappings (see Exercise 15, page 142). As a matter of fact, it is known that a complex entire function is a polynomial if and only if it is compact.

The following invariance theorem is a generalization of Theorem 91.9. The proof for the generalization is the same as that of Theorem 91.9, except for a few minor details involving justification of certain steps. It is left as an exercise for the reader to adjust the proof to the new situation.

95.10. Theorem. *Let $f: X \to Y$ be a perfect mapping. Then if X is second countable, so is Y.*

We now have the necessary machinery to prove the following easily.

95.11. Theorem. *Let X be a separable metric space and suppose $f: X \to Y$ is a perfect mapping. Then Y is a separable metric space.*

EXERCISES: SOME GENERALIZATIONS OF MAPPINGS DEFINED ON COMPACT SPACES

1. Prove the part of Theorem 95.3 that was left for the reader.

2. Suppose $f: X \to Y$ is a continuous bijection where X and Y are first countable Hausdorff spaces. Show that f is a homeomorphism if and only if f is a compact mapping.

3. (a) Prove Theorem 95.4.
   (b) Prove Theorem 95.6.

4. Point out which details in the proof of Theorem 91.9 need to be changed so that the proof becomes a proof of Theorem 95.10.

5. Prove Theorem 95.11.

6. Let $f: X \to Y$ be an open continuous surjection where X and Y are metric spaces. Suppose there is a positive integer k such that $f^{-1}[y]$ consists of exactly k points for each $y \in Y$.
   (a) Show that f is a local homeomorphism; that is, for each $x \in X$, there are open neighborhoods $U \subset X$ and $V \subset Y$ of x and $f(x)$, respectively, such that $f \mid U : U \to V$ is a topological mapping.
   (b) Show that f is a compact mapping.

7. Suppose $f: X \to Y$ is a continuous surjection where X and Y are metric spaces. Suppose there exists a subset $A \subset X$ such that $f[A] = Y$ and $f \mid A : A \to Y$ is a compact mapping. Prove that A is a closed subset of X.

8. Show that if $f: X \to Y$ is a closed (compact) continuous surjection and A is a closed subset or an inverse set (i.e., $A = f^{-1}[f[A]]$), then $f \mid A : A \to f[A]$ is a closed (compact) mapping.

8

Connectedness and Related Concepts

We have already discussed the notion of connectedness in the setting of metric spaces. In this chapter, this important concept will be studied for more general spaces. Many of the basic theorems concerning connectedness in metric spaces carry over to general topological spaces. For example, if $\mathcal{K}$ is a collection of connected subsets of a topological space X and $\bigcap \mathcal{K} \neq \varnothing$, then $\bigcup \mathcal{K}$ is connected. It is also true that the image of a connected topological space under a continuous mapping is connected. On the other hand, there are some facts about connectedness in metric spaces that depend on specialized properties possessed by the topologies of all metric spaces. For example, in a metric space a decreasing sequence of compact connected sets has a connected intersection. This proposition remains true for Hausdorff spaces but not for topological spaces in general.

Some metric spaces have the important property which states that each point in the space has arbitrarily small connected open neighborhoods. Recall that $\mathbf{R}^n$, ℓ^2, and $\mathscr{C}([a, b])$ are examples for which all ε-neighborhoods are connected. In this chapter we shall study a property called local connectedness which generalizes the property possessed by a point having arbitrarily small open connected neighborhoods. Local connectedness, unlike connectedness, is not invariant under all continuous mappings. However, if f is an open or closed continuous surjection defined on a locally connected space, then the range of f is also locally connected.

We shall also define the notions of limit superior and limit inferior of a sequence of subsets of a topological space. In particular, we shall prove a theorem giving conditions under which the limit superior of a sequence of connected sets is connected.

96. CONNECTEDNESS. DEFINITIONS.

This section contains definitions related to the concept of connectedness. In the next section we shall study some basic theorems concerning these concepts.

96.1. Definition. Connected space. *Let X be a topological space. Then X is connected provided X is not the union of two disjoint nonempty open subsets of X.*

As we have done with other topological properties, we shall say that if S is a subset of a space $(X, \mathcal{T})$, then S is a connected subset of $(X, \mathcal{T})$ if and only if $(S, \mathcal{T} \mid S)$ is a connected space.

We have already studied important examples of connected metric spaces, for example, $\mathbf{R}^n$, ℓ^2, and $\mathscr{C}([a, b])$. As we shall see, some of the examples of nonmetric spaces that we have considered are also connected.

96.2. *Example.* Let X be the set of real numbers and let $\mathcal{T}$ consist of $\varnothing$ and the collection of complements of countable subsets of X as in Example 75.7. Since each pair of nonempty open subsets of X has points in common, X cannot be the union of two nonempty open subsets of X with empty intersection. Hence, X is connected.

Recall that the notion of separation of a set was useful in dealing with *connectedness* of subsets of a space. This notion, defined in 54.3 for metric spaces, is repeated here for topological spaces.

96.3. Definition. Mutually separated sets; separation of a set. *Suppose X is a topological space. Two subsets A and B of X are said to be mutually separated in X provided*

$$\text{cl}\,(A) \cap B = A \cap \text{cl}\,(B) = \varnothing.$$

If $S \subset X$ is the union of a pair of nonempty mutually separated sets A and B, then $\{A, B\}$ is said to be a separation of S in X.

As with metric spaces, it is easy to see that a subset S of a topological space is connected if and only if there exists no separation of S in X. The proof given in 55.3 is nonmetric in nature and carries over to this case. If a set is not connected, we shall say that it is *disconnected*. In other words, a set is disconnected if and only if there exists a separation for the set in the space.

We shall see that much of what we have learned so far about the concept of connectedness for metric spaces carries over to general topological spaces. However, as with other concepts, there are differences. Recall, for example, that in a metric space every countable subset containing more than one point is disconnected (see Exercise 2(d), page 112). In a general topological space such sets can be connected. For example, suppose $X = \{0, 1\}$ and $\mathcal{T}$ is the trivial topology for X. Then, obviously, there can be no separation for $(X, \mathcal{T})$.

96.4. Definition. Components of a space. *Suppose that X is a topological space. A subset A of X is said to be a component of X provided A is a maximal connected subset of X; that is, A is a connected subset of X such that if B is any connected subset of X that contains A, then $B = A$.*

It should be observed that a topological space X is connected if and only if it has exactly one component, X itself. Furthermore, for each $x \in X$, x is a member of exactly one component and, thus, the collection of all components of X is a decomposition of X (see 20.1). Some sets are so badly disconnected that every point in the space is a component. For example, any space with the discrete topology has this property.

96.5.   Definition.   Totally disconnected space.   *A topological space S is totally disconnected provided every subset of S consisting of a single point is a component of S.*

It is easy to show that if S is a countable subset of a *metric* space, then S is totally disconnected. Notice also that the collection of all irrational numbers is a totally disconnected subset of the space **R**.

96.6.   Definition.   Sets that separate.   *If X is a connected space and S is a subset of X such that X − S is disconnected, then S is said to separate X. If $x \in X$ and $\{x\}$ separates X, then x is called a cut point of X.*

96.7.   *Example.*   Every point of **R** is a cut point of the space **R**. On the other hand, if S is a circle in $\mathbf{R}^2$, S has no cut points.

In subsequent chapters we shall have frequent occasion to hypothesize that a space is both compact and connected. For example, closed balls in $\mathbf{R}^n$ have this property.

96.8.   Definition.   Continuum.   *A space that is both compact and connected is called a continuum.*

Recall that every ε-neighborhood in a normed linear space is connected. (see Exercise 3, page 146). This notion generalizes to the setting of a topological space as follows.

96.9.   Definition.   Locally connected space.   *A space X is said to be locally connected at a point x in X provided that for each open neighborhood U of x there is an open connected neighborhood V of x that is contained in U. If X is locally connected at each of its points, then X is called a locally connected space.*

We have already seen examples of locally connected metric spaces, for example, $\mathbf{R}^n$, ℓ^2, and $\mathscr{C}([a, b])$. On the other hand, the example in Exercise 7, page 114, is connected but is not locally connected at any point of A. It is locally connected at each point in B.

Because of the definition of the relative topology, it should be clear that if S is a subset of a space $(X, \mathscr{T})$, then S is locally connected at x in S if and only if for each $\mathscr{T}$-open neighborhood U of x, there is a $\mathscr{T}$-open neighborhood V of x contained in U such that $V \cap S$ is connected. It is also useful to notice that a space X is locally connected at $x \in X$ if and only if the neighborhood system of x has a base, each of whose elements is an open connected neighborhood of x.

EXERCISES: CONNECTEDNESS.   DEFINITIONS

1.   Suppose that X is a topological space with the trivial topology. Is X necessarily connected? Is X locally connected?

2. Suppose that $\mathscr{T}_1$ and $\mathscr{T}_2$ are topologies for a set X and $\mathscr{T}_1 \subset \mathscr{T}_2$. If $(X, \mathscr{T}_1)$ is connected, is $(X, \mathscr{T}_2)$ necessarily connected? If $(X, \mathscr{T}_2)$ is connected, is $(X, \mathscr{T}_1)$ necessarily connected?

3. Suppose that Y is a subset of a space X. Let A and B be subsets of Y. Show that A and B are mutually separated sets in Y if and only if they are mutually separated sets in X.

4. In each of the following determine if the space is connected; also determine if the space is locally connected.
   (a) The space in Example 75.4.
   (b) The space in Example 75.5.
   (c) The space in Example 75.6.

5. In Exercise 4, page 165 , the collection $\mathscr{K}$ is a topology for **P**. Is $(\mathbf{P}, \mathscr{K})$ a connected space? Is this space locally connected?

6. In Exercise 5, page 165, the collection $\mathscr{K}$ is a topology for the set X of all real numbers. Is $(X, \mathscr{K})$ a connected space? Is this space locally connected?

7. Suppose S is a connected subset of a space X. Show that if A is a set such that $S \subset A \subset \mathrm{cl}\,(S)$, then A is a connected subset of X.

8. Are components of a space necessarily closed?

9. Give an example of a space that has a component that is not open.

10. Suppose that a nonempty proper subset of a space is both open and closed. What conclusions can you draw about that space?

97. SOME BASIC THEOREMS CONCERNING CONNECTEDNESS

Some of the theorems given in this section are direct generalizations of theorems verified for metric spaces, and the proofs previously given can be carried over word for word. In such cases reference will be made to the previous proof.

97.1. Theorem. *Suppose $\{A, B\}$ is a separation for a subset S of a topological space. If C is a connected subset of S, then $C \subset A$ or $C \subset B$. (See the proof of Theorem 55.5.)*

The following theorem is a more general form of Theorem 55.6. The proof is left as an exercise.

97.2. Theorem. *Suppose $\mathscr{K}$ is a collection of connected subsets of X such that no two are mutually separated. Then $\bigcup \mathscr{K}$ is connected.*

The following two useful facts can be proved easily by making use of the previous theorem. The proofs are left as exercises.

97.3. Theorem. *Suppose X is a topological space and Z is a connected subset of X. Suppose further that $\mathcal{K}$ is a collection of connected subsets of X each of which intersects Z. Then $[\bigcup \mathcal{K}] \cup Z$ is connected.*

97.4. Theorem. *Suppose $\{A_i : i \in \mathbf{P}\}$ is a countable collection of connected subsets such that $A_i \cap A_{i+1} \neq \varnothing$ for each $i \in \mathbf{P}$. Then $\bigcup \{A_i : i \in \mathbf{P}\}$ is connected.*

97.5. Theorem. *Suppose $f: X \to Y$ is a continuous surjection and X is connected. Then Y is also connected.*

PROOF. The proof given in 56.1 for a metric space carries over without change.

The following theorem gives a useful characterization of local connectedness. The proof is left to the next set of exercises.

97.6. Theorem. *A space X is locally connected if and only if for each open subset U of X, the components of U are open.*

In Section 57 we defined the notion of polygonal connectedness for subsets of $\mathbf{R}^n$; that notion is also meaningful for subsets of normed linear spaces. The reader was asked to show in Exercise 4, page 116, that an open subset of $\mathbf{R}^n$ is connected if and only if it is polygonally connected. This fact is also true in any normed linear space. We next define the notion of *path connectedness*, which is a generalization of polygonal connectedness.

97.7. Definition. Path-connectedness. *A path in a subset S of a topological space is a continuous mapping $f: [0, 1] \to S$. If a and b are points in S and $f: [0, 1] \to S$ is a path such that $f(0) = a$ and $f(1) = b$, we then say that the path joins a to b in S. If for each a and b in S there is a path that joins a to b in S, then S is said to be path-connected.*

Since the image of a real interval under a continuous mapping is connected, each two points in a path-connected space X can be joined by a connected set contained in the space. Hence, by using 97.2, it is easy to show that every path-connected space is also connected.

The following theorem is easy to prove and is left as an exercise.

97.8. Theorem. *Suppose that f is a continuous mapping from a path-connected space X onto a topological space Y. Then Y is also path connected.*

97.9. Theorem. *Suppose that f is a continuous open or a continuous closed mapping from a locally connected space X onto a topological space Y. Then Y is locally connected.*

PROOF. We shall prove the part of the theorem for open mappings and leave the part for closed mappings as an exercise.

Suppose $f: X \to Y$ is a continuous open surjection and X is locally connected. Let U be an open subset of Y, and C be a component of U. In the context of theorem 97.6, the proof will be complete if we show that C is open. Let $W = f^{-1}[U]$ and note that W is open in X; since X is locally connected, each component of W is open. Because of the invariance of connectedness under continuous mappings, $f[R]$ is connected for each component R of W. Hence, for each component R of W, if $f[R] \cap C \neq \varnothing$, then $f[R] \subset C$. (Why?) Let $\mathcal{R} = \{R : R$ is a component of W and $f[R] \cap C \neq \varnothing$}. Since $\mathscr{R}$ is a collection of open subsets of X, $\bigcup \mathscr{R}$ is open. Also note that $f^{-1}[C] = \bigcup \mathscr{R}$ so that $f[\bigcup \mathscr{R}] = C$. Then, since f is an open mapping, C is open and the proof of the part of the theorem for open mappings is complete.

EXERCISES: SOME BASIC THEOREMS CONCERNING CONNECTEDNESS

1. Prove each of the following theorems:
   - (a) 97.2.   (d) 97.6.
   - (b) 97.3.   (e) 97.8.
   - (c) 97.4.   (f) Complete the proof of 97.9.

2. Suppose C is a connected space and S is a connected subset of C. Prove that if $\{A, B\}$ is a separation for $C - S$, then $A \cup S$ and $B \cup S$ are connected sets.

3. Prove that if X is a Hausdorff space and is the range of a continuous mapping whose domain is the closed unit interval, then X is a locally connected continuum that is metrizable. Also prove that X is path connected.

   Note. A Hausdorff space that is the image under a continuous mapping of the real interval $[0, 1]$ is called a *Peano space* or a *continuous curve*. There is a remarkable theorem known as the Hahn-Mazurkiewicz theorem that asserts that any nonempty, locally connected, metrizable continuum is the image under a continuous mapping of the real interval $[0, 1]$. See, for example, [6], [19], or [38].

4. Exercise on arcs. If a space X is topologically equivalent to the real interval $[0, 1]$, then it is called a *topological arc*. From the previous exercise, X is a locally connected metrizable continuum. Show that if X is a topological arc, then it has the following additional property.

   > * There exist two points a and b of X that are not cut points of X. For all other points c of X, $X - \{c\}$ is the union of two mutually separated sets A and B containing a and b, respectively.

   There is an interesting theorem which asserts that any metric continuum that satisfies the conditions in the previous paragraph is a topological arc. This accounts for the fact that sometimes a *simple continuous arc* is defined as a metric continuum that satisfies the property given in *. Background material leading up to a proof of this theorem and a proof of this characterization of a topological arc is given, for example, in [38].

5. The image under a homeomorphism of a circle is called a *simple closed curve*. Prove that if S is a simple closed curve, then S is a locally connected, metrizable continuum such that each subset consisting of exactly two points separates S.

6. Let X be a topological space. Suppose H and K are closed subsets of X such that $H \cup K$ and $H \cap K$ are connected. Prove that H and K are connected.

7. Show that if F is a nonempty proper closed subset of a connected locally connected space X, then every component of $X - F$ has a limit point in F. Is the conclusion still true if X is not locally connected?

8. Exercise on ε-chainable sets. Suppose (X, d) is a metric space. If a and b are elements of X, and $\varepsilon > 0$, an ε-chain joining a to b is a finite sequence $(x_1, x_2, \ldots, x_n)$ of points in X such that $a = x_1$, $b = x_n$ and $d(x_i, x_{i+1}) < \varepsilon$ for $i \in \mathbf{P}_{n-1}$. If for each a and b in X there is an ε-chain joining a to b, we say that X is ε-chainable.
   (a) Show that if (X, d) is connected, then it is ε-chainable for each positive ε.
   (b) Show that if (X, d) is ε-chainable for each positive ε, it need not be connected.
   (c) Show that if (X, d) is compact and is ε-chainable for each positive ε, then it is connected.

9. Suppose X is a Hausdorff space and $\{C_i : i \in \mathbf{P}\}$ is a collection of nonempty continua in X such that $C_{i+1} \subset C_i$. Is

$$\bigcap \{C_i : i \in \mathbf{P}\}$$

   necessarily a continuum?

10. Suppose X and Y are Hausdorff spaces and $f : X \to Y$ is a compact continuous surjection such that for each $y \in Y$, $f^{-1}[y]$ is a continuum. Show that for each continuum $C \subset Y$, $f^{-1}[C]$ is a continuum in X.

11. Suppose the metric space (X, d) is compact. Show that X is locally connected if and only if for each $\varepsilon > 0$, X is the union of a finite collection of continua such that each continuum in the collection has a diameter less than ε.

12. Prove the following propositions:

   Suppose the metric space (X, d) is a locally connected continuum. If a and b are two points in X and $\varepsilon > 0$, then there exist continua $M_1, M_2, \ldots, M_n$ in X such that $a \in M_1$, $b \in M_n$, diam $(M_i) < \varepsilon$ for $i \in \mathbf{P}_n$, and $M_i \cap M_j \neq \emptyset$ if and only if $|i - j| \leq 1$. Thus, a and b can be joined by a "connected chain with small links."

   Note: Actually every two points of a locally connected metric continuum can be joined by an arc. The proof of this interesting fact is rather difficult. A development leading up to a proof and a proof can be found in [19].

13. Prove that if X is a locally connected compact space, then X has only a finite number of components.

14. Let X and Y be topological spaces. Prove that the product space $X \times Y$ is connected if and only if X and Y are each connected. Prove that the product space $X \times Y$ is locally connected if and only if X and Y are each locally connected.

98. LIMIT SUPERIOR AND LIMIT INFERIOR OF SEQUENCES OF SUBSETS OF A SPACE

At this point we shall define certain limit processes for sequences of subsets of a topological space. The notions presented here of limit superior, limit inferior, and limit of a sequence of subsets of a topological space are as defined, for example, in [33]. There is a purely set theoretic concept that goes by the same name. See, for example, [16]. In Exercise 1 of the next set of exercises a relation between these two types of limit superior and limit inferior will be pointed out.

In this section, we shall be particularly interested in the situation in which the elements of the sequence of sets are connected sets.

98.1. Definitions. Limit superior; limit inferior. *Suppose* $(X, \mathcal{T})$ *is a topological space and* (A_i) *is a sequence of subsets of* X. *The limit superior of* (A_i), *abbreviated* lim sup (A_i), *is defined to be the set of all* $x \in X$ *such that each neighborhood of* x *intersects* A_i *for an infinite subset of positive integers* i. *The limit inferior of* (A_i), *abbreviated* lim inf (A_i), *is the set of all* $x \in X$ *such that each neighborhood of* x *intersects* A_i *for all but a finite collection of integers* i. *If* lim sup $(A_i) =$ lim inf $(A_i) = L$ *(possibly the empty set), we say that the sequence* (A_i) *converges to* L.

The proof of the next theorem follows easily from the definition and is left as an exercise.

98.2. Theorem. *Let* (A_i) *be a sequence of subsets of a topological space* $(X, \mathcal{T})$. *Then* lim inf $(A_i) \subset$ lim sup (A_i) *and both are closed subsets of* X.

The following theorem is easy to prove by making use of the previously proved fact that a T_1 space that is B.W. compact has the property that every infinite sequence has a cluster point. The proof is left as an exercise.

98.3. Theorem. *Suppose* $(X, \mathcal{T})$ *is a* T_1-*space. Let* (A_i) *be a sequence of nonempty sets in* X *and let* $C =$ lim sup (A_i). *Suppose, further,* cl $(\bigcup \{A_i : i \in P\})$ *is B.W. compact. Then* C *is a nonempty set, and if* U *is an open set that contains* C, *there is an integer* N *such that for* $i \geq N$, $A_i \subset U$.

98.4. Theorem. *Suppose* $(X, \mathcal{T})$ *is a Hausdorff space. Suppose further* (A_i) *is a sequence of nonempty connected subsets of* X *such that* cl $(\bigcup \{A_i : i \in P\})$ *is compact. If* lim inf (A_i) *is nonempty, then* lim sup (A_i) *is a nonempty continuum.*

PROOF. Let $C =$ lim sup (A_i). Note that $C \neq \emptyset$. Also, since C is closed and is contained in the compact Hausdorff subspace cl $(\bigcup \{A_i : i \in P\})$, C is compact. To prove it is connected, suppose $\{A, B\}$ is a separation of C. Then A and B

are disjoint nonempty compact subsets of C. Then, since X is Hausdorff, there exist disjoint open subsets U and V that contain A and B, respectively. On the basis of Theorem 98.3, there is an $N \in \mathbf{P}$ such that if $i \geq N$, then $A_i \subset U \cup V$. Let a be an element (guaranteed by hypothesis) in lim inf (A_i). We suppose $a \in A$. Since U is a neighborhood of a, there is a positive integer M such that $U \cap A_i$ is nonempty for $i \geq M$. Now, there is a $b \in B$ and V is an open neighborhood of b. Since $b \in$ lim sup (A_i), there is an integer $k \geq$ max (N, M) such that $A_k \cap V$ is nonempty. Since $k \geq N$, $A_k \subset U \cup V$. Hence, $A_k = (A_k \cap U) \cup (A_k \cap V)$. Since $k \geq M$, $A_k \cap U \neq \varnothing$; recall that k was chosen so that $A_k \cap V \neq \varnothing$. Thus, the two disjoint sets $A_k \cap V$ and $A_k \cap U$, both open relative to A_k, form a separation of A_k. Since A_k is connected, we have a contradiction.

Note. For other interesting facts about the notions of limit superior and limit inferior of sequences of subsets of a space, see [22], [33], or [39].

EXERCISES: LIMIT SUPERIOR AND LIMIT INFERIOR OF SEQUENCES OF SUBSETS OF A SPACE

1. Let $(X, \mathscr{T})$ be a space where $\mathscr{T}$ is the discrete topology for X and let (A_i) be a sequence of subsets of X. Show that lim sup $(A_i) = \{x : x \in A_i$ for an infinite collection of positive integers $i\}$ and that lim inf $(A_i) = \{x : x \in A_i$ for all but a finite collection of positive integers $i\}$. The expressions for lim sup (A_i) and lim inf (A_i) given here for the discrete case coincide with the set theoretic definitions for these terms when there are no topologies involved. See, for example, [16].

2. Suppose (X, d) is a metric space and (A_i) is a sequence of nonempty subsets of X such that cl $(\bigcup \{A_i : i \in \mathbf{P}\})$ is compact. Let $C = $ lim sup (A_i). Prove that lim $(d(A_i, C)) = 0$.

3. Suppose that (A_i) is a sequence of continua in a compact metric space X.
    (a) Based on Theorem 98.4, if lim inf $(A_i) \neq \varnothing$, then lim sup (A_i) is a continuum. Give an example to show that if lim inf $(A_i) = \varnothing$, then lim sup (A_i) need not be a continuum.
    (b) Suppose F is a closed subset of X such that $A_i \cap F \neq \varnothing$ for each $i \in \mathbf{P}$. Show that for each component K of lim sup (A_i), $K \cap F \neq \varnothing$. (For a more general version of this proposition, see Theorem 1.14 in Chapter IV of [39].)

99. REVIEW EXERCISES

In subsequent chapters we shall be dealing with topological spaces formed from other topological spaces. For example, if $(X, \mathscr{T}(X))$ is a topological space and R is an equivalence relation in X, then $\mathscr{T}(X)$ can be used to induce a useful topology called the quotient topology on the set of all R-equivalence classes. We shall apply the notion of quotient topology to a further study of continuous mappings. We shall also extend the definition of product topology to include the product of an infinite collection of topological spaces. In order to deal better with some of the more difficult questions that will arise, we shall study some types of convergence that are more general than sequential convergence. In all these discussions frequent use will be made of most of the topics treated so far. As a summary of our previous work we include next a set of exercises, some of which will be a review.

EXERCISES: REVIEW

1. If X is a second countable space, is X necessarily separable?

2. If X is a separable first countable space, is X necessarily second countable?

3. Suppose that X is a topological space. Which of the following properties are possessed by each subspace of X provided X has the property?

   (a) T_1.
   (b) T_2.
   (c) Regular.
   (d) Completely regular.
   (e) Normal.
   (f) Separable.
   (g) First countable.
   (h) Second countable.
   (i) Lindelöf.
   (j) Local compactness.
   (k) Local connectedness.

4. Among the properties listed in the previous question are some that are inherited by all subsets and some that are not. Of those which are not, which are inherited by all open subsets? Which are inherited by all closed subsets?

5. Sometimes the concept of local compactness is defined as follows:

   > A topological space X is locally compact provided that for each x in X there is an open neighborhood U of x such that cl (U) is compact.

   Show that this property implies the definition given in this text but that it is not equivalent to it. Are the two definitions equivalent for Hausdorff spaces?

6. Suppose that $f: X \to Y$ is a continuous surjection. Which of the following topological properties are possessed by Y if X has the property?

   (a) T_1.
   (b) T_2.
   (c) Regular.
   (d) First countable.
   (e) Second countable.
   (f) Separable.
   (g) Sequentially compact.
   (h) B.W. compact.
   (i) Compact.
   (j) Locally connected.
   (k) Locally connected metric continuum.

7. Some of the properties listed in question 6 are not necessarily possessed by Y. Which are necessarily possessed by Y if the continuous surjection $f: X \to Y$ is also open?

8. Is the intersection of two compact subsets necessarily compact? Is the intersection of a closed set and a compact set necessarily compact?

9. State a condition weaker than metrizability for which B.W. compactness implies sequential compactness, but too weak for B.W. compactness to imply compactness.

10. Suppose that X is a Hausdorff space, and let X_z be a one-point compactification of X. Give an example to show that X_z is not necessarily Hausdorff. Is X_z necessarily a T_1 space?

11. Suppose that X and Y are homeomorphic spaces and X_z and Y_w are one-point compactifications of X and Y, respectively. Are X_z and Y_w necessarily homeomorphic?

12. Suppose that X is a connected space and X_z is a one-point compactification of X. Is X_z necessarily a continuum? If the answer is no, state a condition under which it is not a continuum.

13. Let $f: X \to \mathbf{R}$ be a continuous surjection and assume that $\mathbf{R}$ is endowed with the Euclidean topology. In each of the following, determine if the statement is necessarily true.
    (a) X is not compact.
    (b) X is connected.
    (c) For each y in $\mathbf{R}$, $X - f^{-1}[y]$ is not connected.
    (d) If f is a closed mapping and X is not metrizable, then f is not one-to-one.

14. Let X be a circle in $\mathbf{R}^2$ and let Y be a "figure 8" in $\mathbf{R}^2$, both with the relative topology. Prove that there can exist no topological map from X onto Y.

15. Suppose that X is a connected space and c is a cut point of X. Prove that for any separation $\{A, B\}$ of $X - \{c\}$, $A \cup \{c\}$ and $B \cup \{c\}$ are connected sets.

16. Let (X, d) be a bounded metric space and let $\mathscr{H}$ be the collection of all nonempty closed subsets of X. For each nonempty subset A of X and positive ε, we shall use the notation $N(A; \varepsilon)$ to denote the set of all points in X that are within ε distance from A; that is,

$$N(A; \varepsilon) = \{x : d(A, x) < \varepsilon\}.$$

We next define the mapping $d_H : \mathscr{H} \times \mathscr{H} \to \mathbf{R}$ as follows:

$$d_H(A, B) = \text{g.l.b. } \{\varepsilon : A \subset N(B; \varepsilon) \quad \text{and} \quad B \subset N(A; \varepsilon)\}$$

(a) Show that d_H is a metric for $\mathcal{K}$. This metric is known as the *Hausdorff metric*.

(b) Show that for all x and y in (X, d), $d(x, y) = d_H(\{x\}, \{y\})$.

(c) Suppose that (K_i) is a sequence in $\mathcal{K}$ that converges to a point K in the space $(\mathcal{K}, d_H)$. Does (K_i) converge to K in the sense that $\lim \sup (K_i) = \lim \inf (K_i) = K$?

(d) Suppose that (K_i) is a sequence of closed subsets of X and that K is a nonempty subset of X such that $\lim \sup (K_i) = \lim \inf (K_i) = K$. Does the sequence (K_i) converge to K in the space $(\mathcal{K}, d_H)$?

9

Quotient Spaces

Recall that if R is an equivalence relation in a set X, then the collection $\{R[x]: x \in X\}$ of all R-equivalence classes is a decomposition of X (see Section 21). This collection is referred to as the quotient set X over R and is denoted by X/R. Conversely, suppose that $\mathcal{R}$ is a decomposition of X and we define the relation R in X by setting

$a\,R\,b$ if and only if a and b belong to the same element of $\mathcal{R}$.

Then R is an equivalence relation in X such that $\mathcal{R} = \{R[x]: x \in X\}$. Thus, if we deal with a decomposition of a set, we can think of it as being generated by an equivalence relation associated with it. Equivalence relations and decompositions come up in many mathematical investigations. In such cases, it is often useful to introduce the quotient or natural mapping $\Psi: X \to X/R$. This mapping carries each x in X into the R-equivalence class $R[x] \in X/R$.

In this chapter we shall be interested in the situation in which $(X, \mathcal{T})$ is a topological space and $\mathcal{R}$ is a decomposition of X, or R is an equivalence relation in X. We shall look at a decomposition of a space $(X, \mathcal{T})$ from two points of view. On one hand, the elements of $\mathcal{R}$ are subsets of X and this collection $\mathcal{R}$ of subsets may have a useful property relative to the topology $\mathcal{T}$. For example, a decomposition $\mathcal{R}$ of a space X is said to be upper semicontinuous provided that for each closed subset F of X the union of all elements of $\mathcal{R}$ that intersect F is closed in X. On the other hand, we can look at the decomposition or quotient set as a set on which we wish to define a useful topology. We shall define such a topology $\mathcal{T}/R$ for X/R and refer to it as the quotient topology. As we shall see, there are strong relations between the two points of view mentioned. For example, it turns out that a decomposition of a space is upper semicontinuous if and only if the associated quotient mapping is a closed mapping.

We shall apply the information that we obtain about the quotient space to a further study of open, closed, and compact mappings. We shall end the chapter with a proof of one form of the well known Eilenberg-Whyburn factorization theorem for compact mappings. This theorem provides a technique which allows one to express a certain type of mapping as a composition of two mappings, each of which has particularly nice properties. The first factor guaranteed by the theorem has the property that the inverse of each point is a continuum. The second factor has totally disconnected point inverses. The technique of factoring a mapping in this manner has been a valuable tool in mapping theory.

100. DECOMPOSITION OF A TOPOLOGICAL SPACE

If $(X, \mathcal{T})$ is a topological space and $\mathcal{R}$ is a decomposition of X (see Section 20), then it is natural to consider the properties of $\mathcal{R}$ that are related to $\mathcal{T}$. In this section we shall consider two types of decompositions that have been especially important in the study of open and closed mappings.

100.1. Definitions. Upper and lower semicontinuous decompositions of a space. *Suppose that $(X, \mathcal{T})$ is a topological space and $\mathcal{R}$ is a decomposition of X.*

100.1(a). *$\mathcal{R}$ is called an upper semicontinuous (u.s.c.) decomposition of $(X, \mathcal{T})$ provided that for each closed subset F in $(X, \mathcal{T})$ the union of the collection of all elements of $\mathcal{R}$ that intersect F is closed in $(X, \mathcal{T})$; that is, $\bigcup \{A : A \in \mathcal{R}, \text{ and } A \cap F \neq \varnothing\}$ is closed if F is closed.*

100.1(b). *$\mathcal{R}$ is called a lower semicontinuous (l.s.c.) decomposition of $(X, \mathcal{T})$ provided that for each open set U in $(X, \mathcal{T})$ the union of the collection of all elements in $\mathcal{R}$ that intersect U is open in $(X, \mathcal{T})$; that is, $\bigcup \{A : A \in \mathcal{R} \text{ and } A \cap U \neq \varnothing\}$ is open if U is open.*

Recall from Definition 11.3 that if R is a relation in X and $U \subset X$, then $R[U] = \{y : u \, R \, y \text{ for at least one } u \in U\}$. Also, $R[U] = \bigcup \{R[u] : u \in U\}$. Furthermore, if R is an equivalence relation in X, then $\{R[x] : x \in X\}$ is precisely the collection of all R-equivalence classes and is, hence, a decomposition of X. (See sections 20 and 21.) Conversely, if $\mathcal{R}$ is any decomposition of X, then $\mathcal{R}$ generates an equivalence relation R in X given by

$$x \, R \, y \text{ if and only if } x \text{ and } y \text{ are in the same element of } \mathcal{R}.$$

These remarks suggest the following useful connection between a decomposition of a space and the equivalence relation associated with it.

100.2. Theorem. *Let $(X, \mathcal{T})$ be a topological space. Suppose that R is an equivalence relation in X and $\mathcal{R}$ is the decomposition of X consisting of all R-equivalence classes. Then,*

100.2(a). *$\mathcal{R}$ is an u.s.c. decomposition of X if and only if $R[F]$ is closed for each closed subset F of X.*

100.2(b). *$\mathcal{R}$ is a l.s.c. decomposition of X if and only if $R[U]$ is open for each open subset U of X.*

The proof of 100.2 is left as an exercise for the reader. The reader should also verify that the following characterization of upper semicontinuity is correct.

100.3. **Theorem.** *Suppose that* $(X, \mathcal{T})$ *is a topological space and $\mathcal{R}$ is a decomposition of X. Then $\mathcal{R}$ is an u.s.c. decomposition if and only if for each open set $U \subset X$, the union of the collection of all elements of $\mathcal{R}$ contained in U (i.e., $\bigcup \{A : A \in \mathcal{R} \text{ and } A \subset U\}$) is open.*

Suppose that $f : X \to Y$ is a mapping. It follows easily from the single-valuedness of a function that $\mathcal{R} = \{f^{-1}[y] : y \in f[X]\}$ is a decomposition of X. We speak of each $f^{-1}[y]$ as a *point inverse* of f and, accordingly, we call $\mathcal{R}$ the *decomposition of X into point inverses of f*. Note that the equivalence relation R associated with $\mathcal{R}$ is given by $x_1 \, R \, x_2$ if and only if $f(x_1) = f(x_2)$. Suppose next that $f : (X, \mathcal{T}(X)) \to (Y, \mathcal{T}(Y))$ is a continuous mapping. As indicated by the next theorem, the decomposition of X into point inverses of f is intimately related to the action of f on the open and closed subsets of X. This fact motivated much of the interest in the study of upper and lower semicontinuous decompositions.

100.4. **Theorem.** *Let $f : X \to Y$ be a closed (an open) continuous surjection and let $\mathcal{R} = \{f^{-1}[y] : y \in Y\}$. Then $\mathcal{R}$ is an upper (a lower) semicontinuous decomposition of X.*

PROOF. We assume that f is a closed (an open) continuous mapping. Let A be a closed (an open) subset of X and let A^* be the union of the collection of all elements of $\mathcal{R}$ that intersect A. Note that $A^* = f^{-1}[f[A]]$. Since A is closed (open) and f is a closed (an open) mapping, then $f[A]$ is closed (open) in Y. Because f is continuous, $f^{-1}[f[A]]$ is closed (open) in X and, thus, $\mathcal{R}$ is an u.s.c. (a l.s.c.) decomposition of X.

Suppose that $f : X \to Y$ is a continuous surjection and $\mathcal{R}$ is the decomposition of X into point inverses of f. On the basis of the previous theorem one might wonder whether f is closed (open) if $\mathcal{R}$ is an upper (a lower) semicontinuous decomposition. The following example gives a negative answer. However, in the next section we shall study a useful class of mappings for which the answer is affirmative.

100.5. *Example.* Let X be the set of all real numbers and let $f(x) = x$ for each x in X. Consider the mapping $f : (X, \mathcal{T}_1) \to (X, \mathcal{T}_2)$ where $\mathcal{T}_1$ is the discrete topology for X and $\mathcal{T}_2$ is the trivial topology for X. It is easy to verify that f is continuous and that the decomposition $\mathcal{R}$ of X into point inverses of f is both upper and lower semicontinuous. However, $f : (X, \mathcal{T}_1) \to (X, \mathcal{T}_2)$ is neither an open nor a closed mapping.

EXERCISES: DECOMPOSITION OF A TOPOLOGICAL SPACE

1. Verify Theorem 100.2.

2. Verify Theorem 100.3.

3. Let $[0, 1]$ be given the relative topology from **R**. Define the relation R in $[0, 1]$ as follows: $x \, R \, y$ if and only if $(x = y)$ or $(x = 1$ and $y = 0)$ or $(x = 0$ and $y = 1)$. Let $\mathcal{R}$ be the set of

all R-equivalence classes. Is $\mathscr{R}$ an u.s.c. decomposition of $[0, 1]$? a l.s.c. decomposition of $[0, 1]$?

4. Let R be the equivalence relation in $\mathbf{R}^2$ given by $a \, R \, b$ if $|a| \leq 1$ and $|b| \leq 1$; otherwise, $a \, R \, b$ if and only if $a = b$. Let $\mathscr{R}$ be the collection of all R-equivalence classes. Is $\mathscr{R}$ an u.s.c. decomposition of $\mathbf{R}^2$? Is $\mathscr{R}$ a l.s.c. decomposition of $\mathbf{R}^2$?

5. Suppose $f: X \to Y$ is an open continuous mapping from a compact space X onto a Hausdorff space Y. Let

$$\mathscr{R} = \{f^{-1}[y] : y \in Y\}.$$

By Theorem 100.4, $\mathscr{R}$ is a l.s.c. decomposition of X. Is $\mathscr{R}$ also an u.s.c. decomposition of X?

101. QUASI-COMPACT MAPPINGS

Suppose $f: X \to Y$ is a mapping. If A is a subset of X such that $A = f^{-1}[C]$ for some subset $C \subset f[X]$, then we shall refer to A as an *f-inverse set* (or simply an inverse set, when omitting the reference to f is not likely to cause any confusion). The following remarks about inverse sets will be useful. In each case, the verification of the remark is left as an exercise.

101.1. *Remark.* If $f: X \to Y$ is a mapping then a subset A of X is an f-inverse set if and only if $f^{-1}[f[A]] = A$.

101.2. *Remark.* Suppose that X and Y are topological spaces and that $f: X \to Y$ is a surjection. Then the image of every open inverse set is open if and only if the image of every closed inverse set is closed.

Mappings for which the image of closed inverse sets are closed have been studied under the name of *quasi-compact* mappings (for example, see [34] and [35]).

101.3. **Definition. Quasi-compact mapping.** *Let $f: X \to Y$ be a surjection such that for each closed inverse set F, $f[F]$ is closed (or, equivalently, for each open inverse set U, $f[U]$ is open). Then f is said to be a quasi-compact mapping.*

Because of Remark 101.2, we have the following.

101.4. *Remark.* Surjections that are open or closed are quasi-compact.

In Theorem 97.9 it is pointed out that local connectedness is invariant under continuous surjections that are either open or closed. The proof shows it is necessary to know only that the image of each open inverse set is open. It is instructive to note that in the proof of Theorem 97.9 the only place that the openness of f is used is to show that $f[\bigcup \mathscr{R}]$ is open. Notice, however, that $\bigcup \mathscr{R}$ is an inverse set. Thus, the proof carries over for continuous quasi-compact mappings. This fact as well as some other information about quasi-compact mappings is contained in the statement of the following theorem, whose proof is left as an exercise.

101.5.   Theorem.   *Let $f:(X, \mathcal{T}) \to (Y, \mathcal{T}')$ be a continuous quasi-compact surjection.*

101.5(a).   *If $A \subset X$ is an open or closed inverse set, then $f \mid A:A \to f[A]$ is quasi-compact (continuity of f not needed for this part).*

101.5(b).   *If $(X, \mathcal{T})$ is locally connected, then so is $(Y, \mathcal{T}')$.*

101.5(c).   *$(Y, \mathcal{T}')$ is a T_1-space if and only if $f^{-1}[y]$ is closed for each $y \in Y$.*

101.5(d).   *If for each $y \in Y$, $f^{-1}[y]$ is connected, and if U is a connected open or closed subset of Y, then $f^{-1}[U]$ is a connected subset of X.*

101.6.   Theorem.   *Suppose that X and Y are topological spaces and $f:X \to Y$ is a quasi-compact surjection. Let $\mathcal{R} = \{f^{-1}[y]:y \in Y\}$. If $\mathcal{R}$ is an u.s.c. (a l.s.c.) decomposition of X, then f is a closed (an open) mapping.*

PROOF.   Keeping in mind that $f:X \to Y$ is quasi-compact and that $\mathcal{R}$ is upper (lower) semicontinuous, we wish to show that f is a closed (an open) mapping. To do this, let S be a closed (an open) subset of X. Notice that $f[S] = f[f^{-1}[f[S]]]$. From the definition of $\mathcal{R}$, $f^{-1}[f[S]]$ is the union of the collection of all elements of $\mathcal{R}$ that intersect S. Then, since $\mathcal{R}$ is an u.s.c. (a l.s.c.) decomposition, $f^{-1}[f[S]]$ is a closed (an open) subset of X. Notice, however, that $f^{-1}[f[S]]$ is an f-inverse set; since it is closed (open) and f is quasi-compact, then

$$f[f^{-1}[f[S]]] = f[S]$$

is closed (open) in Y. Thus, we have shown that f is a closed (an open) mapping.

From 100.4, 101.4, and 101.6, we obtain the following.

101.7.   Theorem.   *Suppose that $f: X \to Y$ is a continuous surjection. Let $\mathcal{R} = \{f^{-1}[y]:y \in Y\}$. Then $f:X \to Y$ is a closed (open) mapping if and only if f is quasi-compact and $\mathcal{R}$ is an u.s.c. (a l.s.c.) decomposition of X.*

EXERCISES: QUASI-COMPACT MAPPINGS

1.   Prove Remarks 101.1 and 101.2.

2.   Prove parts (a), (c), and (d) of Theorem 101.5.

3.   Show that if a continuous surjection is one-to-one, then it is a homeomorphism if and only if it is quasi-compact.

4.   If f is a continuous compact mapping from one metric space onto another, is it necessarily a quasi-compact mapping?

5.   Prove the following proposition:

If $f:X \to Y$ is a continuous quasi-compact surjection, then $g:Y \to Z$ is continuous if and only if $g \circ f:X \to Z$ is continuous.

6.   Prove that the composition of two continuous quasi-compact surjections is a continuous quasi-compact surjection.

7.   We have previously proved that normality is preserved under a closed continuous surjection (see 95.5). Give an alternate proof

of this fact by making use of 100.3 and 100.4. (See [35] for a proof of this proposition and for other theorems about open, closed, and quasi-compact mappings.)

102. THE QUOTIENT TOPOLOGY

Suppose that X is a set and R is an equivalence relation in X. Then the set $\{R[x] : x \in X\}$ of all R-equivalence classes is called the *quotient set* X *by* R and is denoted by X/R.

102.1. Definition. The quotient or natural mapping. *Let* X *be a set and* R *an equivalence relation in* X. *The mapping* $\Psi : X \to X/R$ *given by* $\Psi(x) = R[x]$ *is called the quotient mapping or the natural mapping of* X *onto* X/R.

We make the following observation about Ψ.

102.2. Remark. For each subset $U \subset X$,
$$\Psi^{-1}[\Psi[U]] = R[U] = \{x : x \in X \text{ and } u\,R\,x \text{ for some } u \in U\}.$$
Suppose that $(X, \mathscr{T})$ is a topological space and R is an equivalence relation in X. We shall be able to define a useful topology for X/R that inherits various properties from $(X, \mathscr{T})$. Our study of this problem will make use of the quotient mapping $\Psi : X \to X/R$. To do this we can put the problem in a more general setting as follows:

Suppose that $(X, \mathscr{T})$ is a topological space, Y is a set and $f : X \to Y$ is a surjection. Can we define a useful topology $\mathscr{T}_f$ for Y such that $f : (X, \mathscr{T}) \to (Y, \mathscr{T}_f)$ is continuous? Now suppose that $\mathscr{T}_1$ is any topology for Y for which $f : (X, \mathscr{T}) \to (Y, \mathscr{T}_1)$ is continuous. Since $(Y, \mathscr{T}_1)$ is the image of $(X, \mathscr{T})$ under a continuous mapping, $(Y, \mathscr{T}_1)$ will possess certain properties if $(X, \mathscr{T})$ possesses those properties. Connectedness and compactness are examples of such properties. Furthermore, if $f : (X, \mathscr{T}) \to (Y, \mathscr{T}_1)$ is continuous and if $\mathscr{T}_2$ is smaller than $\mathscr{T}_1$, then $f : (X, \mathscr{T}) \to (Y, \mathscr{T}_2)$ is also continuous. Our strategy will be to choose a topology $\mathscr{T}_f$ for Y that will turn out to be the largest topology for Y with respect to which f is continuous on $(X, \mathscr{T})$. The name "quotient topology" in what follows is motivated by the application of that topology to the quotient set X/R.

102.3. Theorem. *Let* $(X, \mathscr{T})$ *be a topological space,* Y *a set, and* $f : X \to Y$ *a surjection. Let* $\mathscr{T}_f = \{U : U \subset Y \text{ and } f^{-1}[U] \text{ is open}\}$. *Then,*

102.3(a). $\mathscr{T}_f$ *is a topology for* Y.

102.3(b). $f : (X, \mathscr{T}) \to (Y, \mathscr{T}_f)$ *is continuous.*

102.3(c). *If* $\mathscr{T}_1$ *is any topology for* Y *for which* $f : (X, \mathscr{T}) \to (Y, \mathscr{T}_1)$ *is continuous, then* $\mathscr{T}_1 \subset \mathscr{T}_f$.

The proof of Theorem 102.3 is left as an exercise. On the basis of part (a), we make the following definition.

102.4. Definition. Quotient topology. *Let* $(X, \mathscr{T})$ *be a topological space,* Y *a set and* $f : X \to Y$ *a surjection. Then the topology* $\mathscr{T}_f$ *for* Y *as defined in* 102.3 *is called the quotient topology for* Y *determined by* f *and* $(X, \mathscr{T})$.

The next theorem gives a connection between the quotient topology and quasi-compact mappings.

102.5.   Theorem.   *Let $f:(X,\mathscr{T}(X)) \to (Y,\mathscr{T}(Y))$ be a surjection. Then $\mathscr{T}(Y)$ is the quotient topology for Y determined by $f:X \to Y$ and $(X,\mathscr{T}(X))$ if and only if $f:(X,\mathscr{T}(X)) \to (Y,\mathscr{T}(Y))$ is continuous and quasi-compact.*

Proof.   Suppose first that $\mathscr{T}(Y)$ is the quotient topology for Y determined by $f:X \to Y$ and $(X,\mathscr{T}(X))$. We know from 102.3 that $f:(X,\mathscr{T}(X)) \to (Y,\mathscr{T}(Y))$ is continuous and we shall show further that it is quasi-compact. Let U be an open f-inverse set. We shall show that $f[U]$ is open in $(Y,\mathscr{T}(Y))$. Since U is an inverse set, $f^{-1}[f[U]] = U$. Thus, since U is open so is $f^{-1}[f[U]]$. From the fact that $\mathscr{T}(Y)$ is the quotient topology we can conclude that $f[U]$ is open. Hence, we have shown that f is a quasi-compact mapping. We assume next that $f:(X,\mathscr{T}(X)) \to (Y,\mathscr{T}(Y))$ is a continuous quasi-compact surjection and show that $\mathscr{T}(Y)$ is the quotient topology. Let $\mathscr{T}_f$ be the quotient topology for Y determined by $f:X \to Y$ and $(X,\mathscr{T}(X))$. Then $\mathscr{T}_f = \{V:V \subset Y \text{ and } f^{-1}[V] \text{ is open}\}$. We need to show that $\mathscr{T}(Y) = \mathscr{T}_f$. To see this, let $W \in \mathscr{T}(Y)$; then, since

$$f:(X,\mathscr{T}(X)) \to (Y,\mathscr{T}(Y))$$

is continuous, $f^{-1}[W]$ is open and so $W \in \mathscr{T}_f$. Next let $W \in \mathscr{T}_f$. Then $f^{-1}[W]$ is open in X and, hence, is an open inverse set. Consequently, since we are assuming that $f:(X,\mathscr{T}(X)) \to (Y,\mathscr{T}(Y))$ is quasi-compact, $f[f^{-1}[W]] = W$ is open in $(Y,\mathscr{T}(Y))$; that is, $W \in \mathscr{T}(Y)$. Thus, we have shown that $\mathscr{T}(Y) = \mathscr{T}_f$.

We now apply our results concerning the quotient topology to the quotient set X/R. Suppose $(X,\mathscr{T})$ is a topological space and R is an equivalence relation in X. Let $\Psi:X \to X/R$ be the quotient mapping as defined in 102.1. We shall use the notation $\mathscr{T}/R$ to denote the quotient topology for X/R determined by Ψ and $(X,\mathscr{T})$.

102.6.   Definition.   Quotient space.   *Let $(X,\mathscr{T})$, R, and $\mathscr{T}/R$ be as in the previous paragraph. The space $(X/R,\mathscr{T}/R)$ is called the quotient space determined by $(X,\mathscr{T})$ and R. When there is no chance for confusion, we shall use X/R to denote the quotient space.*

Observe that with R an equivalence relation in X and $\mathscr{R}$ the corresponding decomposition, the set X/R is the same as $\mathscr{R}$. Often the space that we are referring to as a quotient space is referred to as a decomposition space, especially when the point of view involves the decomposition $\mathscr{R}$ rather than the corresponding equivalence relation R.

If the decomposition of a space is upper or lower semicontinuous, the decomposition space inherits from X various other properties in addition to those it inherits because Ψ is continuous and quasi-compact. The following theorem is a useful tool for the investigation of such properties.

102.7.   Theorem.   *Suppose $\mathscr{R}$ is a decomposition of $(X,\mathscr{T})$. Then the quotient mapping Ψ is closed (open) if and only if $\mathscr{R}$ is an u.s.c. (a l.s.c.) decomposition of X.*

PROOF. Let $\mathscr{R}$ be a decomposition of $(X, \mathscr{T})$ and let R be the corresponding equivalence relation in X. Consider the quotient mapping $\Psi: (X, \mathscr{T}) \to (X/R, \mathscr{T}/R)$. Note that $\mathscr{R}$ is precisely the decomposition of X into point inverses of Ψ. Since Ψ is quasi-compact and continuous, the theorem follows from 101.7.

The following useful fact is also helpful in the study of decomposition of spaces. It follows at once from the previous theorem and Theorem 95.7.

102.8. Theorem. *Suppose $\mathscr{R}$ is an u.s.c. decomposition of a space $(X, \mathscr{T})$. Let R be the equivalence relation corresponding to $\mathscr{R}$. If each element of $\mathscr{R}$ is a compact subset of X, then $\Psi: (X, \mathscr{T}) \to (X/R, \mathscr{T}/R)$ is a compact mapping.*

Now that we know certain relationships between types of decompositions and the natural mapping, we can deduce information about the decomposition space. We do so by making use of the various theorems that we have proved. Some of the information about the decomposition or quotient space that we can deduce is tabulated in 102.9. The reader should verify the information contained there.

102.9. Properties inherited by quotient spaces

Assumptions on $(X, \mathscr{T})$	Assumptions on decomposition	Properties inherited by $(X/R, \mathscr{T}/R)$
compact		compact
connected		connected
Lindelöf		Lindelöf
separable		separable
locally connected		locally connected
normal	u.s.c.	normal
Hausdorff	u.s.c.; each element compact	Hausdorff
regular	u.s.c.; each element compact	regular
locally compact	u.s.c.; each element compact	locally compact
second countable	u.s.c.; each element compact	second countable
separable metric	u.s.c.; each element compact	separable metric
first countable	l.s.c.	first countable
second countable	l.s.c.	second countable
locally compact	l.s.c.	locally compact

EXERCISES: THE QUOTIENT TOPOLOGY

1. Prove Theorem 102.3.

2. Verify the information contained in 102.9.

3. Exercise on pseudometric spaces:

> Let X be a set. A mapping $d: X \times X \to \mathbf{R}$ is called a *pseudometric* for X provided it satisfies the following conditions:
> (i) $d(x, x) = 0$ for all x in X.
> (ii) $d(x, y) = d(y, x)$ for all x and y in X.
> (iii) $d(x, y) + d(y, z) \geqq d(x, z)$ for all x, y, and z in X.

Observe that pseudometrics like metrics are nonnegative functions. For each x in X and positive number ε, the ε-neighborhood $N(x;\ \varepsilon)$ of x is defined to be the set $\{y : y \in X$ and $d(y, x) < \varepsilon\}$.

(a) Let $\mathscr{B} = \{N(x;\ \varepsilon) : x \in X,\ \varepsilon > 0\}$. Show that $\mathscr{B}$ is a base for a topology $\mathscr{T}(d)$ for X.

   (X, d) is called a *pseudometric* space, and $\mathscr{T}(d)$ as defined in (a) is called the topology for (X, d) generated by d. When we speak of (X, d) as having a topological property, we shall mean that the corresponding space $(X, \mathscr{T}(d))$ has that property. Furthermore, if $(X, \mathscr{T})$ is a topological space for which there exists a pseudo-metric d for which $\mathscr{T}$ is the same as the topology $\mathscr{T}(d)$ that is generated by d, then $(X, \mathscr{T})$ is said to be a *pseudometrizable* space.

(b) Prove that pseudometrizability is a topological property.

(c) Give an example of a psuedometric space that is not a T_1 space.

(d) Prove that a pseudometric space is a metric space if and only if it is T_1.

(e) Prove that every pseudometric space is a first countable normal space. Prove further that a pseudometric space is second countable if and only if it is separable.

(f) Give an example of a B.W. compact pseudometric space that is not compact.

(g) Prove that for pseudometric spaces the properties of compactness, countable compactness, and sequential compactness are equivalent.

(h) Let (X, d) be a pseudometric space. Show that for each $x \in X$, $y \in \mathrm{cl}\,(\{x\})$ if and only if $d(y, x) = 0$. Is $\mathrm{cl}\,(\{x\})$ a compact set?

(i) Let (X, d) be a pseudometric space and let R be the relation defined in X as follows:

$$x \mathrel{R} y \text{ if and only if } d(x, y) = 0.$$

Show that R is an equivalence relation in X. Show also that for each $x \in X$, $R[x] = \mathrm{cl}\,(\{x\})$. Prove, furthermore, that the quotient mapping $\Psi : X \to X/R$ is both open and closed.

(j) Let (X, d) be a pseudometric space. Let R and Ψ be as in part (i). Prove that $(X/R, \mathscr{T}(d)/R)$ is metrizable. Furthermore, show that there is a metric d^* for $(X/R, \mathscr{T}(d)/R)$ such that for all x and y in X, $d(x, y) = d^*(\Psi(x), \Psi(y))$.

NOTE: It should be clear to the reader that there are many metric concepts that can be extended directly to pseudometric spaces. For example, total boundedness, completeness, uniform continuity, and uniform convergence can all be extended to the setting of a pseudometric space and to an even more general type of space known as a *uniform* space. For more information about pseudometric and uniform spaces see [7], [26], or [32].

103. DECOMPOSITION OF A DOMAIN SPACE INTO POINT INVERSES

In this section we shall give further consideration to the decomposition of a domain space into point inverses of f. In particular we shall show that if $f: X \to Y$ is a continuous quasi-compact surjection and $\mathscr{R}$ is the decomposition of X into the point inverses of f, then the associated decomposition or quotient space is homeomorphic to Y.

Suppose that X and Y are topological spaces and $\mathscr{R} = \{f^{-1}[y] : y \in Y\}$. Then the corresponding equivalence relation R in X is given by

$$x_1 \, R \, x_2 \quad \text{if and only if} \quad f(x_1) = f(x_2).$$

In what follows it is important to keep in mind that if $R[x_1] = R[x_2]$, then $f(x_1) = f(x_2)$.

Next let $\Psi: X \to X/R$ be the quotient mapping. By making use of Ψ, we define the mapping $h: X/R \to Y$ as follows: For each $\rho \in X/R$, $\Psi^{-1}[\rho] = R[x] \subset X$ for some $x \in X$. Notice that $f[R[x]] = \{f(x)\}$. We then define $h(\rho) = f(x)$.

It should be clear from the previous remarks that h is well defined; that is, the value of $h(\rho)$ is independent of the particular representation of $R[x]$. In other words, if $\Psi^{-1}[\rho] = R[x] = R[y]$, then $f[R[x]] = f[R[y]] = \{f(x)\} = \{f(y)\}$. The following remarks will also be useful.

103.1. Remarks. Let $f: X \to Y$, $\Psi: X \to X/R$, and $h: X/R \to Y$ be as in the previous discussion. Then

103.1(a). $h: X/R \to Y$ is a bijection.

103.1(b). $h(\Psi(x)) = f(x)$ for each $x \in X$. Hence, $f = h \circ \Psi$.

103.1(c). Because of (b), we have the following commutative diagram:

103.1(d). $h[U] = f[\Psi^{-1}[U]]$ for each subset $U \subset X/R$.

103.1(e). If A is a subset of X, then A is an f-inverse set if and only if A is a Ψ-inverse set.

We leave the proofs of parts (a) through (d) as exercises for the reader and prove part (e).

PROOF. Let A be a subset of X. Then

$$f^{-1}[f[A]] = (h \circ \Psi)^{-1}[h \circ \Psi[A]] = \Psi^{-1}[h^{-1}[h[\Psi[A]]]].$$

From (a), h is a bijection. Hence, the previous set identity becomes

$$f^{-1}[f[A]] = \Psi^{-1}[\Psi[A]].$$

From this and 101.1 we see that A is an f-inverse set if and only if A is a Ψ-inverse set.

103.2. Theorem. *Let $f: X \to Y$ be a continuous surjection. Let $\mathscr{R}$ be the decomposition of X into point inverses. Then the bijection $h: X/R \to Y$ in 103.1 is a homeomorphism if and only if f is quasi-compact.*

PROOF. Let Ψ and h be as in the previous discussion and in the diagram in 103.1(c). We first assume that f is quasi-compact and show that h is a homeomorphism. Toward that end, let U be an open subset of X/R. Then $\Psi^{-1}[U]$ is open in X. Notice that by 103.1(e) $\Psi^{-1}[U]$ is also an f-inverse set. Hence, since f is quasi-compact, $f[\Psi^{-1}[U]]$ is open in Y. But $h[U] = f[\Psi^{-1}[U]]$. Hence, $h[U]$ is open in Y and h is an open mapping. We complete the proof that h is a homeomorphism by proving that h is continuous. To do this, let U be open in Y. Then, since f is continuous, $f^{-1}[U]$ is open. But $f^{-1}[U]$ is also a Ψ-inverse set. Hence, since Ψ is quasi-compact, $\Psi[f^{-1}[U]]$ is open in X/R. But $\Psi[f^{-1}[U]] = h^{-1}[U]$ and, hence, $h^{-1}[U]$ is open. We may thus conclude that h is continuous and, because it is also open and a bijection, h is a homeomorphism.

Next we assume that h is a homeomorphism and show that f is quasi-compact. To do this, let U be an open f-inverse set. We show that $f[U]$ is open in Y. We note first that U also is an open Ψ-inverse set and, since Ψ is quasi-compact, $\Psi[U]$ is open in X/R. Since h is a homeomorphism, $h[\Psi[U]]$ is open in Y. Then because $f = h \circ \Psi$, $f[U] = h[\Psi[U]]$ and, hence, is open in Y. Thus, f is quasi-compact.

103.3. *Example.* Let S be the unit circle in $\mathbf{R}^2$ and let $f: [0, 1] \to S$ be given by $f(x) = (\cos (2\pi x), \sin (2\pi x))$. Since f is a continuous mapping defined on a compact space and S is Hausdorff, f is a closed mapping. The mapping f is therefore quasi-compact. Let R be the equivalence relation in $[0, 1]$ given by $x_1 R x_2$ if and only if $(x_1 = x_2)$ or $(x_1 = 1$ and $x_2 = 0)$ or $(x_1 = 0$ and $x_2 = 1)$. By 103.2, X/R is homeomorphic to S; that is, X/R is a simple closed curve.

103.4. *Example.* Let I be the closed real interval $[0, 1]$ and let R be the equivalence relation in $I \times I$ defined by: $(a_1, a_2) R (b_1, b_2)$ if and only if

(i) $(a_1, a_2) = (b_1, b_2)$ or
(ii) $a_2 = b_2$ and $a_1 = 1, b_1 = 0$ or
(iii) $a_2 = b_2$ and $a_1 = 0, b_1 = 1$.

We can think of forming the quotient space by "pasting together" the two vertical edges of the rectangle shown in the accompanying diagram. By making use of Theorem 103.2, we shall show that the quotient space is topologically equivalent to a cylinder. In order to do this let

$$B = \{(x_1, x_2, x_3): x_1^2 + x_2^2 = 1 \quad \text{and} \quad 0 \leq x_3 \leq 1\} \subset \mathbf{R}^3$$

and note that B is a cylinder. Next let $f: I \times I \to B$ be the continuous surjection given by:

$$f(t_1, t_2) = (\cos (2\pi t_1), \sin (2\pi t_1), t_2) \text{ for every } (t_1, t_2) \text{ in } I \times I.$$

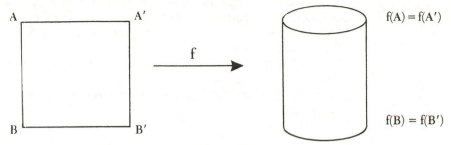

Figure 22.

Observe that the decomposition $\mathscr{R}$ of X into R-equivalence classes is the same as the decomposition of X into point inverses of f. The mapping f is a continuous closed surjection and, hence, is quasi-compact. By Theorem 103.2, X/R is topologically equivalent to B and is, therefore, topologically equivalent to a cylinder.

A quotient space X/R can be thought of as the space obtained by *identifying* the points that are to go into the same R-equivalence class. For example, in 103.3 we identified 0 and 1 as belonging to one equivalence class and any other point x as belonging to an equivalence class with x as its only member. Furthermore, what we have been calling the quotient topology is often called the *identification* topology. Moreover, if $f : (X, \mathscr{T}(X)) \rightarrow (Y, \mathscr{T}(Y))$ is a continuous surjection such that $\mathscr{T}(Y)$ is the same as the quotient or identification topology for Y determined by f and $\mathscr{T}(X)$, then f is called an *identification* mapping. From 102.5 we see that a mapping is an identification mapping if and only if it is a continuous quasi-compact surjection.

EXERCISES: DECOMPOSITION OF A DOMAIN SPACE INTO POINT INVERSES

1. Complete the proof of 103.1.

2. Let R be the equivalence relation for $\mathbf{R}^2$ as given in Exercise 4, page 233. We can visualize the quotient space as being obtained by collapsing the unit closed ball into one point. We might, therefore, guess that $\mathbf{R}^2/R$ is topological plane. Prove that this is correct.

3. (Torus). Let I be the closed real interval $[0, 1]$. We define an equivalence relation R for $I \times I$ by listing the R-equivalence classes as follows:

    For $0 < a_1 < 1, 0 < a_2 < 1$,
    $$R[(a_1, a_2)] = \{(a_1, a_2)\}.$$
    For $a_1 = 0$ or $a_1 = 1, 0 < a_2 < 1$,
    $$R[(a_1, a_2)] = \{(a_1, a_2), (1 - a_1, a_2)\}.$$
    For $0 < a_1 < 1, a_2 = 0$ or $a_2 = 1$,
    $$R[(a_1, a_2)] = \{(a_1, a_2), (a_1, 1 - a_2)\}.$$
    For $a_1 = 0$ or $a_1 = 1, a_2 = 0$ or $a_2 = 1$,
    $$R[(a_1, a_2)] = \{(0, 0), (0, 1), (1, 0), (1, 1)\}.$$

One can visualize the quotient space $(I \times I)/R$ as being obtained by first pasting together two opposite edges of $I \times I$ to form a cylinder and then pasting together the top and bottom of the cylinder to form a torus.

Show that $(I \times I)/R$ is topologically equivalent to $S \times S$ where S is the unit circle in $\mathbf{R}^2$.

NOTE: For other examples of this type the reader is referred to [6] and [28].

104.   TOPOLOGICALLY EQUIVALENT MAPPINGS

Refer to the diagram in 103.1(c) and recall from Theorem 103.2 that if f is continuous and quasi-compact, then Y and X/R are topologically equivalent spaces. Because h is a homeomorphism and $f = h \circ \Psi$, it appears that f and Ψ should act very much alike from a topological point of view. The sense in which f and Ψ act alike is termed "topologically equivalent" and is defined next.

104.1. Definition. Topologically equivalent mappings. *Suppose X_1 is topologically equivalent to X_2, and Y_1 is topologically equivalent to Y_2. The mappings $f_1: X_1 \to Y_1$ and $f_2: X_2 \to Y_2$ are said to be topologically equivalent mappings if there exist topological mappings $h_1: X_1 \to X_2$ and $h_2: Y_1 \to Y_2$ such that*

$$f_1 = h_2^{-1} \circ f_2 \circ h_1.$$

We see from the definition that if $f_1: X_1 \to Y_1$ and $f_2: X_2 \to Y_2$ are topologically equivalent mappings, then X_1 and X_2 are topologically equivalent spaces and the spaces Y_1 and Y_2 are also topologically equivalent. Furthermore, there are topological mappings $h_1: X_1 \to X_2$ and $h_2: Y_1 \to Y_2$ for which we have the following commutative diagram:

$$
\begin{array}{ccc}
X_1 & \xrightarrow{\ f_1\ } & Y_1 \\
\downarrow{\scriptstyle h_1} & & \uparrow\downarrow{\scriptstyle h_2 \ \ h_2^{-1}} \\
X_2 & \xrightarrow{\ f_2\ } & Y_2
\end{array}
$$

The commutativity of the diagram should make it apparent that a topological property is invariant under f_1 or f_1^{-1} if and only if it is invariant under f_2 or f_2^{-1}, respectively. It is not hard to verify the fact that the relation *topologically equivalent* for mappings, is an equivalence relation. Thus, the name "topologically equivalent mappings" is an appropriate one for the concept under discussion.

EXERCISES: TOPOLOGICALLY EQUIVALENT MAPPINGS

1. Verify that "topologically equivalent" is an equivalence relation.

2. Show that if f_1 and f_2 are topologically equivalent mappings, then f_1 is continuous, open, closed, or compact, if and only if f_2 has that property.

3. Verify that with f and h as in 103.1, if f is continuous and quasi-compact, then f and Ψ are topologically equivalent.

105. DECOMPOSITION OF A DOMAIN SPACE INTO COMPONENTS OF POINT INVERSES

So far we have considered the decomposition of the domain space of a mapping into *point inverses*. Sometimes this is referred to as the "natural decomposition" induced by the mapping. Consideration of the decomposition of the domain space of certain mappings into *components of point inverses* has been of considerable importance in the study of continuous mappings.

105.1. Theorem. *Suppose X and Y are Hausdorff first countable spaces and $f: X \to Y$ is a compact continuous surjection. Let $\mathscr{R}$ be the decomposition of X into components of point inverses of f (i.e., $\mathscr{R} = \{F : F$ is a component of $f^{-1}[y]$ for some $y \in Y\}$). Then $\mathscr{R}$ is an u.s.c. decomposition of X.*

PROOF. (It is suggested that the reader review Theorem 98.4 before reading the proof.) Let A be a closed subset of X and let A^* be the union of those elements of $\mathscr{R}$ that intersect A. We complete the proof by showing that A^* is a closed subset of X. To prove this, let $p \in \mathrm{cl}\,(A^*)$. We shall show that $p \in A^*$. There is a sequence (x_i) in A^* that converges to p. For each $i \in \mathbf{P}$, there is a $K_i \in \mathscr{R}$ such that $x_i \in K_i$ and $K_i \cap A \neq \varnothing$. Note that each K_i is a continuum. Since f is continuous, $(f(x_i))$ converges to $f(p)$. The set $Z = \{f(x_i) : i \in \mathbf{P}\} \cup \{f(p)\}$ is compact. Then, because f is a compact mapping, $f^{-1}[Z]$ is compact and closed. Since $\bigcup \{K_i : i \in \mathbf{P}\} \subset f^{-1}[Z]$, $\mathrm{cl}\,(\bigcup\{K_i : i \in \mathbf{P}\})$ is contained in $f^{-1}[Z]$ and is compact. Further, since $p \in \lim \inf (K_i)$, we may use Theorem 98.4 to conclude that $\lim \sup (K_i)$ is a continuum.

We show next that $\lim \sup (K_i) \subset f^{-1}[f(p)]$. To see this let $z \in \lim \sup (K_i)$. Then there is a subsequence (K_{N_i}) of (K_i) and a sequence (z_i), with $z_i \in K_{N_i}$ such that z_i converges to z. But $f(z_i) = f(x_{N_i})$ so that $(f(z_i))$ converges to $f(p)$. Hence, $f(z) = f(p)$ and, consequently, $z \in f^{-1}[f(p)]$. We have thus shown that $\lim \sup (K_i) \subset f^{-1}[f(p)]$.

Next choose $y_i \in K_i \cap A$ for each $i \in \mathbf{P}$. Some subsequence (y_{n_i}) of (y_i) converges to a point $y \in \lim \sup (K_i)$. Since A is a closed subset, $y \in A$. Also, recall that $f[\lim \sup (K_i)] = \{f(p)\}$. We now have y and p as elements of $\lim \sup (K_i)$, which is a connected subset of $f^{-1}[f(p)]$. Thus, y and p are elements of the same component of $f^{-1}f[p]$. Then, since $y \in A$, $p \in A^*$. Hence, A^* is closed.

106. FACTORIZATION OF COMPACT MAPPINGS

It is often easier to study a mapping f by writing f as a composition $f_2 \circ f_1$ where the behavior of the factors f_1 and f_2 is known or easier to determine than that of f. (See comments in section 14.) A well known and now widely used factorization theorem was discovered independently by S. Eilenberg [8] and G. T. Whyburn [33]. The theorem asserts that if f is a continuous mapping from one compact metric space onto another, then f can be factored into the form $f_2 \circ f_1$ in which f_1 and f_2 are continuous, each of the point inverses of f_1 is a continuum (*monotone* mapping), and each of the point inverses of f_2 is totally disconnected (*light* mapping). Later the theorem was generalized from the setting of mappings on compact spaces to that of compact mappings ([36]).

106.1. Definitions. Monotone mappings, light mappings. Suppose X and Y *are topological spaces. A mapping* $f: X \to Y$ *is said to be monotone if for each* $y \in f[X], f^{-1}[y]$ *is a continuum. The mapping* $f: X \to Y$ *is said to be light provided that for each* $y \in f[X], f^{-1}[y]$ *is a totally disconnected set.*

Before proving that certain compact mappings have monotone and light factors, we list a few preliminary results which will be useful. Since the first factor in the factorization will be a natural mapping into a decomposition space, we include some results for factorizations in which the first factor is continuous and quasi-compact.

106.2. Theorem. *Suppose X, Y, and Z are topological spaces and $f: X \to Y$, $f_1: X \to Z$, and $f_2: Z \to Y$ are surjections such that $f = f_2 \circ f_1$. The following conclusions hold:*

106.2(a). *Suppose f_1 is continuous and quasi-compact. Then f is continuous if and only if f_2 is continuous. (See Exercise 5, page 2 3 4.)*

106.2(b). *Suppose f_1 and f_2 are compact mappings. Then f is a compact mapping.*

106.2(c). *Suppose f_1 and f_2 are continuous and f is compact. Then f_2 is compact. If Z is Hausdorff, f_1 is also compact.*

The proofs of (a) and (b) are left as exercises. We prove (c).

PROOF. We first prove that f_2 is compact under the given assumptions. Let K be a compact subset of Y. We wish to show that $f_2^{-1}[K]$ is compact. Now f is compact, so $f^{-1}[K]$ is compact; hence, $f_1^{-1}[f_2^{-1}[K]]$ is compact. Since f_1 is continuous, the image of $f_1^{-1}[f_2^{-1}[K]]$ under f_1 is compact. Hence,

$$f_1[f_1^{-1}[f_2^{-1}[K]]] = f_2^{-1}[K]$$

is compact.

We next assume that Z is Hausdorff in addition to the other assumptions stated in (c). Let H be a compact subset of Z. We complete the proof by showing that $f_1^{-1}[H]$ is compact. Notice first that, since Z is Hausdorff, H is closed and, hence, so is $f_1^{-1}[H]$. Also note that

$$H \subset f_2^{-1}[f_2[H]]$$

and, thus,

$$f_1^{-1}[H] \subset f_1^{-1}[f_2^{-1}[f_2[H]]] = f^{-1}[f_2[H]].$$

Since f_2 is continuous, $f_2[H]$ is compact, and since f is a compact mapping, $f^{-1}[f_2[H]]$ is compact. Thus, $f_1^{-1}[H]$ is compact since it is a closed subset of the compact set $f^{-1}[f_2[H]]$.

We next state a form of the monotone light factorization theorem for compact mappings [36].

106.3. **Theorem.** *Suppose X and Y are Hausdorff first countable spaces and $f: X \to Y$ is a compact continuous surjection (and by 95.8, closed). Then there is a space Z and there are compact closed continuous surjections $f_1: X \to Z$ and $f_2: Z \to Y$ such that f_1 is monotone, f_2 is light, and $f = f_2 \circ f_1$.*

PROOF. The proof of Theorem 106.3 will be broken down into a series of steps. In each of the parts listed the reader should provide a proof if no proof is given in the text.

Let $f: X \to Y$ be as in the statement of Theorem 106.3. Let $\mathscr{R}$ be the decomposition of X into the components of point inverses of f and let $f_1: X \to X/R$ be the corresponding natural mapping. For each $z \in X/R$, note that $f[f_1^{-1}[z]]$ is a set with exactly one element. Call that uniquely determined element $f_2(z)$. Thus, there is defined a mapping $f_2: X/R \to Y$. Each of the following then holds:

106.3(a). $f = f_2 \circ f_1$.

106.3(b). $\mathscr{R}$ is an u.s.c. decomposition of X.

106.3(c). f_1 and f_2 are continuous closed mappings.

106.3(d). f_1 and f_2 are compact mappings.

106.3(e) f_1 is a monotone mapping.

106.3(f). f_2 is a light mapping.

Proof of (c). The mapping f_1 is continuous because it is the natural mapping and, by Theorem 102.7, it is closed, because $\mathscr{R}$ is an u.s.c. decomposition of X. Moreover, it follows from 106.2 (a) that f_2 is continuous. We complete the proof of this part by showing that f_2 is closed. To see this, let L be a closed subset of X/R. Since $f_2[L] = f[f_1^{-1}[L]]$, it follows from the continuity of f_1 and the closedness of f that $f_2[L]$ is a closed subset of Y.

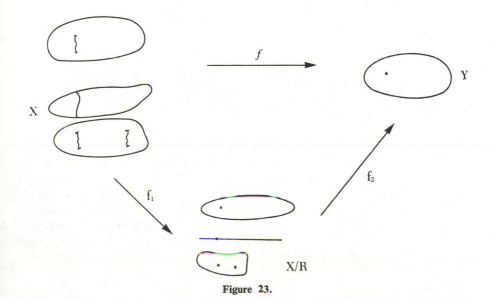

Figure 23.

Proof of (f). Let $y \in Y$, and suppose K is a component of $f_2^{-1}[y]$. We need to show that there is exactly one point in K. Since f_1 is monotone, by 101.5(d), $f_1^{-1}[K]$ is connected. Notice also that $f[f_1^{-1}[K]] = f_2[K] = \{y\}$. Thus, the connected set $f_1^{-1}[K]$ is contained in a single component Q of $f^{-1}[y]$. Hence, $K = f_1[f_1^{-1}[K]] \subseteq f_1[Q]$. But since there is exactly one point in $f_1[Q]$, there is exactly one point in K. Hence, f_2 is light.

The proof is complete if note is taken of the fact that f_1 and f_2 satisfy the requirements for f_1 and f_2 in the statement of the theorem.

EXERCISES: FACTORIZATION OF COMPACT MAPPINGS

1. Verify parts (a), (b), (d), and (e) of 106.3.

2. Let $f : X \to Y$ be a continuous surjection where X and Y are simple closed curves. Let D be the set of points $y \in Y$ for which $f^{-1}[y]$ consists of a single point. Prove that f is monotone if and only if D is a dense subset of Y. Show that $Y - D$ is countable if f is monotone.

3. Suppose $f : X \to Y$ is as in Theorem 106.3. Prove that if f is open, then the second factor obtained in the conclusion is open.

4. Suppose $f : X \to Y$ is a continuous surjection and $f_2 \circ f_1$ and $f_2^* \circ f_1^*$ are two factorizations of f. Whyburn, in [36], calls the two factorizations strictly topologically equivalent provided there is a homeomorphism h from the range space of f_1 onto the range space of f_1^* such that $h \circ f_1 = f_1^*$ and $f_2 = f_2^* \circ h$.

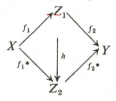

Let $f : X \to Y$ be as in Theorem 106.3. Suppose $f_2 \circ f_1$ and $f_2^* \circ f_1^*$ are two factorizations of f, each of which satisfies the properties of the factorizations guaranteed by Theorem 106.3. Prove that the factorization $f_2 \circ f_1$ is strictly topologically equivalent to the factorization $f_2^* \circ f_1^*$.

HINT: Use the monotone property of f_1 and the continuity property of f_1^* to show that for each z in the range of f_1, $f_1^*[f_1^{-1}[z]]$ is connected. Then use not only the fact that f_2^* is light but whatever else is necessary to show that there is exactly one element in $f_1^*[f_1^{-1}[z]]$. This should now suggest a suitable candidate for a homeomorphism h that satisfies the required properties. The various given properties of the factors f_1, f_2, f_1^*, f_2^* will be needed to show that h is indeed a homeomorphism. (See page 142 in [33].)

5. Give as many properties as you can that must be possessed by Z in Theorem 106.3. Then determine what additional properties, if any, must be possessed by Z in each of the following situations:

    (a) X is compact.

    (b) X is a separable metric space.

    (c) X is normal.

    (d) X is second countable.

    (e) X is locally connected.

10

Net and Filter Convergence

The notion of convergent sequence plays an important role in the investigation of first countable spaces and mappings on first countable spaces. Recall, for example, that if X is a first countable space, then x is a limit point of a subset S of X if and only if there is a sequence in $S - \{x\}$ that converges to x. Of course, since limit points can be characterized in terms of sequences, so can closed sets. We showed that for the class of first countable spaces, a space is Hausdorff if and only if each convergent sequence has a unique limit. If in addition to being first countable a space is Lindelöf, then compactness is equivalent to sequential compactness. Recall, finally, that if a mapping is defined on a first countable space, then it is continuous if and only if it preserves convergent sequences and their limits. That is, if f is a mapping from a first countable space X into a space Y, then f is continuous at x in X if and only if f satisfies the following condition: if a sequence (x_i) in X converges to x, then $(f(x_i))$ converges to $f(x)$ in Y.

The characterization theorems that we have just reviewed do not hold for general topological spaces. However, there are two types of convergence, net and filter convergence, with respect to which we can obtain analogous characterization theorems. In this chapter we shall make a study of these two types of convergence.

107. NETS AND SUBNETS

Recall that a sequence in a set X is a function from the set $\mathbf{P}$ of all positive integers into a set X. It is obvious that the relation $\geq$ in $\mathbf{P}$ plays a critical role in proofs involving convergence of sequences in a topological space. However, in many such proofs involving convergence of sequences in topological spaces it is not the fact that $\mathbf{P}$ is totally ordered (see 22.5) that is essential, rather it is that $\mathbf{P}$ satisfies the somewhat weakened ordering properties stated in the next definition.

107.1. Definition. Directed set. *Suppose that D is a nonempty set and $\geq$ is a relation in D. Then $(D, \geq)$ is called a directed set or a directed system provided that $\geq$ satisfies the following properties:*

107.1(a). $\geq$ *is transitive in D (see 19.2).*

107.1(b). $\geq$ *is reflexive in D (see 19.2).*

107.1(c). *If $a \in D$ and $b \in D$, then there is a $c \in D$ such that $c \geq a$ and $c \geq b$.*

If $(D, \geq)$ is a directed set and $a \geq b$, then we say that *a follows b* or *b precedes a*. Also, we say that $\geq$ is a *direction* for D.

Following are a number of examples of directed sets. The verifications that they are indeed directed sets are left as exercises.

107.2. *Example.* Let X be a set. Recall that the power set $\mathscr{P}(X)$ is the collection of all subsets of X. The relations $\supset$ and $\subset$ direct $\mathscr{P}(X)$. Thus, $(\mathscr{P}(X), \supset)$ and $(\mathscr{P}(X), \subset)$ are directed sets.

107.3. *Example.* Let $(X, \mathscr{T})$ be a topological space. Let $\mathscr{F}$ be the collection of all closed subsets of $(X, \mathscr{T})$. Then $(\mathscr{F}, \subset)$, $(\mathscr{F}, \supset)$, $(\mathscr{T}, \subset)$, and $(\mathscr{T}, \supset)$ are directed sets.

107.4. *Example.* Let $(X, \mathscr{T})$ be a topological space. Let $x \in X$ and suppose $\mathscr{N}_x$ is the neighborhood system of x. Then $(\mathscr{N}_x, \subset)$ and $(\mathscr{N}_x, \supset)$ are directed sets.

107.5. *Example.* Let $[a, b]$ be a closed interval in **R**. By a partition of $[a, b]$, we mean a finite set of points $\{x_0, x_1, x_2, \ldots, x_n\}$ such that $a = x_0 < x_1 < x_2 \cdots < x_n = b$. Let $\mathscr{P}[a, b]$ be the collection of all partitions of $[a, b]$. $(\mathscr{P}[a, b], \supset)$ is a directed set.

107.6. *Example.* (Filter.) Suppose X is a nonempty set and $\mathscr{F}$ is a nonempty collection of nonempty subsets of X such that

107.6(a). if $R \in \mathscr{F}$ and $S \in \mathscr{F}$, then $R \cap S \in \mathscr{F}$ and

107.6(b). if $R \in \mathscr{F}$, then each subset of X that contains R is also an element of $\mathscr{F}$.

$\mathscr{F}$ is then called a *filter* in X.

If $\mathscr{F}$ is a filter in X, then $(\mathscr{F}, \subset)$ and $(\mathscr{F}, \supset)$ are directed sets.

107.7. *Example.* Let X be a set and $\mathscr{C}$ be a nonempty nested collection (see 22.6) of subsets of X. Then $(\mathscr{C}, \supset)$ and $(\mathscr{C}, \subset)$ are directed sets.

107.8. *Example.* Let $(A_1, \geq)$ and $(A_2, \geq^*)$ be directed sets. The following defines a relation $\geqslant$ which directs $A_1 \times A_2$:

$$(a_1, a_2) \geqslant (b_1, b_2) \text{ if and only if } a_1 \geq b_1 \text{ and } a_2 \geq^* b_2$$

Since the set **P** of all positive integers is directed by the relation $\geq$ (given the usual meaning), $(\mathbf{P}, \geq)$ is a directed set. Thus, we may regard a sequence as a function defined on a certain directed set. The notion of a sequence is generalized to that of a net by allowing the domain to be an arbitrary directed set.

107.9.   Definition. Net.   *Let X be a set. A net s in X is a map $s:(D, \geq) \to X$ where $(D, \geq)$ is a directed set. Nets are also called generalized sequences.*

Analogous to the notation used for sequences, if $s:(D, \geq) \to X$ is a net, then for $\alpha \in D$ we shall sometimes write s_α for $s(\alpha)$. The notation $(s(\alpha), \alpha \in D, \geq)$ or the even more simple notation $(s(\alpha))$ will be used to designate a net defined on the directed set $(D, \geq)$.   Even though there is more than one directed set involved in a discussion, it usually will cause no confusion if the same symbol (e.g , $\geq$) is used for the various relations involved. Of course, a very specific relation with meaning of its own is sometimes used. In that case a standard notation for that relation should be used.

107.10.   *Example.*   Let X be a set and let $\mathscr{F}$ be a filter in X (see 107.6). Let s be a choice function for $\mathscr{F}$ guaranteed by the axiom of choice (24.1). Thus, $s(U) \in U$ for each $U \in \mathscr{F}$. The map $s:(\mathscr{F}, \subset) \to X$ is a net in X.

107.11.   *Example.*   Let f be a bounded real-valued function defined on a closed interval $[a, b]$. Let $\mathscr{P}[a, b]$ be the collection of all partitions of $[a, b]$ (see Example 107.5) and let $U:(\mathscr{P}[a, b], \supset) \to \mathbf{R}$ and $L:(\mathscr{P}[a, b], \supset) \to \mathbf{R}$ be given by

$$U(P) = \sum_{i=1}^{n}(\text{l.u.b.} \ (f[[x_{i-1}, x_i]]))(x_i - x_{i-1})$$

and

$$L(P) = \sum_{i=1}^{n}(\text{g.l.b.} \ (f[[x_{i-1}, x_i]]))(x_i - x_{i-1})$$

for each partition $P = \{x_0, x_1, \ldots, x_n\} \in \mathscr{P}[a, b]$. U and L are nets in $\mathbf{R}$. They are the familiar upper and lower sums from Riemann integration theory.

Recall that if $s:\mathbf{P} \to X$ is a sequence in X, then $s^*:\mathbf{P} \to X$ is a subsequence of s provided there exists a strictly increasing map $N:\mathbf{P} \to \mathbf{P}$ such that $s^* = s \circ N$. An analogous notion for nets is that of a subnet as defined next.

107.12.   Definition. Subnet of a net.   *Let $s:(D, \geq) \to X$ be a net in X. Suppose $N:(D^*, \geq^*) \to D$ is a net such that for each $p \in D$ there is a $p^* \in D^*$ such that*

$$d^* \geq^* p^* \text{ implies } N(d^*) \geq p.$$

Then the net $s \circ N:(D^*, \geq^*) \to X$ is called a subnet of s.

As with sequences, the notation s_{N_α} is sometimes used instead of $s(N(\alpha))$ for a term in the subnet.

Note that the "strictly increasing" feature of the function N in the definition of subsequence does not appear in the definition of subnet. Thus, each subsequence of a sequence is a subnet but not every subnet of a sequence is necessarily a subsequence.

EXERCISES: NETS AND SUBNETS

1.   Verify that the examples in 107.2 through 107.8 are directed sets.

2.   Let s^* be the sequence $(1, 2, 3, 3, 4, 5, 6, 6, 7, 8, 9, 9, \ldots)$ formed from the sequence $s = (i)$ by repeating every third term

of the original sequence. Is $s*$ a subsequence of s? Is $s*$ a subnet of s?

3. Suppose $s*$ is a subnet of s and $s**$ is a subnet of $s*$. Show that $s**$ is a subnet of s.

108. CONVERGENCE OF NETS

The notion of convergent sequences in topological spaces can be generalized to convergent nets. As with sequences, unless the space is Hausdorff, the convergence need not be unique.

108.1. Definition. *Let X be a topological space. Let $(x_\alpha, \alpha \in D, \geq)$ be a net in X and $p \in X$. Then (x_α) converges to p provided that for each open neighborhood U of p there exists an $\alpha_0 \in D$, such that for $\alpha \geq \alpha_0$, $x_\alpha \in U$.*

108.2. Example. Let X be a topological space, $x \in X$, and $\mathcal{N}_x$ the neighborhood system of x. For each $U \in \mathcal{N}_x$, let $x(U) \in U$. Then the net $(x(U), U \in \mathcal{N}_x, \subset)$ converges to x.

108.3. Example. In Example 107.11, $(U(P), P \in \mathscr{P}[a, b], \supset)$ converges to $\overline{\int}_a^b f(x)\, dx$, the upper Riemann integral of f on $[a, b]$. $(L(P), P \in \mathscr{P}[a, b], \supset)$ converges to $\underline{\int}_a^b f(x)\, dx$, the lower Riemann integral of f on $[a, b]$, where

$$\overline{\int}_a^b f(x)\, dx = \text{g.l.b. } \{U(P) : P \in \mathscr{P}[a, b]\}$$

and

$$\underline{\int}_a^b f(x)\, dx = \text{l.u.b. } \{L(P) : P \in \mathscr{P}[a, b]\}.$$

108.4. Definition. Cofinal subset. *Suppose that $(D, \geq)$ is a directed set. A subset $D*$ of D is said to be a cofinal subset of D provided that for each $d \in D$ there is a $d* \in D*$ such that $d* \geq d$.*

108.5. Remark. Let $(D, \geq)$ be a directed set. If $D*$ is a cofinal subset of D, then $\geq$ restricted to $D*$ is a direction for $D*$ and, thus, $(D*, \geq)$ is a directed set.

PROOF. Suppose that $a*$ and $b*$ are elements of $D*$. Then there is an element c in D such that c follows both $a*$ and $b*$. Then, since $D*$ is a cofinal subset of D, there is a $c*$ that follows c. From the transitivity of $\geq$, $c*$ follows both $a*$ and $b*$.

The following theorem gives some useful properties of nets. Some of the properties are suggested by analogous ones for sequences. The proofs are left as exercises.

108.6. Theorem. *Suppose that X is a topological space and $(x_\alpha, \alpha \in D, \geq)$ is a net in X. Then*

108.6(a). *If $x_\alpha = x$ for each $\alpha \in D$, then the net (x_α) converges to x.*

108.6(b). *If D^* is a cofinal subset of D, then $(x_\alpha, \alpha \in D^*, \geq)$ is a subnet of $(x_\alpha, \alpha \in D, \geq)$.*

108.6(c) *If the net (x_α) converges to a point p, then every subnet of (x_α) also converges to p.*

108.6(d). *Let $\tilde\alpha \in D$ and $D_{\tilde\alpha} = \{\alpha : \alpha \in D$ and $\alpha \geq \tilde\alpha\}$. Then the relation $\geq$, restricted to $D_{\tilde\alpha}$, directs $D_{\tilde\alpha}$, and $(x_\alpha, \alpha \in D_{\tilde\alpha}, \geq)$ is a subnet of (x_α).*

Recall that in first countable spaces we were able to characterize limit points and points in the closure of a set in terms of convergent sequences. (See 87.3 and Exercise 4, page 197.) The next theorem is an analogous characterization in terms of nets for general topological spaces.

108.7. Theorem. *Let X be a topological space. Then p is a limit point of a subset S if and only if there exists a net (x_α) in $S - \{p\}$ that converges to p. Also, p is in the closure of S if and only if there is a net (x_α) in S that converges to p.*

Theorem 108.7 is easy to prove by making use of 108.2. The proof is left as an exercise.

Every convergent sequence in a Hausdorff space has a unique limit. Among first countable spaces those that are Hausdorff can be characterized by this property, although general Hausdorff spaces cannot be. However, we are able to characterize all Hausdorff spaces as those for which convergent nets have unique limits. This fact is the content of the next theorem.

108.8. Theorem. *A topological space is Hausdorff if and only if each net converges to at most one point.*

PROOF. Let X be a Hausdorff space. Suppose a net $(x_\alpha, \alpha \in D, \geq)$ in X converges to p and q. If $p \neq q$, then there are disjoint neighborhoods N_p and N_q of p and q, respectively. There are elements $\alpha(p)$ and $\alpha(q)$ in D such that if $\alpha \geq \alpha(p)$ and $\beta \geq \alpha(q)$, then $x_\alpha \in N_p$ and $x_\beta \in N_q$. Since D is a directed set, there is a $\gamma \in D$ such that $\gamma \geq \alpha(p)$ and $\gamma \geq \alpha(q)$. But now $x(\gamma) \in N_p \cap N_q$, a contradiction to the fact that N_p and N_q are disjoint.

Next assume that X is not Hausdorff. Then there exist two points x and y in X such that the neighborhood systems $\mathscr{N}_x$ and $\mathscr{N}_y$ of x and y, respectively, satisfy

$$U \cap V \neq \varnothing \quad \text{for each} \quad U \in \mathscr{N}_x \text{ and } V \in \mathscr{N}_y.$$

Hence, for each $(U, V) \in \mathscr{N}_x \times \mathscr{N}_y$, we may choose a $z(U, V) \in U \cap V$. Define the relation $\geq$ on $\mathscr{N}_x \times \mathscr{N}_y$ by

$$(U_1, V_1) \geq (U_2, V_2)$$

if and only if $U_1 \subset U_2$ and $V_1 \subset V_2$. The relation $\geq$ directs $\mathscr{N}_x \times \mathscr{N}_y$. We finish the proof by showing that the net $(z(U, V), (U, V) \in \mathscr{N}_x \times \mathscr{N}_y, \geq)$ converges to x and to y. Let U_0 be a neighborhood of x. Choose $V_0 \in \mathscr{N}_y$. Then consider any $(U, V) \in \mathscr{N}_x \times \mathscr{N}_y$ such that $(U, V) \geq (U_0, V_0)$. Then $z(U, V) \in U \cap V \subset U_0$. Hence, the net $(z(U, V))$ converges to x. By a completely analogous argument, $(z(U, V))$ converges to y also.

Since the previous theorem shows that a net (x_α) in a Hausdorff space has at most one limit, we may make the following definition.

108.9. Definition. Limit of a net. *Suppose X is a Hausdorff space and $(x_\alpha, \alpha \in D, \geqq)$ is a net that converges to a point p in X. We then call p the limit of the net (x_α) and we write*

$$\lim (x_\alpha) = p \qquad \text{or} \qquad \lim (x_\alpha, \alpha \in D, \geqq) = p.$$

Recall that for second countable spaces we were able to characterize compactness by sequential compactness. All topological spaces can be characterized by an analogous "net-type" compactness.

108.10. Theorem. *Let X be a topological space. Then X is compact if and only if each net (x_α) in X has a convergent subnet.*

PROOF. Suppose first that each net in X has a convergent subnet. We will show that X is compact by showing that every collection of closed subsets of X with the finite intersection property has a nonempty intersection (see 90.2).

Let $\mathscr{F}$ be a collection of closed subsets of X, every finite subcollection of which has a nonempty intersection. We wish to show that $\bigcap \mathscr{F}$ is nonempty. Let $\mathscr{F}^*$ be the collection of all finite intersections of elements of $\mathscr{F}$. Notice that the relation $\subset$ directs $\mathscr{F}^*$. For each $F \in \mathscr{F}^*$, choose an $x(F) \in F$. Then $(x(F), F \in \mathscr{F}^*, \subset)$ is a net in X. Thus, some subnet $(x(N(\beta)), \beta \in D, \geqq)$ converges to a point $x \in X$. We complete the proof by showing that $x \in \bigcap \mathscr{F}$. Suppose $x \notin F_0$ for some $F_0 \in \mathscr{F}$. Now $F_0 \in \mathscr{F}^*$. Since $(x(N(\beta)), \beta \in D, \geqq)$ is a subnet of $(x(F), F \in \mathscr{F}^*, \subset)$, there is a $\beta_0 \in D$ such that if $\beta \geqq \beta_0$, then $N(\beta) \subset F_0$. Notice that $X - F_0$ is a neighborhood of x. Since $(x(N(\beta)))$ converges to x, there is a $\beta \geqq \beta_0$ such that $x(N(\beta)) \in (X - F_0)$. But $N(\beta) \subset F_0$ so that $x(N(\beta)) \in X - N(\beta)$, contrary to the way in which $x(N(\beta))$ was chosen. Thus, $\bigcap \mathscr{F} \neq \varnothing$ and X is compact.

Next suppose that X is compact. We assume $(x_\alpha, \alpha \in D, \geqq)$ is a net in X and proceed to find a convergent subnet. To do this, for each $\tilde{\alpha} \in D$ let

$$D_{\tilde{\alpha}} = \{x(\alpha) : \alpha \in D \quad \text{and} \quad \alpha \geqq \tilde{\alpha}\}.$$

Let

$$\mathscr{F} = \{\text{cl}(D_\alpha) : \alpha \in D\}.$$

Because D is a directed set and because of the definition of the D_α, it is easy to show that every finite subcollection of $\mathscr{F}$ has a nonempty intersection. Then since X is compact by 90.2, $\bigcap \mathscr{F} \neq \varnothing$. Let $p \in \bigcap \mathscr{F}$. We shall complete the proof by showing that some subnet of (x_α) converges to p. We form a subnet as follows: Let $\mathscr{N}_p$ be the neighborhood system for p. For every (U_1, α_1) and (U_2, α_2) in $\mathscr{N}_p \times D$, let $(U_2, \alpha_2) \geqslant (U_1, \alpha_1)$ if and only if $U_2 \subset U_1$ and $\alpha_2 \geqq \alpha_1$. This relation $\geqslant$ directs $\mathscr{N}_p \times D$. Next, for each $(U, \alpha) \in \mathscr{N}_p \times D$ we may choose an element $N(U, \alpha) \in D$ such that $N(U, \alpha) \geqq \alpha$ and $x(N(U, \alpha)) \in U \cap D_\alpha$. This is possible since $p \in \text{cl}(D_\alpha)$ and $U \in \mathscr{N}_p$. If we note that $N(U, \alpha) \geqq \alpha$, it is easy to verify that

$$(x(N(U, \alpha)), (U, \alpha) \in \mathscr{N}_p \times D, \geqslant)$$

is a subnet of $(x(\alpha))$. We complete the proof by showing that $(x(N(U, \alpha)))$ converges to p. To see this let $U_0 \in \mathscr{N}_p$. Choose an $\alpha_0 \in D$. Next, suppose $(U, \alpha) \geqslant (U_0, \alpha_0)$. Then $x(N(U, \alpha)) \in U \cap D_\alpha \subset U_0$. Hence, we have shown that the net $(x(N(U, \alpha)))$ converges to p.

We have seen that arguments in which sequences play a part lend themselves quite well to first countable spaces and especially to Hausdorff first countable spaces. In this section we have studied theorems that characterize several properties in terms of nets. By means of these theorems, techniques very similar to those used in sequential arguments can often be employed. We illustrate this fact in the proof of the following theorem.

108.11. Theorem. *Suppose $\mathscr{C}$ is a nonempty nested collection of nonempty continua in a Hausdorff space X. Then $\bigcap \mathscr{C}$ is a nonempty continuum.*

PROOF. We first show that $\bigcap \mathscr{C}$ is a nonempty compact set. To see this let $F_0 \in \mathscr{C}$ and let $\mathscr{C}_0 = \{F:F \in \mathscr{C} \text{ and } F \subset F_0\}$. We first observe that $\bigcap \mathscr{C} = \bigcap \mathscr{C}_0$. Because X is a Hausdorff space, each $F \in \mathscr{C}_0$ is closed, and because $\mathscr{C}_0$ is nested, each finite subcollection of $\mathscr{C}_0$ has a nonempty intersection. Hence, it follows from the compactness of F_0 that $\bigcap \mathscr{C}_0 \neq \varnothing$. Since each of the elements in $\mathscr{C}$ is closed, $\bigcap \mathscr{C}$ is closed, and since $\bigcap \mathscr{C}$ is a subset of the compact Hausdorff subspace F_0, $\bigcap \mathscr{C}$ is compact.

We complete the proof by showing that $\bigcap \mathscr{C}$ is connected. If $\bigcap \mathscr{C}$ is not connected, then there is a separation $\{A, B\}$ of $\bigcap \mathscr{C}$ where A and B are nonempty compact sets. Since X is Hausdorff, there exist disjoint open sets U and V such that $A \subset U$ and $B \subset V$. With $\mathscr{C}_0$, as in the previous paragraph, we note that $(\mathscr{C}_0, \subset)$ is a directed set. For each $F \in \mathscr{C}_0$, $F \cap U \neq \varnothing$, $F \cap V \neq \varnothing$. Hence, because of the connectedness of each element of $\mathscr{C}_0$, each element in $\mathscr{C}_0$ must intersect $X - (U \cup V)$. Then for each $F \in \mathscr{C}_0$, we may choose an element

$$x(F) \in [X - (U \cup V)] \cap F.$$

Notice that $(x(F), F \in \mathscr{C}_0, \subset)$ is a net in $[X - (U \cup V)] \cap F_0$. Since

$$[X - (U \cup V)] \cap F_0$$

is compact, by 108.10, there is a subnet $(x(N(\alpha)))$ of $(x(F))$ that converges to a point $z \in [X - (U \cup V)] \cap F_0$. We shall show that this leads to a contradiction by showing that $z \in \bigcap \mathscr{C} \subset U \cup V$. To see this suppose $z \notin F_1$ for some $F_1 \in \mathscr{C}$. There is no loss in generality in supposing that $F_1 \subset F_0$. Then z is in the open set $X - F_1$. Since $(x(N(\alpha)))$ converges to z, there is an α_1 such that for $\alpha \geq \alpha_1$, $x(N(\alpha)) \in X - F_1$. Since $(x(N(\alpha)))$ is a subnet of $(x(F), F \in \mathscr{C}_0, \subset)$ and $F_1 \in \mathscr{C}_0$, there is an $\alpha_2 \in D$ such that for $\alpha \geq \alpha_2$, $N(\alpha) \subset F_1$. Let $\alpha \geq \alpha_1$ and $\alpha \geq \alpha_2$. Then $x(N(\alpha)) \in N(\alpha) \subset F_1$ and $x(N(\alpha)) \in X - F_1$. We have thus reached a contradiction and the proof is complete.

We shall use the last theorem together with Zorn's lemma (see 24.3 and the remaining part of Section 24) to obtain the following interesting result.

108.12. Theorem. *Let $(X, \mathscr{T})$ be a Hausdorff space. Suppose that A is a subset (possibly empty) of X and that C is a nonempty continuum in X that contains A. Then there is a minimal nonempty subcontinuum K of C that contains A (i.e., K is a nonempty subcontinuum of C that contains A and there does not exist a nonempty proper subcontinuum of K that contains A).*

PROOF. Let $\mathscr{F}$ be the family of all nonempty subcontinua of C that contain A. Order $\mathscr{F}$ by inclusion ($\subset$) and consider the partially ordered set $(\mathscr{F}, \subset)$. Let $(\mathscr{F}^*, \subset)$ be a linearly ordered subset of $\mathscr{F}$. Note that $\bigcap \mathscr{F}^*$ is a nonempty subcontinuum of C by the previous theorem. Also note that $\bigcap \mathscr{F}^* \supset A$ so that

$\bigcap \mathscr{F}^* \in \mathscr{F}$. Thus, every linearly ordered subset of $\mathscr{F}$ has a lower bound. Hence, by Zorn's lemma, $(\mathscr{F}, \subset)$ has a minimal element K. It is easy to see that K satisfies the K in the conclusion of the statement of the theorem that we are proving.

EXERCISES: CONVERGENCE OF NETS

1. Prove Theorems 108.6 and 108.7.

2. Recall that in a topological space if $(x(i))$ converges to p, then $\{x(i): i \in \mathbf{P}\} \cup \{p\}$ is a compact set. Show by an example that the analogous statement for nets does not necessarily hold.

3. Suppose $(f(\alpha), \alpha \in D \geq)$ and $(g(\alpha), \alpha \in D \geq)$ are two real-valued nets defined on a directed set D. For each $\alpha \in D$, let $s(\alpha) = f(\alpha) + g(\alpha), p(\alpha) = f(\alpha)g(\alpha)$, and $q(\alpha) = \dfrac{f(\alpha)}{g(\alpha)}$ provided $g(\alpha) \neq 0$. Prove each of the following:
    (a) If $\lim (f(\alpha)) = F$ and $\lim (g(\alpha)) = G$, then $\lim (p(\alpha)) = FG$ and $\lim (s(\alpha)) = F + G$.
    (b) If $g(\alpha) \neq 0$ for each $\alpha \in D$, $\lim (f(\alpha)) = F$, and $\lim (g(\alpha)) = G \neq 0$, then $\lim (q(\alpha)) = \dfrac{F}{G}$.

4. Let $(f_\alpha : (X, \mathscr{T}) \to (Y, d))$ be a net of mappings from a topological space into a metric space.
    (a) Extend the notion of pointwise convergence and uniform convergence of sequences of mappings to nets of mappings. Then try out your definition on the following proposition.
    (b) Suppose $(f_\alpha : (X, \mathscr{T}) \to (Y, d))$ is a net of continuous mappings where (Y, d) is metric. If the net (f_α) converges uniformly to a mapping $f : (X, \mathscr{T}) \to (Y, d)$, then f is continuous.

5. A continuous surjection $f : X \to Y$ is said to be strongly irreducible provided that for no closed proper subset X^* of X is it true that $f[X^*] = Y$. Prove the following proposition: Let X and Y be Hausdorff spaces and assume that X is compact. If $f : X \to Y$ is a continuous surjection, then there is a closed subset X^* of X such that $f[X^*] = Y$ and the restriction $f \mid X^* : X^* \to Y$ is a strongly irreducible mapping. (See [33] for other theorems about irreducible mappings.)

109. FILTERS

Recall that in Example 107.6, a filter was defined in connection with our considerations of directed sets. A theory of convergence has been developed in terms of filters that is analogous to the theory of convergence in terms of nets. In this section we shall make a brief study of filter convergence theory. In the next chapter we shall have occasion to apply results from the theory of nets and filters in the discussion of product spaces.

109.1. Definition. Filter on a set. *Let X be a set. A nonempty collection $\mathscr{F}$ of nonempty subsets of X is called a filter on X provided that*

$$\text{if} \quad A \in \mathscr{F} \quad \text{and} \quad B \in \mathscr{F}, \quad \text{then} \quad A \cap B \in \mathscr{F}$$

and

$$\text{if} \quad A \in \mathscr{F} \quad \text{and} \quad A \subset B, \quad \text{then} \quad B \in \mathscr{F}.$$

We emphasize the fact that in the previous definition $\mathscr{F}$ is a *nonempty* collection of subsets of X and that each element of $\mathscr{F}$ is a *nonempty* subset of X.

109.2. *Example.* Let X be a topological space and let $x \in X$. Then the neighborhood system $\mathscr{N}_x$ of x is a filter on X.

The fact that the neighborhood system of each point is a filter will play an important role in our study of filter convergence.

109.3. Definition. Neighborhood filter of a point. *Let X be a topological space. For each $x \in X$, the neighborhood system $\mathscr{N}_x$ of x is called the neighborhood filter of x.*

109.4. Definitions. Filter base and filter subbase. *Let $\mathscr{B}$ and $\mathscr{S}$ be nonempty collections of nonempty subsets of X.*
$\mathscr{B}$ is said to be a filter base on X provided that if $A \in \mathscr{B}$ and $B \in \mathscr{B}$, then there is a $C \in \mathscr{B}$ such that $C \subset A \cap B$.
$\mathscr{S}$ is said to be a filter subbase on X provided for every nonempty finite subcollection $\mathscr{S}^*$ of $\mathscr{S}$, $\bigcap \mathscr{S}^* \neq \varnothing$.

The terms base and subbase with reference to filters can be justified on the basis of the following theorem, the proof of which is left as an exercise.

109.5. Theorem. *Suppose that X is a set.*

109.5(a). *If $\mathscr{B}$ is a filter base on X, then the collection*

$$\{U : U \subset X \quad \text{and} \quad B \subset U \quad \text{for some} \ B \in \mathscr{B}\}$$

is a filter on X.

109.5(b). *If $\mathscr{S}$ is a filter subbase on X, then the collection*

$$\{U : U = \bigcap \mathscr{S}^* \text{ for some finite subcollection } \mathscr{S}^* \text{ of } \mathscr{S}\}$$

is a filter base on X.

In what follows the filter defined in 109.5(a) will be called the *filter on X generated by $\mathscr{B}$* and the filter generated by the base defined in 109.5(b) will be called the *filter on X generated by $\mathscr{S}$*.

109.6. *Example.* Suppose that X is a topological space and $x \in X$. Let $\mathscr{B}$ be the collection of all open neighborhoods of the point x. Then $\mathscr{B}$ is a filter base on X, and $\mathscr{B}$ generates the neighborhood filter $\mathscr{N}_x$ of x.

109.7. Definition. Convergent filters. *Let $(X, \mathcal{T})$ be a topological space, let $x \in X$, and let $\mathcal{F}$ be a filter on X. Then $\mathcal{F}$ is said to converge to x provided $\mathcal{F}$ contains the neighborhood filter of x. In particular, the neighborhood filter of x converges to x.*

The next theorem follows immediately from the previous definition.

109.8. Theorem. *If in a topological space X a filter $\mathcal{F}$ converges to x, then any filter $\mathcal{F}^*$ on X that contains $\mathcal{F}$ also converges to x.*

The previous theorem suggests that if $\mathcal{F}$ and $\mathcal{F}^*$ are filters in a space and s and s^* are nets, then "$\mathcal{F}^* \supset \mathcal{F}$" is a statement for filter convergence theory that is analogous to "s^* is a subnet of s" in net convergence theory.

The next theorem is the analogue for filters of Theorem 108.8.

109.9. Theorem. *A topological space X is Hausdorff if and only if each filter on X converges to at most one point in X.*

PROOF. Suppose that X is Hausdorff and a filter $\mathcal{F}$ on X converges to a point x in X. Suppose that $y \in X$ and $y \neq x$. Then there exist neighborhoods N_x and N_y of x and y, respectively, such that $N_x \cap N_y = \varnothing$. Since $\mathcal{F}$ converges to x, $N_x \in \mathcal{F}$, and since $N_x \cap N_y = \varnothing$, $N_y \notin \mathcal{F}$. But then $\mathcal{F}$ cannot converge to y.

Next suppose that X is not Hausdorff. Then there exist points p and q in $X, p \neq q$, such that if we let $\mathcal{N}_p$ and $\mathcal{N}_q$ be the neighborhood systems for p and q, respectively, then

$$(*) \quad N_p \cap N_q \neq \varnothing \quad \text{for each } N_p \text{ in } \mathcal{N}_p \quad \text{and} \quad N_q \text{ in } \mathcal{N}_q.$$

Next let

$$\mathcal{B} = \{U : U = N_p \cap N_q \text{ for some } N_p \in \mathcal{N}_p \text{ and } N_q \in \mathcal{N}_q\}.$$

Notice that for all B_1 and B_2 in $\mathcal{B}$, $B_1 \cap B_2 \in \mathcal{B}$. Then because of $(*)$, $\mathcal{B}$ is a filter base. Let $\mathcal{F}$ be the filter generated by $\mathcal{B}$. We prove next that $\mathcal{F} \supset \mathcal{N}_p$ and $\mathcal{F} \supset \mathcal{N}_q$. To see this notice that if $N_p \in \mathcal{N}_p$ and $N_q \in \mathcal{N}_q$, then $N_p \supset N_p \cap N_q$ and $N_q \supset N_p \cap N_q$. From this it follows that N_p and N_q are elements of the filter $\mathcal{F}$. We have shown that the filter $\mathcal{F}$ converges to both p and q and the proof is complete.

A useful tool in connection with filters is the notion of a *maximal filter*. We shall make use of it in some of our considerations of compactness in the next chapter.

109.10. Definition. Maximal filter or ultrafilter. *Let $\mathcal{F}$ be a filter on a set X. $\mathcal{F}$ is said to be a maximal filter or an ultrafilter provided that no other filter on X properly contains $\mathcal{F}$.*

109.11. Theorem. *If X is a set and $\mathcal{F}_0$ is a filter on X, then there exists a maximal filter $\mathcal{M}$ on X that contains $\mathcal{F}_0$.*

PROOF. (The proof will make use of Zorn's lemma.)

Let $\mathcal{F}_0$ be a filter on a set X. Let $\mathcal{C}^*$ be the collection of all filters on X that contain $\mathcal{F}_0$. Consider the partially ordered set $(\mathcal{C}^*, \subset)$. Suppose $\mathcal{H}^*$ is a linearly ordered subset of $\mathcal{C}^*$. Let $\mathcal{H} = \bigcup \mathcal{H}^*$. We shall show that $\mathcal{H}$ is a filter on X

and that $\mathcal{H}$ is an upper bound for $\mathcal{H}^*$. Since for each $A \in \mathcal{H}$, $A \in \mathcal{F}$ for some $\mathcal{F} \in \mathcal{H}^*$, A must be nonempty. Also if $A \in \mathcal{H}$ and $B \in \mathcal{H}$, then $A \in \mathcal{F}_1$, $B \in \mathcal{F}_2$ for some $\mathcal{F}_1$ and $\mathcal{F}_2$ in $\mathcal{H}^*$. Since $\mathcal{H}^*$ is linearly ordered, A and B must be in $\mathcal{F}_1$ or A and B must be in $\mathcal{F}_2$. Hence, $A \cap B \in \mathcal{F}_1$ or $A \cap B \in \mathcal{F}_2$. Thus, $A \cap B \in \mathcal{H}$. Finally, suppose $A \in \mathcal{H}$ and $A \subset B$. Then, since $A \in \mathcal{F}_1$ for some $\mathcal{F}_1 \in \mathcal{H}^*$, $B \in \mathcal{F}_1$ also. Hence, $B \in \mathcal{H}$. We have thus shown that $\mathcal{H}$ is a filter on X. Furthermore, $\mathcal{F}_0 \subset \mathcal{F} \subset \mathcal{H}$ for each $\mathcal{F} \in \mathcal{H}^*$. Thus, $\mathcal{H} \in \mathcal{C}^*$ and $\mathcal{H}$ is an upper bound in $\mathcal{C}^*$ for $\mathcal{H}^*$. Hence, by Zorn's lemma, $\mathcal{C}^*$ has a maximal element and any such maximal element is an ultrafilter on X that contains $\mathcal{F}_0$.

In the next theorem, we characterize maximal filters in an interesting and useful way. This theorem will play an important role in the applications of filters to compactness.

109.12. Theorem. *Let $\mathcal{M}$ be a filter on a set X. Then $\mathcal{M}$ is a maximal filter on X if and only if it has the following property: For each subset A of X, either $A \in \mathcal{M}$ or $X - A \in \mathcal{M}$.*

PROOF. We first suppose that $\mathcal{M}$ is a maximal filter on X. Let A be a subset of X and assume that $X - A \notin \mathcal{M}$. We shall show that in that case $A \in \mathcal{M}$. Since we are assuming that $X - A \notin \mathcal{M}$, it follows that for each $F \in \mathcal{M}$, F could not be contained in $X - A$. For, otherwise, $X - A \in \mathcal{M}$. Hence, $F \cap A \neq \varnothing$ for each $F \in \mathcal{M}$. (Using this fact, we shall construct a filter $\mathcal{F}$ that has A as an element. We shall then show that $\mathcal{F}$ contains the maximal filter $\mathcal{M}$ and, hence, is the same as $\mathcal{M}$.) Let

$$\mathcal{B} = \{B : B = F \cap A \qquad \text{for some } F \in \mathcal{M}\}.$$

Notice that each element in $\mathcal{B}$ is nonempty. Furthermore, the intersection of each two elements of $\mathcal{B}$ is itself an element of $\mathcal{B}$. Hence, $\mathcal{B}$ is a filter base on X. Let $\mathcal{F}$ be the filter on X generated by $\mathcal{B}$. We shall show $\mathcal{M} \subset \mathcal{F}$. Suppose that $F \in \mathcal{M}$. Then $F \cap A \subset F$. Hence, since $A \cap F \in \mathcal{F}$, $F \in \mathcal{F}$. However, since $\mathcal{M}$ is a maximal filter, $\mathcal{M} = \mathcal{F}$. We show finally that $A \in \mathcal{M}$. Let $F \in \mathcal{M}$. Then $F \cap A \in \mathcal{F}$ and, hence, $F \cap A \in \mathcal{M}$. Moreover, since $F \cap A \subset A$, $A \in \mathcal{M}$. This completes this part of the proof since we have shown that if $X - A \notin \mathcal{M}$, then $A \in \mathcal{M}$.

Next, assume that $\mathcal{M}$ is a filter that satisfies the following property:

For each subset A of X, $A \in \mathcal{M}$ or $(X - A) \in \mathcal{M}$.

We shall show that $\mathcal{M}$ is an ultrafilter. Suppose that $\mathcal{F}$ is a filter on X such that $\mathcal{M} \subset \mathcal{F}$. We complete the proof by showing that $\mathcal{F} \subset \mathcal{M}$. Let $F \in \mathcal{F}$. We wish to show that $F \in \mathcal{M}$. To see this, assume that $F \notin \mathcal{M}$. Then from the property assumed for $\mathcal{M}$, $X - F \in \mathcal{M}$, and since $\mathcal{M} \subset \mathcal{F}$, $X - F \in \mathcal{F}$. From this it follows that $F \notin \mathcal{F}$, a contradiction.

109.13. Theorem. *A topological space X is compact if and only if every maximal filter on X converges.*

PROOF. Suppose first that X is compact. Let $\mathcal{M}$ be a maximal filter on X and let us suppose $\mathcal{M}$ does not converge. Then for each $x \in X$, since $\mathcal{M}$ does not converge to x, we can find a neighborhood N_x of x such that $N_x \notin \mathcal{M}$

and we may assume that N_x is open. By the previous theorem, since $\mathcal{M}$ is maximal and $N_x \notin \mathcal{M}$, $X - N_x \in \mathcal{M}$. The collection $\{N_x : x \in X\}$ of neighborhoods is chosen so that $N_x \notin \mathcal{M}$ is an open covering of X. Since X is compact, some finite subcovering $\{N_{x_i} : i \in \mathbf{P}_n\}$ also covers X. But each $X - N_{x_i} \in \mathcal{M}$. Hence, $\bigcap \{X - N_{x_i} : i \in \mathbf{P}_n\} \neq \varnothing$ since $\mathcal{M}$ is a filter. But

$$\bigcap \{X - N_{x_i} : i \in \mathbf{P}_n\} = X - \bigcup \{N_{x_i} : i \in \mathbf{P}_n\} = X - X = \varnothing.$$

Thus, we have arrived at a contradiction and $\mathcal{M}$ converges.

Assume next that every maximal filter on X converges. Let $\mathcal{F}$ be a collection of closed subsets of X such that the intersection of each finite subcollection of $\mathcal{F}$ has a nonempty intersection. We wish to show that $\bigcap \mathcal{F} \neq \varnothing$. Note that $\mathcal{F}$ is a filter subbase. Let $\mathcal{F}^*$ be the filter generated by $\mathcal{F}$. By Theorem 109.11 there is an ultrafilter $\mathcal{M}$ on X that contains $\mathcal{F}^*$. Then $\mathcal{M}$ converges to some point $p \in X$. We complete the proof by showing that $p \in \bigcap \mathcal{F}$. If not, then there is an $S \in \mathcal{F}$ such that $p \in X - S$. Now $(X - S) \in \mathcal{M}$ since $\mathcal{M}$ converges to p and $X - S$ is a neighborhood of p. But $S \in \mathcal{F}$ and so $S \in \mathcal{M}$. Since S and $X - S$ are both elements of $\mathcal{F}$, $S \cap (X - S) \neq \varnothing$ and we have arrived at an obvious contradiction.

Since each filter on a set is contained in a maximal filter and since compactness on a topological space is equivalent to the property stated in the last theorem, we have for filters the following analogue of Theorem 108.10.

109.14.   Theorem.   *A topological space is compact if and only if each filter on X is contained in a filter that converges.*

NOTE: The type of net convergence that we have been studying is known as Moore-Smith convergence. The reader is referred to pages 72 and 78 in [32] for some interesting historical remarks on the development of nets and filters. The reader is also referred to [27] for an interesting expository account of Moore-Smith convergence.

The next theorem will be useful to us in our consideration of product spaces in the next chapter.

109.15.   Theorem.   *Let $f : X \to Y$ be a surjection and let $\mathcal{F}$ be a filter on X. Then $\{f[F] : F \in \mathcal{F}\}$ is a filter on Y.*

PROOF.   Let $\mathcal{G} = \{f[F] : F \in \mathcal{F}\}$. It is clear that each element of $\mathcal{G}$ is nonempty. Suppose first that $G \in \mathcal{G}$ and $G \subset H$. We shall show that $H \in \mathcal{G}$. Since $G \in \mathcal{G}$, there is an $F \in \mathcal{F}$ such that $f[F] = G$. Then $F \subset f^{-1}[H]$ and, hence $f^{-1}[H] \in \mathcal{F}$. Consequently $f[f^{-1}[H]] = H \in \mathcal{G}$. Next suppose that G_1 and G_2 are elements of $\mathcal{G}$. Then there are elements F_1 and F_2 of $\mathcal{F}$ such that $f[F_1] = G_1$ and $f[F_2] = G_2$. Since $F_1 \cap F_2 \in \mathcal{F}$, it follows that $f[F_1 \cap F_2] \in \mathcal{G}$. Further, since $f[F_1 \cap F_2] \subset f[F_1] \cap f[F_2]$, it follows from the first part of the proof that $f[F_1] \cap f[F_2] \in \mathcal{G}$. This completes the proof.

If $f : X \to Y$ is a map that is not surjective and $\mathcal{F}$ is a filter on X, then we can no longer conclude that $\mathcal{G} = \{f[F] : F \in \mathcal{F}\}$ is a filter. However, $\mathcal{G}$ is a filter base. The verification of this fact is left as an exercise for the reader.

EXERCISES: FILTERS

1. Prove Theorem 109.5.

2. Suppose X is a topological space and A is a nonempty subset of X. Show that $p \in X$ is a limit point of A if and only if $A - \{p\}$ is an element of a filter on X that converges to p.

3. Suppose $f : X \to Y$ is a map and $\mathscr{F}$ is a filter on X. Show that $\mathscr{B}$ is a filter base on Y where

    $$\mathscr{B} = \{f[F] : F \in \mathscr{F}\}.$$

4. Prove that a mapping $f : (X, \mathscr{T}) \to (Y, \mathscr{T}^*)$ is continuous at $x \in X$ if and only if the following condition holds:

    For each filter $\mathscr{F}$ on X, if $\mathscr{F}$ converges to x, then the filter $\mathscr{F}^*$ on Y, generated by the filter base $\{f[F] : F \in \mathscr{F}\}$, converges to $f(x)$.

5. Suppose $X = A \cup B$ and $\mathscr{M}$ is an ultrafilter on X. Prove that either $A \in \mathscr{M}$ or $B \in \mathscr{M}$.

6. Let $f : X \to Y$ be a map. Suppose $\mathscr{M}$ is a maximal filter on X. Then the filter $\mathscr{M}_f$ on Y generated by the filter base

    $$\{f[F] : F \in \mathscr{M}\}$$

    is a maximal filter on Y.

7. The proposition in Exercise 4 states a characterization of continuity at a point in terms of filters. State and prove an analogous proposition that uses nets instead of filters.

11

Product Spaces

In this chapter the notion of *product space* will be extended to include the product of infinite collections of spaces. As in the finite case, the projection mappings play an essential role both in the definition and in the investigation of infinite product spaces. We shall show that the projection mapping π_α from a nonempty product space into the coordinate space X_α is an open continuous surjection. Furthermore, each coordinate space has a homeomorphic copy of itself in the product space. Thus, each coordinate space inherits from the product space every property that is invariant under a continuous open map and every property that is inherited by each subspace of a topological space. There are also properties that are inherited by the product space provided each coordinate space has that property. For example, the well known and important Tychonoff theorem states that the product space for a collection of compact spaces is compact. The Hausdorff property is another property that is inherited by the product space from the coordinate spaces. Some properties are inherited by countable product spaces but not by arbitrary products. For example, the product of a countable collection of metric spaces is metrizable. However, the product space for an uncountable collection of metric spaces cannot be metrizable unless all but a countable number of the coordinate spaces are trivial metric spaces (contain no more than one point).

110. CARTESIAN PRODUCTS

Suppose that $\{X_i : i \in \mathbf{P}_n\}$ is a collection of sets indexed by $\mathbf{P}_n$ for some positive integer n. In 15.5 the Cartesian product $\mathsf{X}\,\{X_i : i \in \mathbf{P}_n\}$ is defined as the set of all finite-sequences $(x_1, x_2, x_3, \ldots, x_n)$ such that $x_i \in X_i$ for $i \in \mathbf{P}_n$. Similarly, if $\{X_i : i \in \mathbf{P}\}$ is a collection of sets indexed by $\mathbf{P}$, then $\mathsf{X}\,\{X_i : i \in \mathbf{P}\}$ is defined as the collection of all infinite sequences (x_i) such that $x_i \in X_i$. Now recall from 15.1 and 15.2 that a finite-sequence $(x_1, x_2, x_3, \ldots, x_n)$ is a function defined on $\mathbf{P}_n$, whereas an infinite sequence (x_i) is a function defined on $\mathbf{P}$. We see, therefore, that if $\mathbf{P}^*$ is the set $\mathbf{P}_n$ or $\mathbf{P}$, then $\mathsf{X}\,\{X_i : i \in \mathbf{P}^*\}$ is the set of all *functions* x defined on $\mathbf{P}^*$ such that $x(i) \in X_i$ for each $i \in \mathbf{P}^*$. This formulation suggests the extension of the notion of Cartesian products to include products of arbitrary indexed collections of sets.

110.1. Definition. Cartesian product. *Suppose* $\{X_\alpha : \alpha \in A\}$ *is an indexed collection of sets. Then by the Cartesian product* $\mathsf{X}\,\{X_\alpha : \alpha \in A\}$ *of the collection* $\{X_\alpha : \alpha \in A\}$, *we shall mean the collection of all functions* x *defined on* A *such that* $x(\alpha) \in X_\alpha$. *For each* $\alpha \in A$, X_α *is called the* α-*coordinate space, and for each* $x \in \mathsf{X}\,\{X_\alpha : \alpha \in A\}$, $x(\alpha)$, *also written as* x_α, *is called the* α-*coordinate of* x.

It is to be noted that if $A \neq \varnothing$ and each $X_\alpha \in \{X_\alpha : \alpha \in A\}$ is nonempty, then $\mathsf{X}\,\{X_\alpha : \alpha \in A\}$ is nonempty by the axiom of choice. As a matter of fact, each $x \in \mathsf{X}\,\{X_\alpha : \alpha \in A\}$ is what we have previously referred to in 24.1 as a choice function for the collection $\{X_\alpha : \alpha \in A\}$.

110.2. Definition. The projection mappings. *Let* $\mathsf{X}\,\{X_\alpha : \alpha \in A\}$ *be a Cartesian product. For each* $\alpha \in A$, *let* $\pi_\alpha : \mathsf{X}\,\{X_\alpha : \alpha \in A\} \to X_\alpha$ *be defined by*

$$\pi_\alpha(x) = x_\alpha.$$

The mapping π_α *is called the projection mapping of* $\mathsf{X}\,\{X_\alpha : \alpha \in A\}$ *into the* α-*coordinate space* X_α.

EXERCISES: CARTESIAN PRODUCTS

1. Let $X = \mathsf{X}\,\{X_\alpha : \alpha \in A\}$ be a nonempty Cartesian product and suppose π_β is the projection mapping into X_β. Let

$$\tilde{x} \in \mathsf{X}\,\{X_\alpha : \alpha \in A\}$$

   and let

$$\tilde{X}_\beta = \{x : x \in X \text{ and } x_\alpha = \tilde{x}_\alpha \text{ for } \alpha \neq \beta\}.$$

   Show that

$$\pi_\beta \,|\, \tilde{X}_\beta : \tilde{X}_\beta \to X_\beta$$

   is a bijection.

2.   (a) Let X be the closed real interval $[0, 1]$ and $\mathscr{F}$ the collection of all real-valued functions defined on X. For each $t \in X$, let Y_t be the set $\mathbf{R}$ of all real numbers. Show that $\mathscr{F} = \mathsf{X}\,\{Y_t : t \in X\}$.
     (b) Let $\mathscr{S}$ be the collection of all real-valued infinite sequences. Express $\mathscr{S}$ as a Cartesian product.

111. THE PRODUCT TOPOLOGY

Suppose that $\{(X_\alpha, \mathscr{T}_\alpha) : \alpha \in A\}$ is an indexed collection of topological spaces. In this section we shall define a topology $\mathscr{T}$ for the set $\mathsf{X}\,\{X_\alpha : \alpha \in A\}$. The topology to be introduced is a straightforward generalization of the one introduced in 76.8 for finite collections of topological spaces.

111.1. Definition. Product space. *Let* $\{(X_\alpha, \mathcal{T}_\alpha): \alpha \in A\}$ *be an indexed collection of topological spaces. Let* $\mathcal{S}$ *be the following collection of subsets of* $\mathsf{X}\{X_\alpha : \alpha \in A\}$:

$$\mathcal{S} = \{U : U = \pi_\alpha^{-1}[U_\alpha] \text{ for an } \alpha \in A \text{ and open set } U_\alpha \subset X_\alpha\}$$

The topology $\mathcal{T}$ *for* $X = \mathsf{X}\{X_\alpha : \alpha \in A\}$ *that has* $\mathcal{S}$ *as a subbase is called the product topology and the space* $(X, \mathcal{T})$ *is known as the product space.*

If $\{(X_\alpha, \mathcal{T}_\alpha): \alpha \in A\}$ is a collection of topological spaces, we shall usually denote the corresponding product space by $\mathsf{X}\{X_\alpha : \alpha \in A\}$ with no specific reference to the topology.

111.2. Remark. A base $\mathcal{B}$ for the product topology defined in 111.1 is the collection of all sets of the form

$$\bigcap \{\pi_\alpha^{-1}[U_\alpha] : \alpha \in F\}$$

where F is a *finite* subset of the indexing set A and, for each α in F, U_α is open in X_α. Furthermore, $\bigcap \{\pi_\alpha^{-1}[U_\alpha] : \alpha \in F\} = \mathsf{X}\{W_\alpha : \alpha \in A\}$, where $W_\alpha = U_\alpha$ for each $\alpha \in F$ and $W_\alpha = X_\alpha$ for $\alpha \in A - F$.

It is interesting to note that for each nonempty open set W in $\mathsf{X}\{X_\alpha : \alpha \in A\}$, $\pi_\alpha[W] = X_\alpha$ for all but a finite number of $\alpha \in A$. This is easily seen as follows: Let $x \in W$; then there is a base element U, as given in 111.2, such that $x \in U \subset W$. But $U = \mathsf{X}\{U_\alpha : \alpha \in A\}$ where $U_\alpha = X_\alpha$ for $\alpha \in A - F$ for some finite subset F of A. Now let $\beta \in A - F$. Then $\pi_\beta[U] \subset \pi_\beta[W]$. Hence, $U_\beta = X_\beta \subset \pi_\beta[W]$ and, consequently, $\pi_\beta[W] = X_\beta$.

111.3. Theorem. *Let* $\mathsf{X}\{X_\alpha : \alpha \in A\}$ *be a nonempty product space. Then for each* $\beta \in A$, *the projection mapping*

$$\pi_\beta : \mathsf{X}\{X_\alpha : \alpha \in A\} \to X_\beta$$

is an open continuous surjection.

PROOF. It follows at once from the definition of the product that π_β is a surjection. That π_β is continuous follows from the fact that for each U_β open in X_β, $\pi_\beta^{-1}[U_\beta]$ is an element of the defining subbase for the product topology. To show that π_β is open it is sufficient to show that $\pi_\beta[W]$ is open where W is a base element as in 111.2. Let $W = \bigcap \{\pi_\alpha^{-1}[U_\alpha] : \alpha \in F\}$, where F is finite and U_α is open in X_α. But then $\pi_\beta[W] = U_\beta$ if $\beta \in F$ or $\pi_\beta[W] = X_\beta$ if $\beta \in A - F$. In either case $\pi_\beta[W]$ is open in X_β.

Suppose that $\{(X_i, d_i) : i \in \mathbf{P}_n\}$ is a finite collection of metric spaces and (X, d) is the corresponding product metric space. In 51.5 we proved that a sequence (x_i) in X converges to a point x in X if and only if for each fixed $j \in \mathbf{P}_n$, $(\pi_j(x_i))$ converges to $\pi_j(x)$. This theorem suggests the following analogous statements for nets and filters, the proofs of which are left as exercises.

111.4. Theorem.

111.4(a). *Suppose that* $(x(\beta))$ *is a net in a nonempty product space*

$$\mathsf{X}\,\{X_\alpha : \alpha \in A\}.$$

Then $(x(\beta))$ *converges to a point* $\tilde{x} \in \mathsf{X}\,\{X_\alpha : \alpha \in A\}$ *if and only if for each fixed* $\alpha \in A$, *the net* $(\pi_\alpha(x(\beta)))$ *converges to* $\pi_\alpha(\tilde{x})$.

111.4(b). *Suppose* $\mathscr{F}$ *is a filter on the product space* $\mathsf{X}\,\{X_\alpha : \alpha \in A\}$. *For each* $\alpha \in A$, *let* $\mathscr{F}_\alpha$ *be the filter* $\{\pi_\alpha[F] : F \in \mathscr{F}\}$ *(See* Theorem 109.15.*). Then* $\mathscr{F}$ *converges to* $x \in \mathsf{X}\,\{X_\alpha : \alpha \in A\}$ *if and only if* $\mathscr{F}_\alpha$ *converges to* x_α *for each* $\alpha \in A$.

The next theorem is a useful tool in dealing with product spaces. It will be reminiscent of a well known situation in $\mathbf{R}^n$.

111.5.   Theorem.   *Let* $\mathsf{X}\,\{X_\alpha : \alpha \in A\}$ *be a nonempty product space. Then if* $c \in \mathsf{X}\,\{X_\alpha : \alpha \in A\}$ *and* $\beta \in A$, *there is a subspace* X_β^c *such that* $c \in X_\beta^c$ *and* X_β^c *is homeomorphic to* X_β.

PROOF.   Let $X = \mathsf{X}\,\{X_\alpha : \alpha \in A\}$, $c \in X$, and $\beta \in A$. Let X_β^c be the following subset of $\mathsf{X}\,\{X_\alpha : \alpha \in A\}$:

$$X_\beta^c = \{x : x_\alpha = c_\alpha \text{ if } \alpha \neq \beta\}.$$

Consider the mapping $\pi_\beta \,|\, X_\beta^c : X_\beta^c \to X_\beta$. This mapping is a continuous bijection. We complete the proof by showing that $\pi_\beta \,|\, X_\beta^c$ is an open mapping. Let $\mathscr{B}$ be a base for the product topology as given in 111.1. Then $\{W \cap X_\beta^c : W \in \mathscr{B}\}$ is a base $\mathscr{B}_\beta^c$ for the relative topology for X_β^c. Note that if $W_\beta^c \in \mathscr{B}_\beta^c$, then $W_\beta^c = \mathsf{X}\,\{W_\alpha : \alpha \in A\}$, where W_α contains only c_α if $\alpha \neq \beta$ and W_α is an open subset (possibly X_α) if $\alpha = \beta$. Hence, $\pi_\beta[W_\beta^c]$ is open in X_β. Since we have shown that elements of the base $\mathscr{B}_\beta^c$ are carried onto open subsets of X_β by the mapping $\pi_\beta \,|\, X_\beta^c$, it follows that $\pi_\beta \,|\, X_\beta^c : X_\beta^c \to X_\beta$ is an open mapping. This completes the proof.

We can now deduce that each coordinate space inherits many properties from the product space. In this respect, the following observation is useful.

111.6.   *Remark.*   Any property that is invariant under continuous open surjections is inherited by each coordinate space from the product space (111.3). Any property that is inherited by a subspace from a topological space that contains it is inherited by all coordinate spaces of a product space. (This follows from 111.5.)

Thus, it is relatively easy to find properties that the product space induces upon the coordinate spaces. It is usually a more important and difficult problem to deduce information about the product space from the coordinate spaces.

111.7.   Theorem.   *Suppose* $\mathsf{X}\,\{X_\alpha : \alpha \in A\}$ *is a nonempty product space. Then* $\mathsf{X}\,\{X_\alpha : \alpha \in A\}$ *is Hausdorff if and only if each coordinate space* X_α *is Hausdorff.*

PROOF.   If the product space is Hausdorff, it follows from Remark 111.6 that each coordinate space is Hausdorff. Next, suppose each X_α is Hausdorff. Suppose x and y are distinct elements in $\mathsf{X}\,\{X_\alpha : \alpha \in A\}$. Then for some $\alpha \in A$, $x_\alpha \neq y_\alpha$. Since X_α is Hausdorff, there exists a pair of open disjoint neighborhoods U_α and V_α of x_α and y_α, respectively. Then $\pi_\alpha^{-1}[U_\alpha]$ and $\pi_\alpha^{-1}[V_\alpha]$ are disjoint open neighborhoods of x and y. Hence, the product space is Hausdorff.

The important Tychonoff product theorem states that the product space for a collection of compact spaces is compact. The proof which we shall give for that theorem makes use of the fact that a space is compact if and only if every ultra-filter (maximal filter) on the space converges (109.13). Recall also that a filter $\mathscr{F}$ on a set X is an ultrafilter if and only if for each subset A of X, either $A \in \mathscr{F}$ or $X - A \in \mathscr{F}$ (109.12). In addition, we will need 109.15 and the following lemma.

111.8. Lemma. Let $f: X \to Y$ be a surjection. Suppose $\mathscr{M}$ is an ultrafilter on X. Then $\{f[F]: F \in \mathscr{M}\}$ is an ultrafilter on Y.

PROOF. Let $\mathscr{M}_f = \{f[F]: F \in \mathscr{M}\}$. It follows from Theorem 109.15 that $\mathscr{M}_f$ is a filter on Y. We shall show that $\mathscr{M}_f$ is an ultrafilter by showing that for every subset A of Y either $A \in \mathscr{M}_f$ or $Y - A \in \mathscr{M}_f$ (109.12). To see this, let $A \subset Y$. Suppose $A \notin \mathscr{M}_f$. Then $f^{-1}[A] \notin \mathscr{M}$ because, otherwise, $f[f^{-1}[A]] = A \in \mathscr{M}_f$. Since $\mathscr{M}$ is an ultrafilter and $f^{-1}[A] \notin \mathscr{M}$, it follows that $X - f^{-1}[A] \in \mathscr{M}$. Hence $f[X - f^{-1}[A]] = Y - A \in \mathscr{M}_f$. We have thus shown that for every $A \subset Y$, $A \in \mathscr{M}_f$ or $Y - A \in \mathscr{M}_f$. This shows that $\mathscr{M}_f$ is an ultra-filter.

111.9. Theorem. The Tychonoff product theorem. *If* $\{X_\alpha: \alpha \in A\}$ *is a collection of compact spaces, then* $X = \underset{}{\times} \{X_\alpha: \alpha \in A\}$ *is compact.*

PROOF. We shall prove the product space X is compact by showing that every ultrafilter on it converges (109.13). Let $\mathscr{M}$ be an ultrafilter on X. For each $\alpha \in A$, let $\mathscr{M}_\alpha$ be the filter $\{\pi_\alpha[F]: F \in \mathscr{M}\}$. By Lemma 111.8, each $\mathscr{M}_\alpha$ is an ultrafilter in X_α. Each X_α is compact; hence, by 109.13, $\mathscr{M}_\alpha$ converges to a point $x_\alpha \in X_\alpha$. Hence, by 111.4(b), $\mathscr{M}$ converges to the point $x \in \underset{}{\times} \{X_\alpha: \alpha \in A\}$, where $\pi_\alpha(x) = x_\alpha$ for each α in A.

The proof of the Tychonoff theorem presented here uses a method as found, for example, in [3] and [6]. In net convergence theory there is a notion of uni-versal net that is analogous to the notion of ultrafilter. The development of the notion of universal net and a proof of the Tychonoff product theorem that is analogous to the one in 111.9 can be found, for example, in [6]. Still other proofs may be found in [24] and [26].

111.10. Theorem. *Let* $X = \underset{}{\times} \{X_\alpha: \alpha \in A\}$ *be a nonempty product space. Then* X *is compact if and only if each coordinate space* X_α *is compact.*

PROOF. The proof follows from Remark 111.6, and the Tychonoff theorem.

The next theorem is a generalization of the proposition stated in Exercise 14, page 225.

111.11. Theorem. *Let* $\underset{}{\times} \{X_\alpha: \alpha \in A\}$ *be a product space and suppose each coordinate space* X_α *is connected. Then the product space is connected.*

PROOF. Suppose each X_α is connected and $X = \underset{}{\times} \{X_\alpha: \alpha \in A\}$ is not con-nected. Then there is a separation $\{U, V\}$ for X. The sets U and V are nonempty open subsets of X. Let $p \in U$ and $q \in V$. (We shall get a contradiction by showing that there is a point $z \in V$ such that z can be "chained" to p by a finite number of connected sets $X_{\alpha_1}^*, X_{\alpha_2}^*, \ldots, X_{\alpha_n}^*$ such that $X_{\alpha_i}^* \cap X_{\alpha_{i+1}}^* \neq \varnothing$ for $i = 1, 2, \ldots,$ $n - 1$.) Since $q \in V$ and V is open, there is an open set of the form

$$W = \bigcap \{\pi_{\alpha_i}^{-1}[W_{\alpha_i}]: i = 1, 2, \ldots, n\}$$

such that $q \in W \subset V$ and W_{α_i} is open in X_{α_i}. Let

$$X_{\alpha_1}^* = \{x : x_\alpha = p_\alpha \text{ for } \alpha \neq \alpha_1\}.$$

Notice that $X_{\alpha_1}^*$ is a connected subset of X since it is a homeomorphic copy of X_{α_1}. This follows from the proof of 111.5. Next define

$$X_{\alpha_2}^* = \{x : x_{\alpha_1} = q_{\alpha_1} \text{ and } x_\alpha = p_\alpha \text{ for } \alpha \in A - \{\alpha_1, \alpha_2\}\}.$$

Observe that the α_2-coordinate is the only "free" coordinate. The set $X_{\alpha_2}^*$ is a homeomorphic copy of X_{α_2} and, hence, is a connected subset of X. Note that $p \in X_{\alpha_1}^*$ and $z^1 \in X_{\alpha_1}^* \cap X_{\alpha_2}^*$ where $z_{\alpha_1}^1 = q_{\alpha_1}$ and $z_\alpha^1 = p_\alpha$ for $\alpha \in A - \{a_1\}$. More generally, define $X_{\alpha_k}^* = \{x : x_\alpha = q_\alpha \text{ for } \alpha \in \{\alpha_1, \alpha_2, \ldots, \alpha_{k-1}\} \text{ and } x_\alpha = p_\alpha \text{ for } \alpha \in A - \{\alpha_1, \alpha_2, \ldots, \alpha_k\}\}$. Notice that $\alpha = \alpha_k$ is the only "free" index. For $k = 1, 2, \ldots, n$, $X_{\alpha_k}^*$ is a homeomorphic copy of X_{α_k} and, hence, is connected. Next for $k = 1, 2, \ldots, n - 1$, there is a point z^k such that $z^k \in X_{\alpha_k}^* \cap X_{\alpha_{k+1}}^*$. To see this, for $k \in \{1, 2, 3, \ldots, n\}$ consider the point z^k given by

$$z_\alpha^k = q_\alpha \quad \text{for} \quad \alpha \in \{\alpha_1, \alpha_2, \ldots, \alpha_k\}$$

and

$$z_\alpha^k = p_\alpha \quad \text{for} \quad \alpha \in A - \{\alpha_1, \alpha_2, \ldots, \alpha_k\}.$$

Observe that $z^k \in X_{\alpha_k}^* \cap X_{\alpha_{k+1}}^*$ for $k = 1, 2, 3, \ldots, n - 1$ and that $z^n \in X_{\alpha_n}^* \cap W$. It follows from Theorem 97.4 that $\bigcup \{X_{\alpha_i}^* : i = 1, 2, \ldots, n\}$ is connected. But $\{p, z^n\} \subset \bigcup \{X_{\alpha_i}^* : i = 1, 2, \ldots, n\}$. This is a contradiction, since $p \in U$ and $z^n \in V$. Thus, $\mathsf{X} \{X_\alpha : \alpha \in A\}$ is connected.

The next theorem is useful in considering subspaces and local properties of product spaces. The easy proof is left as an exercise.

111.12. Theorem. *Let* $\mathsf{X} \{X_\alpha : \alpha \in A\}$ *be a product space. Suppose that for each* $\alpha \in A$, S_α *is a subset of* X_α *and is given the relative topology from* X_α. *Then the product topology for* $\mathsf{X} \{S_\alpha : \alpha \in A\}$ *is the same as the relative topology induced on* $\mathsf{X} \{S_\alpha : \alpha \in A\}$ *by the product space* $\mathsf{X} \{X_\alpha : \alpha \in A\}$.

Suppose next that $\mathsf{X} \{X_\alpha : \alpha \in A\}$ is a product space and for each $\alpha \in A$, S_α is a connected subspace of X_α. Then by 111.11, the product space $\mathsf{X} \{S_\alpha : \alpha \in A\}$ is a connected space. Hence, by the previous theorem, $\mathsf{X} \{S_\alpha : \alpha \in A\}$ is also a connected subspace of $\mathsf{X} \{X_\alpha : \alpha \in A\}$. We shall use this observation in the proof of the next theorem.

111.13. Theorem. *Let* $X = \mathsf{X} \{X_\alpha : \alpha \in A\}$ *be a nonempty product space. Then* X *is locally connected if and only if each coordinate space* X_α *is locally connected and, with only a finite number of exceptions,* X_α *is connected.*

PROOF. Suppose first that the product space X is locally connected. Recall that for $\alpha \in A$, the projection mapping $\pi_\alpha : X \to X_\alpha$ is an open continuous surjection. Hence, from 97.9, each X_α is locally connected. Next let $x \in X$. Then since X is locally connected, we can find an open connected neighborhood U of x. This open set U contains an open neighborhood W of x of the form given in 111.2; that is, there is a finite subset F of A and open sets W_α of X_α such that if we let $W_\alpha = X_\alpha$ for $\alpha \in A - F$, then

$$x \in W = \mathsf{X} \{W_\alpha : \alpha \in A\} \subset U.$$

Since U is connected and for each α the projection mapping π_α is continuous, $\pi_\alpha[U]$ is connected. Next observe that for each $\alpha \in A - F$,

$$X_\alpha = \pi_\alpha[W] \subset \pi_\alpha[U].$$

Hence, X_α is connected for each $\alpha \in A - F$; that is, with at most a finite number of exceptions, all the coordinate spaces are connected.

We next assume that for each $\alpha \in A$, X_α is locally connected and that there is a finite set $F \subset A$ such that for $\alpha \in A - F$, X_α is connected. Let $x \in X$ and U be an open neighborhood of x. There is a finite set $F^* \subset A$ and neighborhood $\times \{W_\alpha : \alpha \in A\}$ of x contained in U such that each W_α is open in X_α and $W_\alpha = X_\alpha$ for α in $A - F^*$. For each α in $F \cup F^*$, let V_α be a connected open neighborhood of $\pi_\alpha(x)$ such that $V_\alpha \subset W_\alpha$. For each $\alpha \in A - (F \cup F^*)$, let $V_\alpha = X_\alpha$. Then the set $V = \times \{V_\alpha : \alpha \in A\}$ is an open neighborhood of x that is contained in U. Furthermore, by 111.11 and 111.12, V is a connected subset of the product space X. Hence, X is locally connected.

111.14. Theorem. *Let $X = \times \{X_\alpha : \alpha \in A\}$ be a nonempty product space. Then X is second (first) countable if and only if each X_α is second (first) countable, and with at most a countable number of exceptions all the X_α's have the trivial topology.*

PROOF. (The proofs for the first countable and second countable cases are similar. We give the second countable case only.) Assume that $\times \{X_\alpha : \alpha \in A\}$ is second countable. Then it follows from 111.6 that each X_α is second countable. We show next that there exists a countable subset C of A such that for all $\alpha \in A - C$, X_α has the trivial topology. To show this suppose that $\mathscr{B} = \{B^n : n \in \mathbf{P}\}$ is a countable base for the product topology. We may assume that all the elements of $\mathscr{B}$ are nonempty. Then for each $n \in \mathbf{P}$, there is a finite subset F_n of A such that $\pi_\alpha[B^n] = X_\alpha$ for all $\alpha \in A - F_n$. (See paragraph after 111.2.) Let C be the countable set $\bigcup \{F_n : n \in \mathbf{P}\}$. We shall show that for each $\alpha \in A - C$, X_α has the trivial topology. Let $\alpha \in A - C$ and let U_α be a nonempty open subset of X_α. Choose an $x_\alpha \in U_\alpha$ and a point x in X such that $\pi_\alpha(x) = x_\alpha$ and let $U = \pi_\alpha^{-1}[U_\alpha]$. For some $n \in \mathbf{P}$, there is a B^n such that $x \in B^n \subset U$. Now $\pi_\beta[B^n] = X_\beta$ for all $\beta \in A - F_n$. Since $\alpha \in A - C \subset A - F_n$, $\pi_\alpha[B^n] = X_\alpha$. Then because $B^n \subset U$, we have $X_\alpha = \pi_\alpha[B^n] \subset \pi_\alpha[U] = U_\alpha$. Hence, $U_\alpha = X_\alpha$. Thus we have shown that there are no nonempty proper open subsets of X_α and, hence, X_α has the trivial topology.

To prove the converse, assume that each X_α is second countable and that there is a countable subset $C \subset A$ such that X_α has the trivial topology for each $\alpha \in A - C$. There is no loss in assuming that $C \neq \varnothing$. Now for each $\alpha \in C$, let $\mathscr{B}_\alpha$ be a countable base for X_α and let $\mathscr{S}_\alpha = \{\pi_\alpha^{-1}[B] : B \in \mathscr{B}_\alpha\}$. Next let $\mathscr{S} = \bigcup \{\mathscr{S}_\alpha : \alpha \in C\}$. Finally, let $\mathscr{B}$ be the base for X formed by taking all finite intersections of members of $\mathscr{S}$. (See 111.1.) It is left as an exercise for the reader to show that $\mathscr{B}$ is a countable base for the product topology.

A metric space consisting of more than one point cannot have the trivial topology. Hence, from the previous theorem we see that the product of an uncountable collection of nontrivial metric spaces cannot be metrizable for, if it were, it would have to be first countable. It is true, however, that the product space for a countable collection of metrizable spaces is metrizable. We already have this result for finite products (see 51.1 and 51.2) and we next obtain it for countably infinite products.

111.15.   Theorem.   *Let* $\{(X_i, \mathscr{T}_i): i \in \mathbf{P}\}$ *be a countably infinite collection of metrizable spaces. Then the product topology for* $\times \{X_i: i \in \mathbf{P}\}$ *is metrizable.*

PROOF.   For each $i \in \mathbf{P}$, let d_i be a metric for $(X_i, \mathscr{T}_i)$ such that $d_i(x_i, y_i) \leq 1$ for all (x_i, y_i) in $X_i \times X_i$. Let X be the product set $\times \{X_i: i \in \mathbf{P}\}$. It is easy to show that the following formula defines a metric d for X:

$$d(x, y) = \sum_{i=1}^{\infty} 2^{-i} d_i(x_i, y_i).$$

We shall show that the topology $\mathscr{T}(d)$ generated by d is the product topology. Let $\mathscr{B}(d)$ be the base for $\mathscr{T}(d)$ consisting of all open d-spheres. Let $\mathscr{S}$ be the usual subbase for the product topology $\mathscr{T}$. It will be sufficient to show that each element of $\mathscr{S}$ is open in $(X, \mathscr{T}(d))$ and that each element in $\mathscr{B}(d)$ is open in the product space. Let $U \in \mathscr{S}$ and $x \in U$. Then $U = \pi_k^{-1}[U_k]$ for some open set $U_k \subset X_k$. Since $\pi_k(x) = x_k \in U_k$, there is a positive ε such that $N_{d_k}(x_k, \varepsilon) \subset U_k$. Now let $\delta = 2^{-k}\varepsilon$. We shall show that $N_d(x; \delta) \subset U$. To do this, let $y \in N_d(x; \delta)$. Then

$$d(x, y) = \sum_{i=1}^{\infty} 2^{-i} d_i(x_i, y_i) < \delta.$$

But then $2^{-k} d_k(x_k, y_k) < \delta = 2^{-k}\varepsilon$ and, hence, $d_k(x_k, y_k) < \varepsilon$ Thus,

$$y_k \in N_{d_k}(x_k, \varepsilon)$$

and, consequently, $y_k \in U_k$, from which it follows that $y \in U$. Thus we have shown that $N_d(x; \delta) \subset U$ and, hence, that U is in $\mathscr{T}(d)$. We next let $N_d(x; \varepsilon)$ be an open d-sphere and we show that it is open in the product space. To do this let $y \in N_d(x; \varepsilon)$. Choose a positive δ such that $N_d(y; \delta) \subset N_d(x; \varepsilon)$; for example, $\delta = \varepsilon - d(y, x)$ will do. We shall finish the proof by finding a neighborhood W of y in the product space such that $W \subset N_d(y; \delta)$ and, hence, such that $W \subset N_d(x; \varepsilon)$. This will imply that $N_d(x; \varepsilon)$ is open in the product space. To accomplish this let N be a positive integer sufficiently large so that $2^{-N} < \tfrac{1}{2}\delta$ and, consequently, such that $\sum_{i=N+1}^{\infty} 2^{-i} < \tfrac{1}{2}\delta$. Next let

$$W = \bigcap \{\pi_i^{-1}[N_{d_i}(y_i; \tfrac{1}{2}\delta)]: i \in \mathbf{P}_N\}.$$

We shall complete the proof by showing $W \subset N_d(y; \delta)$. Let $z \in W$. Then $d_i(z_i, y_i) < \tfrac{1}{2}\delta$ for $i \in \mathbf{P}_N$. To show that $z \in N_d(y; \delta)$, we calculate an estimate for $d(z, y)$ as follows:

$$d(z, y) \leq \sum_{i=1}^{N} 2^{-i} d_i(z_i, y_i) + \sum_{i=N+1}^{\infty} 2^{-i}$$

$$< \sum_{i=1}^{N} 2^{-i}(\tfrac{1}{2}\delta) + \tfrac{1}{2}\delta < \tfrac{1}{2}\delta + \tfrac{1}{2}\delta = \delta.$$

This completes the proof.

Suppose that $\{(X_i, d_i): i \in \mathbf{P}\}$ is a countably infinite collection of metric spaces. Then for each $i \in \mathbf{P}$, the function $d_i^*: X_i \times X_i \to \mathbf{R}$ given by

$$d_i^*(x_i, y_i) = d_i(x_i, y_i)[1 + d_i(x_i, y_i)]^{-1}$$

is a metric for X_i that is equivalent to d_i and is bounded by 1. Hence, from the proof of the previous theorem, an example of an admissible metric for the product space is given by

$$d(x, y) = \sum_{i=1}^{\infty} 2^{-i} d_i(x_i, y_i)[1 + d_i(x_i, y_i)]^{-1}.$$

EXERCISES: THE PRODUCT TOPOLOGY

1. Prove Theorem 111.4.

2. In Theorem 111.5 show that X_β^c need not be closed. If we suppose that $\mathsf{X}\{X_\alpha : \alpha \in A\}$ is Hausdorff is X_β^c closed in $\mathsf{X}\{X_\alpha : \alpha \in A\}$?

3. Prove, without using the results of filters, that the Cartesian product space of two compact spaces is compact.

4. Prove that the product topology is the smallest topology for which each of the projection mappings is continuous.

5. By making use of Remark 111.6, deduce as many properties as you can that each of the coordinate spaces inherits from the product space when the product space has the property.

6. Prove Theorem 111.12.

7. Complete the proof of 111.14.

112. MAPPINGS INTO PRODUCT SPACES

The notion of representing a function f into $\mathbf{R}^n$ in terms of its coordinate functions (i.e., $f = (f_1, f_2, \ldots, f_n)$) is probably already familiar to the reader from the calculus. The reader is probably also familiar with the fact that it is often useful to represent a complex-valued function f in terms of its real and imaginary parts u and v. These are special instances of the representation of a mapping into a product space in terms of its coordinate functions. In this section this notion will be extended to the setting of general product spaces. We shall prove, for example, that a mapping into a product space is continuous if and only if each of its coordinate functions is continuous.

112.1. Definition. Coordinates of a mapping. *Suppose that f is a mapping from a set X into a product set $\mathsf{X}\{Y_\alpha : \alpha \in A\}$. For each $\alpha \in A$, let*

$$\pi_\alpha : \mathsf{X}\{Y_\alpha : \alpha \in A\} \to Y_\alpha$$

be the α-projection mapping and let $f_\alpha = \pi_\alpha \circ f$. Then for each α in A, $f_\alpha : X \to Y_\alpha$ is called the α-coordinate of f.

112.2. Theorem. *Suppose that f is a mapping from a topological space X into a product space $\mathsf{X}\{Y_\alpha : \alpha \in A\}$. Then f is continuous if and only if for each $\alpha \in A$ the coordinate mapping $f_\alpha : X \to Y_\alpha$ is continuous.*

PROOF. For each $\alpha \in A$, π_α is continuous and $f_\alpha = \pi_\alpha \circ f$. Hence, if we assume that f is continuous, then f_α is continuous. Next assume that for each $\alpha \in A$, f_α is continuous. To show that f is continuous, it is sufficient to show that $f^{-1}[U]$ is open for each element U of a subbase for the product space. (See 82.1(e).) To show this, we let $\mathscr{S}$ be the usual subbase for the product topology as given in 111.1. Let $U \in \mathscr{S}$. Then $U = \pi_\alpha^{-1}[U_\alpha]$ for some $\alpha \in A$ and open subset $U_\alpha \subset Y_\alpha$. Then $f^{-1}[U] = f^{-1}[\pi_\alpha^{-1}[U_\alpha]]$. Next observe that since $f_\alpha = \pi_\alpha \circ f$,

$$f_\alpha^{-1}[U_\alpha] = f^{-1}[\pi_\alpha^{-1}[U_\alpha]].$$

Hence,

$$f_\alpha^{-1}[U_\alpha] = f^{-1}[U].$$

Then because f_α is continuous, $f_\alpha^{-1}[U_\alpha]$ is open and, consequently, $f^{-1}[U]$ is open. This completes the proof that f is continuous if each f_α is continuous.

If for each $\alpha \in A$, X_α and Y_α are topologically equivalent spaces, then one would expect that the product space $\mathbf{X}\{X_\alpha : \alpha \in A\}$ is topologically equivalent to the product space $\mathbf{X}\{Y_\alpha : \alpha \in A\}$. That such is indeed the case follows immediately from the next theorem.

112.3. Theorem. *Suppose that $X = \mathbf{X}\{X_\alpha : \alpha \in A\}$ and $Y = \mathbf{X}\{Y_\alpha : \alpha \in A\}$ are product spaces and for each $\alpha \in A$, $f_\alpha : X_\alpha \to Y_\alpha$ is a mapping. If we let $F : X \to Y$ be the mapping whose α-coordinate function F_α for each α in A is given by*

$$F_\alpha(x) = f_\alpha(x_\alpha),$$

then the following conclusions hold for F:

112.3(a). *If each $f_\alpha : X_\alpha \to Y_\alpha$ is continuous, then $F : X \to Y$ is continuous.*

112.3(b). *If each $f_\alpha : X_\alpha \to Y_\alpha$ is a surjection, then $F : X \to Y$ is a surjection.*

112.3(c). *If each $f_\alpha : X_\alpha \to Y_\alpha$ is a topological mapping, then $F : X \to Y$ is a topological mapping.*

PROOF. We shall prove 112.3(a) and leave the other two parts as exercises. By the previous theorem, F is continuous if, for each α in A, the coordinate function F_α is continuous. For each α in A let π_α be the α-projection mapping defined on $\mathbf{X}\{X_\alpha : \alpha \in A\}$. Then for each $x \in X$,

$$F_\alpha(x) = f_\alpha(x_\alpha) = f_\alpha(\pi_\alpha(x)).$$

From this we see that for each α in A, $F_\alpha = f_\alpha \circ \pi_\alpha$. Since each f_α and π_α is continuous, it follows that each F_α is continuous. Since each F_α is continuous, it follows from the previous theorem that F is continuous.

The reader has seen various instances in which information about a function can be obtained by representing the function as a composition of simpler functions. Often one of the previous two theorems can be used to good advantage to accomplish this.

112.4. Example. Let E be a real normed linear space. Suppose that $f_1 : E \to E$ and $f_2 : E \to E$ are continuous mappings. Let α and β be real numbers. Define the mappings $G : E \times E \to E$ and $H : E \to E$ by:

$$G((x, y)) = \alpha f_1(x) + \beta f_2(y) \quad \text{for} \quad (x, y) \text{ in } E \times E$$

and

$$H(x) = \alpha f_1(x) + \beta f_2(x) \quad \text{for} \quad x \text{ in } E.$$

We shall show that G and H are continuous. In order to do this, we first define the mappings $s: E \times E \to E$, $r: E \times E \to E \times E$ and $q: E \to E \times E$ as follows:

$$s((x, y)) = x + y \quad \text{for} \quad (x, y) \text{ in } E \times E,$$

$$r((x, y)) = (\alpha\, f_1(x), \beta f_2(y)) \quad \text{for} \quad (x, y) \text{ in } E \times E,$$

$$q(x) = (x, x) \quad \text{for} \quad x \text{ in } E.$$

It is easy to prove that s is continuous. That q is continuous follows from 112.2, since the identity mapping $i: E \to E$ is continuous. It is easy to see that αf_1 and βf_2 are continuous. Hence, by 112.3(a), r is continuous. Observe that $G = s \circ r$. Hence, G is continuous since s and r are continuous. Next note that $H = G \circ q$ and H is continuous. The situation in this example is summarized in the following commutative mapping diagram:

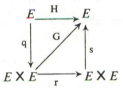

EXERCISES: MISCELLANEOUS

I. **1.** Prove 112.3(b) and 112.3(c).

2. Suppose that $F: X \to \bigtimes \{Y_\alpha : \alpha \in A\}$ is an open mapping. For each α in A, let $F_\alpha: X \to Y_\alpha$ be the α-coordinate of F. Prove that each coordinate $F_\alpha: X \to Y_\alpha$ of F is an open mapping.

3. Let $I: \mathbf{R}^n \times \mathbf{R}^n \to \mathbf{R}$ be the inner product mapping for $\mathbf{R}^n$ as given in Section 34.
   (a) By making use of the properties of the inner product, prove that I is continuous.
   (b) For each $i \in P_n$, assume that $f_i: \mathbf{R} \to \mathbf{R}$ and $g_i: \mathbf{R} \to \mathbf{R}$ are continuous mappings. Let $H: \mathbf{R} \to \mathbf{R}$ be the function given by

$$H(x) = \sum_{i=1}^{n} f_i(x) g_i(x) \quad \text{for} \quad x \text{ in } \mathbf{R}$$

   Use the results of part (a) and appropriate theorems in §112 to show that H is continuous.

4. Suppose that $\mathscr{F}$ is a family of continuous mappings $f: X \to Y_f$. Note that each $f \in \mathscr{F}$ is defined on the same topological space X, but that we are not assuming that the spaces Y_f are the same. For each point x in X, let $e(x)$ be the point in the product space $\bigtimes \{Y_f : f \in \mathscr{F}\}$ whose f-coordinate is $f(x)$. This defines a mapping $e:X \to X\{Y_f:f\in\mathscr{F}\}$ which is known as the *evaluation mapping* for the family $\mathscr{F}$.

(a) Prove that the evaluation mapping is continuous.

(b) If for each pair of distinct points x and y in X, there is an $f\in\mathscr{F}$ such that $f(x)\neq f(y)$, then the family is said to *distinguish points*. Prove that the evaluation mapping e is an injection if and only if the family $\mathscr{F}$ distinguishes points.

(c) Suppose that X is a compact space and for each $f\in\mathscr{F}$, Y_f is a Hausdorff space. Suppose further that the family $\mathscr{F}$ distinguishes points. Prove that the evaluation mapping e is a homeomorphism from X onto $e[X]$, where $e[X]$ is given the relative topology from the product space $X\{Y_f:f\in\mathscr{F}\}$.

(For more information concerning the evaluation map, see, for example, [26] or [32].)

II.   We assume that $\mathbf{R}$ is given the usual topology and that R is the relation in $\mathbf{R}$ defined as follows:

$$R = \{(x, x):x\in\mathbf{R}\} \cup \{(m, n):m\in\mathbf{Z} \quad \text{and} \quad n\in\mathbf{Z}\}.$$

1. Is the quotient mapping $p:\mathbf{R} \to \mathbf{R}/R$ an open mapping? Is it a closed mapping? Is it a compact mapping?

2. Is the quotient space $\mathbf{R}/R$ metrizable? Is it locally connected?

3. Is the product space $\mathbf{R} \times (\mathbf{R}/R)$ a metric space?

III.   1. Suppose that $\mathscr{F}$ is an ultrafilter on a set X. Can $\bigcap\mathscr{F}$ contain more than one point?

2. Suppose that $\mathscr{F}$ is a filter on a set X and $\bigcap\mathscr{F} = \varnothing$. Prove that X is necessarily an infinite set.

3. Let X be the set of all real numbers and let $\mathscr{T}$ be the topology for X as in Example 75.6. Is $\mathscr{T}$ a filter on X? Is $\mathscr{T}-\{\varnothing\}$ a filter on X?

4. Let $\mathscr{F}$ be a filter on a set X. Prove that $\mathscr{F}$ is the intersection of all ultrafilters that contain it.

IV.   This set of exercises deals with the notion of composition of relations. The results will be needed in V.

Let X be a set and let U and V be two relations in X. By the composition $U \circ V$, we shall mean the relation defined as follows:

$$U \circ V = \{(v, u):(v, x)\in V \quad \text{and} \quad (x, u)\in U$$

for some $x\in X\}$.

1. Show that if the relations U and V happen to be functions from X into X, then the definition of composition as defined here is consistent with the definition for composition of functions.

2. Prove the associative law for composition of relations; that is, prove that if U, V, and W are relations in X, then

$$(U \circ V) \circ W = U \circ (V \circ W)$$

3. Recall from 11.4 the definition of R^{-1} for a relation R. Prove that if U and V are relations in a set X, then $(U \circ V)^{-1} = V^{-1} \circ U^{-1}$.

NOTE: In view of the associativity of composition of relations, parentheses will not be used for $(U \circ V) \circ W$ in the exercises in V.

V. In this sequence of exercises we shall be dealing with the notion of *uniform space*. As the reader will see, such concepts as uniform continuity, uniform convergence, and other metric type properties can be extended to the setting of uniform spaces.

In these exercises the reader will need to make use of the results in IV. Also, extensive use will be made of the notations in Section 11, especially 11.2, 11.3 and 11.4.

Let X be a nonempty set. In what follows we shall refer to the set $\{(x, x):x \in X\}$ as the diagonal of $X \times X$ and denote it by Δ. A uniformity for the set X is a nonempty collection $\mathscr{U}$ of subsets of $X \times X$ (and, hence, a nonempty collection of relations in X) that satisfies each of the following properties:

(i) $\Delta \subset U$ for each $U \in \mathscr{U}$.
(ii) If $U \in \mathscr{U}$ and $U \subset W \subset X \times X$, then $W \in \mathscr{U}$.
(iii) If U and V are elements of $\mathscr{U}$, then $U \cap V \in \mathscr{U}$.
(iv) If $U \in \mathscr{U}$, then $U^{-1} \in \mathscr{U}$.
(v) If $U \in \mathscr{U}$, then there is a $V \in \mathscr{U}$ such that $V \circ V \subset U$.

If X is a nonempty set and $\mathscr{U}$ is a uniformity for X, then $(X, \mathscr{U})$ is called a *uniform space*.

In each of the following, we assume that $(X, \mathscr{U})$ is a uniform space.

1. Prove that $\mathscr{U}$ is a filter on $X \times X$.

2. Show that for each $U \in \mathscr{U}$, $U \subset U \circ U \subset U \circ U \circ U$.

Hence, for example, if $U \in \mathscr{U}$, then $U \circ U$ and $U \circ U \circ U$ are also elements of $\mathscr{U}$.

3. Let $U \in \mathscr{U}$. Then there is a $V \in \mathscr{U}$ such that $V \circ V \circ V \subset U$.

4. Recall that a relation is symmetric if and only if $U = U^{-1}$. Prove that if $U \in \mathscr{U}$, then there is a symmetric $V \in \mathscr{U}$ such that $V \circ V \subset U$.

HINT: Prove first that if $W \in \mathscr{U}$, then $W \cap W^{-1}$ is a symmetric element of $\mathscr{U}$.

5. Let (X, d) be a metric space. For each $\varepsilon > 0$, let

$$U_\varepsilon = \{(x, y):d(x, y) < \varepsilon\} \subset X \times X.$$

Let $\mathscr{U}$ be the following collection of subsets of $X \times X$:

$$\mathscr{U} = \{U : U_\varepsilon \subset U \ \text{ for some } \varepsilon > 0\}.$$

Prove that the collection $\mathscr{U}$ is a uniformity for X.

6. Let $\mathscr{C}$ be the collection of all continuous real-valued functions defined on the real line $\mathbf{R}$. For each compact set $K \subset \mathbf{R}$ and for each $\varepsilon > 0$ let

$$U(\varepsilon, K) = \{(f, g) : f \in \mathscr{C}, g \in \mathscr{C}, |f(x) - g(x)| < \varepsilon$$

for all $x \in K\}$.

Next let $\mathscr{U}$ be the collection of *all* subsets $\mathscr{C} \times \mathscr{C}$ that contain a set of this form; that is,

$$\mathscr{U} = \{U : U \supset U(\varepsilon, K) \text{ for some } \varepsilon > 0$$

$$\text{and compact subset } K \text{ of } \mathbf{R}\}.$$

Prove that $\mathscr{U}$ is a uniformity for $\mathscr{C}$.

7. Let $(X, \mathscr{U})$ be a uniform space. Let $\mathscr{T}(\mathscr{U})$ be the following collection of subsets of X:

$$\mathscr{T}(\mathscr{U}) = \{S : \text{for each } x \in S, \text{ there is a}$$

$$U \in \mathscr{U} \text{ such that } U[x] \subset S\}.$$

Show that $\mathscr{T}(\mathscr{U})$ is a topology for X. This topology is known as the *uniform topology*. In what follows, a uniform space $(X, \mathscr{U})$ will be said to have a topological property provided the topological space $(X, \mathscr{T}(\mathscr{U}))$ has that property. Also, suppose that $(X, \mathscr{U})$ and $(Y, \mathscr{V})$ are uniform spaces. We shall say that a mapping $f : (X, \mathscr{U}) \to (Y, \mathscr{V})$ is continuous provided $f : (X, \mathscr{T}(\mathscr{U})) \to (Y, \mathscr{T}(\mathscr{V}))$ is continuous.

8. Suppose that $(X, \mathscr{U})$ is a uniform space. Let $U \in \mathscr{U}$ and $x \in X$. Prove that the set $U[x]$ is a neighborhood of x (relative to the uniform topology).

HINT: Show that the set

$$N_x = \{y : V[y] \subset U[x] \ \text{ for some } V \in \mathscr{U}\}$$

is an open subset of $U[x]$ that contains x. Property $\cdot$ (v) for uniform spaces will play a critical role in the proof.

9. A mapping f from a uniform space $(X, \mathscr{U})$ into a uniform space $(Y, \mathscr{V})$ is called *uniformly continuous* provided that for each $V \in \mathscr{V}$, there is a $U \in \mathscr{U}$ such that

$$(x, y) \in U \text{ implies that } (f(x), f(y)) \in V.$$

Prove that if f is uniformly continuous, then it is continuous.

10. Let (X, d) be a metric space, and let $\mathscr{U}$ be the uniformity for X as given in Exercise 5. Show that the topology for X generated by d and the uniform topology are the same.

11. Let $f:(X, \mathcal{U}) \to (Y, \mathcal{V})$ be a mapping where $(X, \mathcal{U})$ and $(Y, \mathcal{V})$ are uniform spaces. Prove that if $(X, \mathcal{U})$ is compact and f is continuous, then f is uniformly continuous.

    HINT: Imitate the proof of the metric case with property (v) for uniform spaces used very much like the triangle inequality.

12. Let $(f_\alpha, \alpha \in D, \geq)$ be a net of mappings from a topological space $(X, \mathcal{T})$ into a uniform space $(Y, \mathcal{V})$. Then the net (f_α) is said to converge uniformly to the mapping $f:(X, \mathcal{T}) \to (Y, \mathcal{V})$ provided that for every $V \in \mathcal{V}$ there is an $N \in D$ such that for all $\alpha \geq N$,

$$(f_\alpha(x), f(x)) \in V \quad \text{for all } x \in X.$$

    Prove that if the net (f_α) converges uniformly to f and if each of the f_α's is continuous, then the mapping f is continuous also. (See Exercise 4, page 255 .)(It should be pointed out that the definition of uniform convergence is meaningful if the common domain for the f_α's is a set rather than a space. However, a space was needed for the proposition.)

    The notions of total boundedness, completeness and other metric-like properties can also be extended to the setting of uniform spaces. The reader who wishes to pursue this topic is referred, for example, to [6], [26], or [32].

References

[1] Abian, A.: *The Theory of Sets and Transfinite Arithmetic*. Philadelphia, W. B. Saunders Company, 1965.

[2] Baum, J. D.: *Elements of Point Set Topology*. Englewood Cliffs, New Jersey, Prentice-Hall, Inc., 1964.

[3] Bourbaki, N.: *General Topology, Part 1*. Paris, Hermann; Reading, Massachusetts, Addison-Wesley Publishing Company, 1966.

[4] Bushaw, D.: *Elements of General Topology*. New York, John Wiley & Sons, Inc., 1963.

[5] Cronin, J.: *Fixed Points and Topological Degree in Nonlinear Analysis*. Providence, Rhode Island, American Mathematical Society, 1964.

[6] Cullen, H. F.: *Introduction to General Topology*. Boston, D. C. Heath & Company, 1968.

[7] Dugundji, J.: *Topology*. Boston, Allyn & Bacon, Inc., 1966.

[8] Eilenberg, S.: Sur les transformations continués d'espaces métriques compacts. *Fundamenta Mathematicae*, 22:292–296, 1934.

[9] Engelking, R.: *Outline of General Topology*. New York, John Wiley & Sons, Inc., 1968.

[10] Fomin, S. V., and Kolmogorov, A. N.: *Elements of the Theory of Functions and Functional Analysis*. (Translated by Leo F. Boron) Albany, New York, Graylock Press, 1957.

[11] Gemignani, M. C.: *Elementary Topology*. Reading, Massachusetts, Addison-Wesley Publishing Company, 1967.

[12] Gillman, L., and Jerison, M.: *Rings of Continuous Functions*. Princeton, New Jersey, D. Van Nostrand Company, Inc., 1960.

[13] Gleason, A. M.: *Fundamentals of Abstract Analysis*. Reading, Massachusetts, Addison-Wesley Publishing Company, 1966.

[14] Goffman, C.: Preliminaries to functional analysis. *Studies in Modern Analysis*. M.A.A. Studies in Mathematics, Mathematical Association of America, 1962.

[15] Goldberg, R. R.: *Methods of Real Analysis*. New York, Blaisdell Publishing Company, 1964.

[16] Graves, L. M.: *The Theory of Functions of Real Variables*. New York, McGraw-Hill Book Company, Inc., 1946.

[17] Greenspan, D.: *Theory and Solution of Ordinary Differential Equations*. New York, The Macmillan Company, 1960.

[18] Greever, J.: *Theory and Examples of Point-Set Topology*. Belmont, California, Brooks/Cole Publishing Company, 1967.

[19] Hall, D. W., and Spencer, G. L.: *Elementary Topology*. New York, John Wiley & Sons, Inc., 1955.

[20] Hayden, S., and Kennison, J. F.: *Zermelo-Fraenkel Set Theory*. Columbus, Ohio, Charles E. Merrill Publishing Company, 1968.

[21] Heider, L. J., and Simpson, J. E.: *Theoretical Analysis*. Philadelphia, W. B. Saunders Company, 1967.

[22] Hocking, J. G., and Young, G. S.: *Topology*. Reading, Massachusetts, Addison-Wesley Publishing Company, Inc., 1961.

[23] Hu, S. T.: *Elements of General Topology*. San Francisco, Holden-Day, Inc., 1964.

[24] Hu, S. T.: *Introduction to General Topology*. San Francisco, Holden-Day, Inc., 1966.

[25] Husain, T.: *Introduction to Topological Groups*. Philadelphia, W. B. Saunders Company, 1966.

[26] Kelly, J. L.: *General Topology*. New York, D. Van Nostrand Company, Inc., 1955.

[27] McShane, E. J.: A theory of limits. *Studies in Modern Analysis.* M.A.A. Studies in Mathematics, Mathematical Association of America, 1962.

[28] Mendelson, B.: *Introduction to Topology.* 2nd ed. Boston, Allyn & Bacon, Inc., 1968.

[29] Pervin, W. J.: *Foundations of General Topology.* New York, Academic Press, Inc., 1964.

[30] Simmons, G. F.: *Introduction to Topology and Modern Analysis.* New York, McGraw Hill Book Company, Inc., 1963.

[31] Thomas, J.: A regular space, not completely regular. Amer. Math. Monthly 76 (2):181–182, 1969.

[32] Thron, W. J.: *Topological Structures.* New York, Holt, Rinehart and Winston, 1966.

[33] Whyburn, G. T.: *Analytic Topology.* American Mathematical Society Colloquium Publications, Vol. 28. New York, American Mathematical Society, 1942.

[34] Whyburn, G. T.: Dynamic topology. Amer. Math. Monthly 77 (6): 556–570, 1970.

[35] Whyburn, G. T.: Open and closed mappings. Duke Math. J. 17: 69–74, 1950.

[36] Whyburn, G. T.: Open mappings on locally compact spaces. Mem. Amer. Math. Soc. 1, 1950.

[37] Whyburn, G. T.: *Topological Analysis.* Rev. ed. Princeton, New Jersey, Princeton University Press, 1964.

[38] Whyburn, G. T.: *Topological Analysis.* Princeton, New Jersey, Princeton University Press, 1958.

[39] Wilder, R. L.: *Topology of Manifolds.* American Mathematical Society Colloquium Publications, Vol. 32. New York, American Mathematical Society, 1949.

Index